Encyclopaedia of Mathematical Sciences

Volume 31

Editor-in-Chief: R.V. Gamkrelidze

Yu. V. Egorov M. A. Shubin (Eds.)

Partial Differential Equations II

Elements of the Modern Theory. Equations with Constant Coefficients

Springer-Verlag
Berlin Heidelberg New York
London Paris Tokyo
Hong Kong Barcelona
Budapest

Title of the Russian edition:
Itogi nauki i tekhniki, Sovremennye problemy matematiki,
Fundamental'nye napravleniya, Vol. 31, Differentsial'nye
uravneniya s chastnymi proizvodnymi 2
Publisher VINITI, Moscow 1988

Mathematics Subject Classification (1991):
35-xx, 35Sxx, 58G15, 35Axx

ISBN 3-540-52001-5 Springer-Verlag Berlin Heidelberg New York
ISBN 0-387-52001-5 Springer-Verlag New York Berlin Heidelberg

CIP data applied for

Printed in Germany
Typesetting: Asco Trade Typesetting Ltd., Hong Kong
SPIN 10008987 41/3140/SPS - 5 4 3 2 1 0 - Printed on acid-free paper

List of Editors and Authors

Editor-in-Chief

R. V. Gamkrelidze, Russian Academy of Sciences, Steklov Mathematical Institute, ul. Vavilova 42, 117966 Moscow, Institute for Scientific Information (VINITI), ul. Usievicha 20a, 125219 Moscow, Russia, e-mail: gam@ips.ac.msk.su

Consulting Editors

Yu. V. Egorov, U.F.R. M.I.G., Université Paul Sabatier, 118, route de Narbonne, 31062 Toulouse, France, e-mail: egorov@cict.fr
M. A. Shubin, Department of Mathematics, Northeastern University, Boston, MA 02115, USA, e-mail: shubin@neu.edu

Authors

Yu. V. Egorov, U.F.R. M.I.G., Université Paul Sabatier, 118, route de Narbonne, 31062 Toulouse, France, e-mail: egorov@cict.fr
A. I. Komech, Department of Mathematics, Moscow State University, 119899 Moscow, Russia, e-mail: kom@compnet.msu.su
M. A. Shubin, Department of Mathematics, Northeastern University, Boston, MA 02115, USA, e-mail: shubin@neu.edu

Translator

P. C. Sinha, Jagat Narayan Road, 800003 Patna, India

Contents

I. Linear Partial Differential Equations. Elements of the Modern Theory

Yu.V. Egorov, M.A. Shubin

Translated from the Russian
by P.C. Sinha

Contents

Preface

In this paper we have made an attempt to present a sketch of certain ideas and methods of the modern theory of linear partial differential equations. It can be regarded as a natural continuation of our paper (Egorov and Shubin [1988], EMS vol. 30) where we dealt with the classical questions, and therefore we quote this paper for necessary definitions and results whenever possible. The present paper is basically devoted to those aspects of the theory that are connected with the direction which originated in the sixties and was later called "*microlocal analysis*". It contains the theory and applications of pseudodifferential operators and Fourier integral operators and also uses the language of wave front sets of distributions. But where necessary we also touch upon important topics connected both with the theory preceding the development of microlocal analysis, and sometimes even totally classical theories. We do not claim that the discussion is complete. This paper should be considered simply as an introduction to a series of more detailed papers by various other authors which are being published in this and subsequent volumes in the present series and which will contain a detailed account of most of the questions raised here.

The bibliographical references given in this paper are in no way complete. We have tried to quote mostly books or review papers whenever possible and have not made any attempt to trace original sources of described ideas or theorems. This will be rectified at least partially in subsequent papers of this series.

We express our sincere gratitude to M.S. Agranovich who went through the manuscript and made a number of useful comments.

Yu.V. Egorov M.A. Shubin

Notation

We shall use the following standard symbols.

$\mathbb{R}$ is the set of all real numbers.

$\mathbb{C}$ is the set of all complex numbers.

$\mathbb{Z}$ is the set of all integers.

$\mathbb{Z}_+$ is the set of all non-negative integers.

$\mathbb{R}^n$ is the standard n-dimensional real vector space.

$\mathbb{C}^n$ is the standard n-dimensional complex vector space.

$\partial/\partial x = (\partial/\partial x_1, \ldots, \partial/\partial x_n)$, where $x = (x_1, \ldots, x_n) \in \mathbb{R}^n$.

$D = i^{-1}\partial/\partial x$, where $i = \sqrt{-1} \in \mathbb{C}$; $D_j = i^{-1}\partial/\partial x_j$.

$D^\alpha = D_1^{\alpha_1} \ldots D_n^{\alpha_n}$, where α is a multi-index, that is, $\alpha = (\alpha_1, \ldots, \alpha_n)$ with $\alpha_j \in \mathbb{Z}_+$.

$\xi^\alpha = \xi_1^\alpha \ldots \xi_n^{\alpha_n}$, where $\xi = (\xi_1, \ldots, \xi_n) \in \mathbb{R}^n$ or $\mathbb{C}^n$ and $\alpha = (\alpha_1, \ldots, \alpha_n)$ is a multi-index.

$x \cdot \xi = x_1\xi_1 + \cdots + x_n\xi_n$ if $x = (x_1, \ldots, x_n) \in \mathbb{R}^n$ and $\xi = (\xi_1, \ldots, \xi_n) \in \mathbb{R}^n$.

$C_0^\infty(\Omega)$ is the space of C^∞-functions having compact support in a domain $\Omega \subset \mathbb{R}^n$.

$\Delta = \Delta_x = \partial^2/\partial x_1^2 + \cdots + \partial^2/\partial x_n^2$ is the standard Laplacian in $\mathbb{R}^n$.

$|x| = (x_1^2 + \cdots + x_n^2)^{1/2}$ for $x = (x_1, \ldots, x_n) \in \mathbb{R}^n$.

$|\alpha| = \alpha_1 + \cdots + \alpha_n$, where α is a multi-index.

$\alpha! = \alpha_1! \ldots \alpha_n!$ for a multi-index α.

$\mathscr{D}'(\Omega)$ is the space of all distributions in Ω.

$\mathscr{E}'(\Omega)$ is the space of all distributions with compact support in Ω.

$L_2(\Omega)$ is the Hilbert space of all square integrable functions in Ω.

$S(\mathbb{R}^n)$ is the Schwartz space of C^∞-functions on $\mathbb{R}^n$ whose derivatives decay faster than any power of $|x|$ as $|x| \to \infty$.

$S'(\mathbb{R}^n)$ is the space of all distributions with temperate growth on $\mathbb{R}^n$.

supp u denotes the support of a function (or distribution) u.

sing supp u is the singular support of a distribution u.

$H^s(\mathbb{R}^n)$ denotes the Sobolev space consisting of those distributions $u \in S'(\mathbb{R}^n)$ for which $(1 + |\xi|^2)^{s/2}\tilde{u}(\xi) \in L_2(\mathbb{R}^n)$; here $\tilde{u}$ is the Fourier transform of u.

$H^s(\Omega)$, where $s \in \mathbb{Z}_+$, is the Hilbert space containing those functions $u \in L_2(\Omega)$ for which $D^\alpha u \in L_2(\Omega)$ with $|\alpha| \leqslant s$.

$H^s_{\mathrm{comp}}(\Omega) = \mathscr{E}'(\Omega) \cap H^s(\mathbb{R}^n)$.

$H^s_{\mathrm{loc}}(\Omega)$ is the space of those $u \in \mathscr{D}'(\Omega)$ such that $\varphi u \in H^s(\mathbb{R}^n)$ for any function $\varphi \in C_0^\infty(\Omega)$.

$\mathring{H}^s(\Omega)$ is the completion of the space $C_0^\infty(\Omega)$ in the topology of $H^s(\Omega)$.

$C^\alpha(\Omega)$, where $\alpha \in (0, 1)$, is the space of functions continuous in Ω such that $\sup\limits_{x, y \in K} |u(x) - u(y)|\,|x - y|^{-\alpha} < \infty$ for each $K \subset\subset \Omega$.

§ 1. Pseudodifferential Operators

1.1. Definition and Simplest Properties (Agranovich [1965], Egorov [1984, 1985], Eskin [1973], Friedrichs [1968], Hörmander [1971, 1983, 1985], Kohn and Nirenberg [1965], Kumano-go [1982], Nirenberg [1970], Palais [1965], Reed and Simon [1972–1978], Rempel and Schulze [1982], Shubin [1978], Taylor [1981], and Trèves [1980]). The theory of pseudodifferential operators, in its present form, appeared in the mid-sixties (Kohn and Nirenberg [1965]). Its principal aim was to extend to operators with variable coefficients the standard application of the Fourier transformation to operators having constant coefficients, in which case this transformation reduces the differentiation D^α to multiplication by ξ^α.

We consider the *differential operator*

$$A = \sum_{|\alpha| \leq m} a_\alpha(x) D^\alpha \tag{1.1}$$

in a domain $\Omega \subset \mathbb{R}^n$, where $a_\alpha \in C^\infty(\Omega)$, $D = i^{-1}\partial/\partial x$ and $\alpha = (\alpha_1, \ldots, \alpha_n)$ is a multi-index with $|\alpha| = \alpha_1 + \cdots + \alpha_n$. We express the function $u \in C_0^\infty(\Omega)$ by means of the formula for the inverse Fourier transform

$$u(x) = (2\pi)^{-n} \int e^{ix\cdot\xi} \tilde{u}(\xi)\, d\xi, \tag{1.2}$$

where

$$\tilde{u}(\xi) = \int e^{-iy\cdot\xi} u(y)\, dy \tag{1.3}$$

and we assume that u has been extended to be zero in $\mathbb{R}^n \backslash \Omega$. Applying the operator A to both sides of (1.2), we have

$$Au(x) = (2\pi)^{-n} \int e^{ix\cdot\xi} a(x, \xi) \tilde{u}(\xi)\, d\xi, \tag{1.4}$$

where

$$a(x, \xi) = \sum_{|\alpha| \leq m} a_\alpha(x) \xi^\alpha. \tag{1.5}$$

The function $a(x, \xi)$ is known as the *symbol* or the *total symbol* of A and the operator itself is often denoted by $a(x, D)$ or $a(x, D_x)$. We see that $a \in C^\infty(\Omega \times \mathbb{R}^n)$ and that $a(x, \xi)$ is a polynomial in ξ with coefficients in $C^\infty(\Omega)$. If we substitute the expression for $\tilde{u}(\xi)$ from (1.3) into (1.4), we can also write A in the form

$$Au(x) = (2\pi)^{-n} \iint e^{i(x-y)\cdot\xi} a(x, \xi) u(y)\, dy\, d\xi, \tag{1.6}$$

where the integral should be understood as a repeated integral.

In the theory of pseudodifferential operators we study operators of the form (1.4) (or (1.6)) with more general symbols $a(x, \xi)$ than (1.5). For example, a convenient class of symbols is obtained if the estimates

$$|\partial_\xi^\alpha \partial_x^\beta a(x, \xi)| \leqslant C_{\alpha\beta K}(1 + |\xi|)^{m-|\alpha|}, \quad x \in K, \quad \xi \in \mathbb{R}^n \tag{1.7}$$

hold, where α and β are multi-indices, K is a compact set in Ω and m is a real number. The class of symbols $a \in C^\infty(\Omega \times \mathbb{R}^n)$ satisfying these estimates is denoted by $S^m(\Omega \times \mathbb{R}^n)$, or simply S^m if it is either clear or irrelevant what domain Ω is involved. Clearly, the symbols (1.5) of differential operators satisfy (1.7) if m is taken to be the order of the operator A or any larger number.

To give an example of a symbol in S^m for any $m \in \mathbb{R}$, we can mention the symbol $(1 + |\xi|^2)^{m/2}$. The corresponding operator in $\mathbb{R}^n$ is denoted by $(1 - \Delta)^{m/2}$, which is consistent with the definition of the powers of the differential operator $1 - \Delta$ when $m/2$ is an integer.

Let $a \in S^m(\Omega \times \mathbb{R}^n)$. We define A by (1.4), or by (1.6) with the integral taken as a repeated integral. It follows easily from (1.7) that the integral in (1.4) converges absolutely if $u \in C_0^\infty(\Omega)$ and that it can be infinitely differentiated with respect to x under the integral sign for $x \in \Omega$. Thus we obtain a continuous linear operator

$$A\colon C_0^\infty(\Omega) \to C^\infty(\Omega), \tag{1.8}$$

which is denoted by $a(x, D)$ or $a(x, D_x)$, as in the case of a differential operator. Operators of the form $a(x, D)$, with symbols $a \in S^m$, are the simplest examples of pseudodifferential operators.

Let us examine the properties of a pseudodifferential operator of the form $a(x, D)$ with symbol $a \in S^m$. We first note that the integral in (1.6) converges absolutely if $m < -n$ and, by changing the order of integration, we can write $A = a(x, D)$ in the form

$$Au(x) = \int K_A(x, y)u(y)\, dy, \tag{1.9}$$

where

$$K_A(x, y) = (2\pi)^{-n} \int e^{i(x-y)\cdot\xi} a(x, \xi)\, d\xi. \tag{1.10}$$

In the present case where $m < -n$, the *kernel* K_A is continuous on $\Omega \times \Omega$. Using the identity

$$e^{i(x-y)\cdot\xi} = |x - y|^{-2N}(-\Delta_\xi)^N e^{i(x-y)\cdot\xi}, \tag{1.11}$$

where N is a non-negative integer, and then integrating (1.10) by parts, we can write, in place of (1.10),

$$K_A(x, y) = (2\pi)^{-n}|x - y|^{-2N} \int e^{i(x-y)\cdot\xi}(-\Delta_\xi)^N a(x, \xi)\, d\xi, \tag{1.12}$$

where $x \neq y$. Since $(-\Delta_\xi)^N a(x, \xi) \in S^{m-2N}$, this integral can be differentiated k times with respect to x and y provided that $m - 2N + k < -n$. Because N is arbitrary, it follows that $K_A \in C^\infty$ for $x \neq y$, that is, off the diagonal in $\Omega \times \Omega$.

In the general case, the kernel K_A of the operator A is a distribution on $\Omega \times \Omega$. This result follows from the Schwartz theorem on the kernel (see Egorov and Shubin [1988; § 1.11, Chap. 2] and Hörmander [1983, 1985; Chap. 5]) or can be established directly as follows. For $u, v \in C_0^\infty(\Omega)$, we write $\langle Au, v\rangle$ as a repeated integral:

$$\langle Au, v\rangle = (2\pi)^{-n} \iiint e^{i(x-y)\cdot\xi} a(x, \xi) u(y) v(x)\, dy\, d\xi\, dx.$$

We integrate this integral by parts and use the identity

$$e^{i(x-y)\cdot\xi} = (1 + |\xi|^2)^{-N} (1 - \Delta_y)^N e^{i(x-y)\cdot\xi} \tag{1.13}$$

to obtain

$$\langle Au, v\rangle = (2\pi)^{-n} \iiint e^{i(x-y)\cdot\xi} (1 + |\xi|^2)^{-N} a(x, \xi) \times (1 - \Delta_y)^N [u(y) v(x)]\, dy\, d\xi\, dx. \tag{1.14}$$

This integral already converges absolutely for sufficiently large N and remains absolutely convergent if $u(y)v(x)$ is replaced by $\varphi = \varphi(x, y) \in C_0^\infty(\Omega \times \Omega)$. This enables us to write

$$\langle Au, v\rangle = \langle K_A, v \otimes u\rangle,$$

where $K_A \in \mathscr{D}'(\Omega \times \Omega)$. We now integrate (1.14) by parts, using (1.11), and find that in the general case too $K_A \in C^\infty(\Omega \times \Omega \backslash \Delta)$, where Δ is the diagonal in $\Omega \times \Omega$. This property is referred to as the *pseudolocality* of a pseudodifferential operator A. It is equivalent to the condition that $Au \in \mathscr{D}'(\Omega) \cap C^\infty(\Omega')$ if $u \in \mathscr{E}'(\Omega) \cap C^\infty(\Omega')$, where Ω' is an open subset of Ω.

The operator $A = a(x, D_x)$ with symbol $a \in S^m$ can be extended uniquely to a continuous map

$$A\colon \mathscr{E}'(\Omega) \to \mathscr{D}'(\Omega). \tag{1.15}$$

To see this, we introduce the *transpose operator* ${}^tA$ by means of the identity

$$\langle Au, v\rangle = \langle u, {}^tAv\rangle, \quad u, v \in C_0^\infty(\Omega). \tag{1.16}$$

It can be seen easily that such an operator can be defined by the formula

$${}^tAv(y) = (2\pi)^{-n} \iint e^{i(x-y)\cdot\xi} a(x, \xi) v(x)\, dx\, d\xi,$$

or, what is the same, by the formula

$${}^tAv(x) = (2\pi)^{-n} \iint e^{i(x-y)\cdot\xi} a(y, -\xi) v(y)\, dy\, d\xi. \tag{1.17}$$

The operator ${}^tA$ defines a map

$$ {}^tA: C_0^\infty(\Omega) \to C^\infty(\Omega), \tag{1.18}$$

which, by duality, yields the continuous map (1.15) that extends the map (1.8) in view of (1.16). The formula (1.16) can be regarded as the definition of Au for $u \in \mathscr{E}'(\Omega)$ by taking any $v \in C_0^\infty(\Omega)$.

In addition to the transpose ${}^tA$, we can also study a *formal adjoint operator* A^*. This operator is defined by the formula

$$(Au, v) = (u, A^*v), \quad u, v \in C_0^\infty(\Omega), \tag{1.19}$$

where $(\cdot, \cdot)$ denotes the scalar product in $L_2(\Omega)$. Such an operator is given by the formula

$$A^*v(x) = (2\pi)^{-n} \iint e^{i(x-y)\cdot\xi} \overline{a(y, \xi)} v(y)\, dy\, d\xi, \tag{1.20}$$

and also maps $C_0^\infty(\Omega)$ into $C^\infty(\Omega)$.

The pseudolocality of the operator A is equivalent to the following property of the extended map (1.15):

$$\text{sing supp}(Au) \subset \text{sing supp}\, u. \tag{1.21}$$

1.2. The Expression for an Operator in Terms of Amplitude. The Connection Between the Amplitude and the Symbol. Symbols of Transpose and Adjoint Operators (Egorov [1984, 1985], Hörmander [1971, 1983, 1985], Kumano-go [1982], Shubin [1978], Taylor [1981], and Trèves [1980]). The formulae (1.17) and (1.20) defining ${}^tA$ and A^* have a slightly different form from the formula (1.6) for the operator $A = a(x, D_x)$. They give us reason to examine more general operators A that are defined by expressions of the form

$$Au(x) = (2\pi)^{-n} \iint e^{i(x-y)\xi} a(x, y, \xi) u(y)\, dy\, d\xi, \tag{1.22}$$

where the function $a = a(x, y, \xi) \in C^\infty(\Omega \times \Omega \times \mathbb{R}^n)$ lies in $S^m = S^m(\Omega \times \Omega \times \mathbb{R}^n)$, that is, it satisfies the estimates

$$|\partial_\xi^\alpha \partial_x^{\beta_1} \partial_y^{\beta_2} a(x, y, \xi)| \leqslant C_{\alpha\beta_1\beta_2 K}(1 + |\xi|)^{m-|\alpha|}, \quad (x, y) \in K,$$

where K is a compact set in $\Omega \times \Omega$. The function $a(x, y, \xi)$ in (1.22) is known as the *amplitude* of A. The class of operators of the form (1.22) with amplitudes $a \in S^m$ is denoted by L^m or $L^m(\Omega)$ and constitutes the simplest class of pseudodifferential operators. It can be seen easily that any pseudodifferential operator $A \in L^m$ defines the continuous maps (1.8) and (1.15). Every $A \in L^m$ has a transpose ${}^tA$ and a formal adjoint operator A^* which also belong to L^m. Their amplitudes ${}^ta$ and a^* are expressed in terms of the amplitude of A by the formulae

$${}^ta(x, y, \xi) = a(y, x, -\xi), \quad a^*(x, y, \xi) = \overline{a(y, x, \xi)}. \tag{1.23}$$

Every operator $A \in L^m$ is pseudolocal, this result being proved in the same way as for the operators $A = a(x, D_x)$.

The expression for $A \in L^m$ in the form (1.22) is not unique. For example, on integrating by parts, we can replace the amplitude $a(x, y, \xi)$ by the amplitude $a_1(x, y, \xi) = (1 + |x - y|^2)^{-N}(1 - \Delta_\xi)^N a(x, y, \xi)$ without changing the operator itself. But the expression for A in the form $a(x, D_x)$ reduces this non-uniqueness significantly. For example, such an expression is, in general, unique if $\Omega = \mathbb{R}^n$, that is, the symbol $a(x, \xi)$ is uniquely determined by A. Therefore it is desirable to simplify (1.22) by moving over, for example, to (1.6) with a suitably chosen symbol. It turns out that this can be done to within operators with smooth kernels, as the following theorem shows.

Theorem 1. *Any operator $A \in L^m(\Omega)$, with amplitude $a \in S^m$, can be expressed in the form $A = \sigma_A(x, D_x) + R$, where $\sigma_A \in S^m$ and R is an operator with kernel $K_R \in C^\infty(\Omega \times \Omega)$. This expression can be so chosen that*

$$\sigma_A(x, \xi) - \sum_{|\alpha| \leqslant N-1} \frac{1}{\alpha!} \partial_\xi^\alpha D_y^\alpha a(x, y, \xi)|_{y=x} \in S^{m-N}(\Omega \times \mathbb{R}^n)\ [1] \tag{1.24}$$

for any integer $N > 0$.

We shall indicate the main points of the proof later, and for the present we note that all the terms in the summation (1.24) depend only on the values of $a(x, y, \xi)$ and its derivatives for $y = x$ (in particular, the principal term in the summation is simply $a(x, y, \xi)$). This means that to within symbols of order $m - N$ (for any N) the symbol σ_A is determined by the values of the amplitude $a(x, y, \xi)$ near $\Delta \times \mathbb{R}^n$, where Δ is the diagonal in $\Omega \times \Omega$. In fact, $K_A \in C^\infty(\Omega \times \Omega)$ if $a(x, y, \xi) = 0$ in a neighbourhood of $\Delta \times \mathbb{R}^n$, because in this case integration by parts, with the aid of (1.11), enables us to replace the amplitude $a(x, y, \xi)$ by the amplitude $|x - y|^{-2N}(-\Delta_\xi)^N a(x, y, \xi) \in S^{m-N}$ without changing the operator. Let us mention, by the way, that any operator R with a smooth kernel K_R can be written in the form (1.22) with amplitude $a_R(x, y, \xi) = (2\pi)^n e^{-i(x-y)\xi} K_R(x, y)\psi(\xi)$, where the function $\psi \in C_0^\infty(\mathbb{R}^n)$ is such that $\int \psi(\xi)\, d\xi = 1$. Clearly, $a_R \in S^{-\infty}$, where

$$S^{-\infty} = \bigcap_{m \in \mathbb{R}} S^m.$$

In what follows, we shall also use the following notation:

$$L^{-\infty} = \bigcap_{m \in \mathbb{R}} L^m, \quad S^\infty = \bigcup_{m \in \mathbb{R}} S^m, \quad L^\infty = \bigcup_{m \in \mathbb{R}} L^m.$$

Clearly, $L^{-\infty}$ is precisely the class of all operators with smooth kernels.

The set of relations (1.24), with $N = 1, 2, \ldots$, will be written below in short in the form of an asymptotic series

$$\sigma_A(x, \xi) \sim \sum_\alpha \frac{1}{\alpha!} \partial_\xi^\alpha D_y^\alpha a(x, y, \xi)|_{y=x}. \tag{1.24'}$$

---

[1] We recall that $\alpha! = \alpha_1! \ldots \alpha_n!$ for any multi-index α.

More generally, if we have a system of functions $a_j = a_j(x, \xi) \in S^{m_j}$, $j = 0, 1, 2, \ldots$, where $m_j \to -\infty$ as $j \to \infty$, and a function $a = a(x, \xi)$, we shall write

$$a \sim \sum_{j=0}^{\infty} a_j \tag{1.25}$$

if

$$a - \sum_{j=0}^{N-1} a_j \in S^{\overline{m}_N} \tag{1.26}$$

for any integer $N \geqslant 0$, where $\overline{m}_N = \max_{j \geqslant N} m_j$. Instead of this last definition of $\overline{m}_N$, it is clearly sufficient to assume that $\overline{m}_N$ are arbitrary numbers for which $\overline{m}_N \to -\infty$ as $N \to \infty$. The function a is obviously defined uniquely up to addition of any function from $S^{-\infty}$. Clearly $a \in S^m$, where $m = \max_{j \geqslant 0} m_j$. It can easily be seen that for any sequence $a_j \in S^{m_j}$ there exists a function a such that (1.25) holds. For this it is sufficient to take

$$a(x, \xi) = \sum_{j=0}^{\infty} \chi\left(\frac{\xi}{t_j}\right) a_j(x, \xi), \tag{1.27}$$

where $\chi \in C^\infty(\mathbb{R}^n)$ and $\chi(\xi) = 1$ for $|\xi| \geqslant 2$ while $\chi(\xi) = 0$ for $|\xi| \leqslant 1$ and the numbers t_j tend to $+\infty$ sufficiently rapidly as $j \to \infty$.

The asymptotic sums for the amplitudes $a(x, y, \xi)$ are defined in an analogous manner.

With the aid of asymptotic summation, it is useful to identify in S^m the class S^m_{cl} of *classical* or *polyhomogeneous symbols*. This class consists of symbols $a \in S^m$ that have a decomposition of the form (1.25) in which $m_j = m - j$ and the function a_j is positive homogeneous in ξ, with $|\xi| \geqslant 1$, and of degree $m_j = m - j$, that is,

$$a_j(x, t\xi) = t^{m-j} a_j(x, \xi), \quad |\xi| \geqslant 1, \quad t \geqslant 1.$$

The *classical amplitudes* can be defined in a similar fashion.

Let $a^0_{m-j}(x, \xi)$ be a positive homogeneous function in ξ (now for all $\xi \neq 0$) on $\Omega \times (\mathbb{R}^n \backslash 0)$ which coincides with $a_j(x, \xi)$ for $|\xi| \geqslant 1$. Such a function is uniquely defined, and for $a \in S^m_{cl}$ we shall write

$$a \sim \sum_{j=0}^{\infty} a^0_{m-j} \tag{1.28}$$

in place of (1.25). This expression is well-defined because the functions a^0_{m-j} also define $a (\operatorname{mod} S^{-\infty})$. The function $a^0_m = a^0_m(x, \xi)$, which is homogeneous of degree m in ξ, is called the *principal symbol* of the operator A.

It is clear that the transition from the amplitude a to the symbol σ_A by means of (1.24′) does not take us out of S^m_{cl}, that is, $\sigma_A \in S^m_{cl}$ if $a \in S^m_{cl}$. The symbols of differential operators are also classical. The results of a number of other operations also remain within the class of classical symbols and amplitudes.

We now present the main points in the proof of Theorem 1.1. First, as we noted earlier, by multiplying the amplitude $a(x, y, \xi)$ by a cut-off function $\chi = \chi(x, y) \in C^\infty(\Omega \times \Omega)$ which is unity in a neighbourhood of the diagonal, we only change the operator A by adding an operator with a smooth kernel, and we can arrange that $a(x, y, \xi) = 0$ for $(x, y) \notin U$, where U is an arbitrary pre-assigned neighbourhood of the diagonal in $\Omega \times \Omega$. We now expand $a(x, y, \xi)$ in y by the Taylor formula for $y = x$:

$$a(x, y, \xi) = \sum_{|\alpha| \leqslant N-1} \frac{1}{\alpha!}[D_y^\alpha a(x, y, \xi)]|_{y=x}[i(y-x)]^\alpha$$
$$+ \sum_{|\alpha|=N} [i(y-x)]^\alpha r_\alpha(x, y, \xi), \tag{1.29}$$

where $r_\alpha \in S^m$. Substituting this expansion into (1.22) and noting that

$$[i(y-x)]^\alpha e^{i(x-y)\xi} = (-\partial_\xi)^\alpha e^{i(x-y)\xi},$$

we obtain, on integrating by parts the terms of the first sum in (1.29), operators with symbols which are equal to terms of the sum in (1.24). The remainder (that is, the second sum in (1.29)) can be transformed in the same manner into an operator with amplitude in S^{m-N}. This implies that if a symbol σ_A is expressed as the asymptotic series (1.24′), then the operator $A - \sigma_A(x, D_x)$ will belong to L^{m-N} for any N and will therefore be an operator with a smooth kernel, as required.

Theorem 1.1 easily yields formulae that express the symbols $\sigma_{{}^tA}$ and σ_{A^*} of the transpose and formally adjoint operators in terms of σ_A (mod $S^{-\infty}$). Indeed, to within operators with smooth kernels, ${}^tA$ and A^* can be taken to be defined by the amplitudes

$${}^ta = {}^ta(x, y, \xi) = \sigma_A(y, -\xi), \quad a^*(x, y, \xi) = \overline{\sigma_A(y, \xi)}$$

(see (1.23)). It now follows from Theorem 1.1 that

$$\sigma_{{}^tA}(x, \xi) \sim \sum_\alpha \frac{1}{\alpha!} \partial_\xi^\alpha D_x^\alpha \sigma_A(x, -\xi), \tag{1.30}$$

$$\sigma_{A^*}(x, \xi) \sim \sum_\alpha \frac{1}{\alpha!} \partial_\xi^\alpha D_x^\alpha \overline{\sigma_A(x, \xi)}. \tag{1.31}$$

In particular, $\sigma_{A^*} - \bar{\sigma}_A \in S^{m-1}$. This implies that $\operatorname{Im} \sigma_A \in S^{m-1}$ if the operator $A \in L^m$ is formally self-adjoint (that is, if it is symmetric on $C_0^\infty(\Omega)$). Further, if the operator is also classical, then the principal homogeneous part $\sigma_A^0(x, \xi)$ (of order m) of its symbol is real valued.

We cite two important examples of pseudodifferential operators which are not differential operators.

Example 1.1 (*One-dimensional singular integral operator*). Let us consider on $\mathbb{R}^1$ an operator A of the form

$$Au(x) = a(x)u(x) + \text{v.p.} \frac{1}{\pi i} \int_{-\infty}^{\infty} \frac{L(x, y)}{x - y} u(y)\, dy,$$

where $a \in C^\infty(\mathbb{R})$, $L \in C^\infty(\mathbb{R} \times \mathbb{R})$ and v.p. denotes the "valeur principale" or the *principal value of the integral*, that is,

$$\text{v.p.}\frac{1}{\pi i}\int_{-\infty}^{\infty} \frac{L(x, y)}{x - y} u(y)\, dy = \lim_{\varepsilon \to +0} \frac{1}{\pi i}\int_{|y-x| \geqslant \varepsilon} \frac{L(x, y)}{x - y} u(y)\, dy.$$

By Hadamard's lemma, we write $L(x, y) = b(x) + (y - x)L_1(x, y)$, where $b(x) = L(x, x)$ and $L_1 \in C^\infty(\mathbb{R} \times \mathbb{R})$, and obtain

$$Au(x) = a(x)u(x) + b(x)Su(x) + R_1 u(x),$$

where $R_1 \in L^{-\infty}$ and S is the *Hilbert transformation* defined by

$$Su(x) = \text{v.p.}\frac{1}{\pi i}\int_{-\infty}^{\infty} \frac{u(y)}{x - y}\, dy.$$

This transformation leads to the multiplication of $\tilde{u}(\xi)$ by $-\operatorname{sgn} \xi$, and therefore, to within an operator with a smooth kernel, A has the form $a(x, D_x)$, where $a(x, \xi) = a(x) - b(x)\chi(\xi) \operatorname{sgn} \xi$, with $\chi \in C^\infty(\mathbb{R})$ such that $\chi(\xi) = 1$ for $|\xi| \geqslant 1$ and $\chi(\xi) = 0$ for $|\xi| < 1/2$. Thus A is a classical pseudodifferential operator of order zero with principal symbol $\sigma(x, \xi) = a(x) - b(x) \operatorname{sgn} \xi$.

Example 1.2 (*Multidimensional singular integral operator*). We consider in $\mathbb{R}^n$ an operator A of the form

$$\begin{aligned} Au(x) &= a(x)u(x) + \text{v.p.}\int \frac{L\left(x, \dfrac{x - y}{|x - y|}\right)}{|x - y|^n} u(y)\, dy \\ &= a(x)u(x) + \lim_{\varepsilon \to +0} \int_{|y-x| \geqslant \varepsilon} \frac{L\left(x, \dfrac{x - y}{|x - y|}\right)}{|x - y|^n} u(y)\, dy, \end{aligned}$$

where $a \in C^\infty(\mathbb{R}^n)$ and $L = L(x, \omega) \in C^\infty(\mathbb{R}^n \times S^{n-1})$ is such that

$$\int_{S^{n-1}} L(x, \omega)\, d\omega = 0, \quad x \in \mathbb{R}^n.$$

Here S^{n-1} denotes the unit sphere in $\mathbb{R}^n$. Then the expression $|z|^{-n} L(x, z)$ defines a homogeneous distribution of order $-n$ on $\mathbb{R}^n_z$ which depends smoothly on x (see Egorov and Shubin [1988, § 1, Chap. 2] and Hörmander [1983, 1985, § 3.2]). The Fourier transform of this function with respect to z is a distribution $g(x, \xi)$ on $\mathbb{R}^n_\xi$ that is homogeneous in ξ of degree zero and smooth for $\xi \neq 0$, and also depends smoothly on $x \in \mathbb{R}^n$. Then it follows easily that, to within an operator with a smooth kernel, A can be written in the form $a(x, D_x)$, where $a(x, \xi) = a(x) + \chi(\xi)g(x, \xi)$, with $\chi \in C^\infty(\mathbb{R}^n)$ such that $\chi(\xi) = 1$ if $|\xi| \geqslant 1$ and $\chi(\xi) = 0$ if $|\xi| \leqslant 1/2$. In particular, A is a classical pseudodifferential operator of order zero with principal symbol $\sigma(x, \xi) = a(x) + g(x, \xi)$.

1.3. The Composition Theorem. The Parametrix of an Elliptic Operator (Egorov [1984, 1985], Hörmander [1983, 1985], Kumano-go [1982], Shubin [1978], Taylor [1981], Trèves [1980]). Let us take two pseudodifferential operators A and B: $C_0^\infty(\Omega) \to C^\infty(\Omega)$. For the composition $A \circ B$ to be defined, it is necessary either that B maps $C_0^\infty(\Omega)$ into $C_0^\infty(\Omega)$ or that A can be extended to a continuous operator acting from $C^\infty(\Omega)$ into $C^\infty(\Omega)$. In fact, it is convenient to impose a slightly stronger condition that operators be properly supported. Namely, we say that an operator $A \in L^m(\Omega)$ is *properly supported* if both the natural projections π_1, π_2: supp $K_A \to \Omega$ are proper maps. We recall that a map $f: X \to Y$ of two locally compact spaces is said to be *proper* if the inverse image $f^{-1}(K)$ of any compact set $K \subset Y$ is compact in X. The property of A being properly supported is equivalent to the following two simultaneous conditions: 1) for any compact set $K \subset \Omega$ there exists a compact set $K_1 \subset \Omega$ such that A maps $C_0^\infty(K)$ into $C_0^\infty(K_1)$; 2) the same is true for the transpose ${}^tA$.

By truncating the kernel K_A near the diagonal, we can obtain the decomposition $A = A_1 + R$ for any pseudodifferential operator $A \in L^m(\Omega)$, where $R \in L^{-\infty}(\Omega)$ and A_1 is a properly supported operator. This remark enables us, without loss of generality, to confine our attention to properly supported operators in a majority of cases.

A properly supported operator $A \in L^m(\Omega)$ defines the following continuous maps:

$$A: C_0^\infty(\Omega) \to C_0^\infty(\Omega),$$

$$A: C^\infty(\Omega) \to C^\infty(\Omega),$$

$$A: \mathscr{E}'(\Omega) \to \mathscr{E}'(\Omega),$$

$$A: \mathscr{D}'(\Omega) \to \mathscr{D}'(\Omega).$$

Thus if one of the $L^\infty(\Omega)$ operators A and B is properly supported, the composition $A \circ B$ is defined.

To describe the symbol of the composition, we first examine the case of differential operators $A = a(x, D_x)$ and $B = b(x, D_x)$. Set $C = A \circ B$. By the Leibniz formula, we have

$$Cu(x) = a(x, D_x + D_y)[b(x, D_y)u(y)]|_{y=x}$$

$$= \sum_\alpha \frac{1}{\alpha!} \partial_\xi^\alpha a(x, D_y) D_x^\alpha b(x, D_y) u(y)|_{y=x},$$

where we have used the Taylor formula to expand $a(x, D_x + D_y)$ as a power series in D_x. This implies that $C = c(x, D_x)$ with

$$c(x, \xi) = \sum_\alpha \frac{1}{\alpha!} \partial_\xi^\alpha a(x, \xi) D_x^\alpha b(x, \xi).$$

The sum here is finite because a is a polynomial in ξ. This sum makes sense as an asymptotic sum if $a \in S^{m_1}$ and $b \in S^{m_2}$.

Theorem 1.2 (*the composition theorem*). *Let $A \in L^{m_1}$ and $B \in L^{m_2}$ be two pseudodifferential operators in Ω one of which is properly supported. Then $C = A \circ B \in L^{m_1+m_2}$ and $C = c(x, D_x) + R$, where $R \in L^{-\infty}$ and $c(x, \xi)$ has the asymptotic expansion*

$$c(x, \xi) \sim \sum_\alpha \frac{1}{\alpha!}[\partial_\xi^\alpha a(x, \xi)].[D_x^\alpha b(x, \xi)]. \tag{1.32}$$

This result can be proved by arguing in the same way as for differential operators but by confining Taylor expansion to a finite sum and estimating the remainder. An alternative way is as follows. Using the formula $B = {}^t({}^tB)$, we represent B by means of the amplitude $\tilde{b}(y, \xi) = \sigma_{{}^tB}(y, -\xi)$, which implies that

$$\widetilde{Bu}(\xi) = \iint e^{-iy\xi}\tilde{b}(y, \xi)u(y)\,dy$$

and

$$Cu(x) = (2\pi)^{-n} \iint e^{i(x-y)\xi}a(x, \xi)\tilde{b}(y, \xi)u(y)\,dy\,d\xi.$$

Thus C is an operator with amplitude $c(x, y, \xi) = a(x, \xi)\tilde{b}(y, \xi)$. Theorem 1.1 then shows that $C \in L^{m_1+m_2}$. Now an application of (1.30), which gives $\tilde{b}(y, \xi)$, together with (1.24) leads to (1.32) after simple algebraic simplifications.

We note that the principal part on the right-hand side of (1.32) is simply the product $a(x, \xi)b(x, \xi)$ and therefore

$$c(x, \xi) - a(x, \xi)b(x, \xi) \in S^{m_1+m_2-1}. \tag{1.33}$$

If A and B are classical pseudodifferential operators of orders m_1 and m_2 respectively, then $C = A \circ B$ is a classical pseudodifferential operator of order $m_1 + m_2$ whose principal symbol is

$$c^0_{m_1+m_2}(x, \xi) = a^0_{m_1}(x, \xi)b^0_{m_2}(x, \xi). \tag{1.34}$$

We now present an important definition.

Definition 1.1. An operator $A = a(x, D_x) \in L^m(\Omega)$ is said to be an *elliptic pseudodifferential operator* of order m if for every compact $K \subset \Omega$ there exist positive constants $R = R(K)$ and $\varepsilon = \varepsilon(K)$ such that

$$|a(x, \xi)| \geqslant \varepsilon|\xi|^m, \quad x \in K, \quad |\xi| \geqslant R, \tag{1.35}$$

for any compact set $K \subset \Omega$. A more general operator $A = a(x, D_x) + R \in L^m(\Omega)$, where $R \in L^{-\infty}$ and $a \in L^m$, is *elliptic* if the operator $a(x, D_x)$ is elliptic. If (1.35) holds, then $a(x, \xi)$ is referred to as an *elliptic symbol.*

If $A \in L^m_{cl}(\Omega)$ and $a^0_m(x, \xi)$ is the principal symbol (homogeneous of degree m) of A, then the ellipticity of A is equivalent to the fact that

$$a^0_m(x, \xi) \neq 0 \quad \text{for } \xi \neq 0, \tag{1.36}$$

and this is consistent with the definition of ellipticity of a differential operator (see Egorov and Shubin [1988, §2, Chap. 1]).

Theorem 1.3. *If A is an elliptic pseudodifferential operator of order m in Ω, then there exists a properly supported pseudodifferential operator $B \in L^{-m}(\Omega)$ such that*

$$B \circ A = I - R_1, \quad A \circ B = I - R_2, \tag{1.37}$$

where $R_j \in L^{-\infty}(\Omega)$, $j = 1, 2$. Such a pseudodifferential operator B is unique up to addition of operators with smooth kernels and is an elliptic pseudodifferential operator of order $-m$. If A is a classical pseudodifferential operator of order m, then B is a classical pseudodifferential operator of order $-m$.

The operator B satisfying the hypotheses of this theorem is called a *parametrix* of A. The fact that the parametrix of an elliptic pseudodifferential operator is also a pseudodifferential operator shows that the choice of the class of pseudodifferential operators is reasonable. In particular, the parametrix of an elliptic differential operator is a classical pseudodifferential operator.

In order to construct the parametrix B of the operator A, it is necessary to take for the first approximation the operator $B_0 \in L^{-m}$ whose symbol is $a^{-1}(x, \xi)$ for large $|\xi|$ (that is, for $x \in K$ and $|\xi| > R(K)$ for any compact set $K \subset \Omega$). Now B_0 can be made a properly supported operator by addition of an operator with a smooth kernel. The composition theorem then implies that

$$B_0 A = I - T_1, \quad AB_0 = I - T_2; \quad T_j \in L^{-1}(\Omega), \quad j = 1, 2.$$

We now construct properly supported operators B_0' and B_0'' such that

$$B_0' \sim I + T_1 + T_1^2 + \cdots, \quad B_0'' \sim I + T_2 + T_2^2 + \cdots,$$

by which we mean that the symbols of B_0' and B_0'' are defined by the corresponding asymptotic sums of symbols of the operators on the right-hand sides. We then set $B_1 = B_0' B_0$ and $B_2 = B_0 B_0''$. This yields

$$B_1 A = I - R_1', \quad AB_2 = I - R_2'; \quad R_j' \in L^{-\infty}, \quad j = 1, 2.$$

On multiplying the first equation by B_2 on the right and using the second equation, we find that $B_1 - B_2 \in L^{-\infty}(\Omega)$. Thus for B we can take either B_j. We have also established at the same time that B is unique up to operators belonging to $L^{-\infty}(\Omega)$.

The existence of a parametrix implies that the solutions of elliptic equations with smooth right-hand sides are regular. To see this, let $u \in \mathscr{D}'(\Omega)$ and $Au = f \in \mathscr{D}'(\Omega) \cap C^\infty(\Omega_1)$, where $\Omega_1 \subset \Omega$. Here we have assumed that Au is meaningful, and for this it is sufficient, for example, that $f \in \mathscr{E}'(\Omega)$ or that A is properly supported. Then $u \in C^\infty(\Omega_1)$ because, by applying B to both sides of the equation $Au = f$, we obtain $u = Bf + R_1 u$. Then, since B is pseudolocal, it follows that $Bf \in C^\infty(\Omega_1)$ and $R_1 u \in C^\infty(\Omega)$ as R_1 is an operator with a smooth kernel. More precise regularity theorems can be formulated in terms of the Sobolev norms, and this will be done below.

We shall describe in some detail the structure of the parametrix B for a classical elliptic pseudodifferential operator A of order m. Suppose that the symbol $a(x, \xi)$ of A has the asymptotic expansion (1.28). Let the symbol $b(x, \xi)$ of the parametrix B have a similar expansion

$$b(x, \xi) \sim \sum_{k=0}^{\infty} b^0_{-m-k}(x, \xi). \tag{1.38}$$

If we use the composition formula for finding $B \circ A - I$, then all the homogeneous components must vanish. We now group the members of the series defining the symbol of $B \circ A - I$ according to the degree of homogeneity and obtain the equations

$$\begin{gathered} b^0_{-m} a^0_m = 1, \\ b^0_{-m-l} a^0_m + \sum_{\substack{|\alpha|+j+k=l \\ k<l}} \frac{1}{\alpha!} [\partial^\alpha_\xi b_{-m-k}] [D^\alpha_x a_{m-j}] = 0 \end{gathered} \tag{1.39}$$

to determine the functions b^0_{m-k}. The first of these equations implies that $b^0_{-m} = (a^0_m)^{-1}$. The other equations enable us to determine by induction all the members of the sum in (1.38), so defining the parametrix to within operators belonging to $L^{-\infty}(\Omega)$.

1.4. Action of Pseudodifferential Operators in Sobolev Spaces and Precise Regularity Theorems for Solutions of Elliptic Equations (Egorov [1984, 1985], Hörmander [1983, 1985], Kumano-go [1982], Shubin [1978], Taylor [1981], Trèves [1980]). A key to the discussion of pseudodifferential operators in Sobolev spaces is the following theorem.

Theorem 1.4. *Suppose that the operator* $A = a(x, D_x) \in L^0(\mathbb{R}^n)$ *has a symbol* $a(x, \xi)$ *that satisfies the estimates*

$$|\partial^\alpha_\xi \partial^\beta_x a(x, \xi)| \leqslant C_{\alpha\beta}(1 + |\xi|)^{-|\alpha|}, \quad x, \xi \in \mathbb{R}^n. \tag{1.40}$$

Then A *can be extended to a continuous operator*

$$A: L_2(\mathbb{R}^n) \to L_2(\mathbb{R}^n).$$

The distinction between the estimates (1.40) and the usual estimates (1.7), which define the class $S^0(\mathbb{R}^n \times \mathbb{R}^n)$ of symbols, consists in the following. The constants $C_{\alpha\beta}$ in (1.40) do not depend on x whereas the estimates (1.7) are satisfied for $x \in K$ with constants $C_{\alpha\beta K}$ which depend on K for any compact set $K \subset \mathbb{R}^n$. In particular, the estimates (1.40) hold for any symbol $a \in S^m(\mathbb{R}^n \times \mathbb{R}^n)$ such that $a(x, \xi) = 0$ for $|x| > R$ or, more generally, $a(x, \xi) = a_\infty(\xi)$ for $|x| > R$; that is, for a symbol that either vanishes or does not depend on x for large x. In particular, the singular integral operators of Examples 1.1 and 1.2 can be extended to bounded operators in $L_2(\mathbb{R}^n)$ provided that the defining functions $a(x)$ and $L(x, y)$ are such that the operator in question reduces to a convolution operator in a neighbourhood of infinity. This means that $a(x) = a_\infty$ for $|x| > R$

and $L(x, y) = L_0$ for $|x| > R$ or $|y| > K$ in Example 1.1 while, in Example 1.2, $L(x, z) = L_\infty(z)$.

One of the possible proofs of Theorem 1.4 is based on the algebraic formalism already developed. To see this, we use the composition theorem, modified for the case where the estimates of symbols are uniform in x, and choose a constant $M > 0$ such that $M > \overline{\lim}_{|\xi|\to\infty} \sup_x |a(x, \xi)|$. We can then construct an operator $B = b(x, D_x) \in L^0(\mathbb{R}^n)$, whose symbol $b(x, \xi)$ also satisfies the estimates of the form (1.40), such that $M^2 = A^*A + B^*B + R$, where $R = r(x, D_x) \in L^{-\infty}$ and the symbol $r(x, \xi)$ also satisfies the estimates of the class $S^{-\infty}(\mathbb{R}^n \times \mathbb{R}^n)$ uniformly in x. It then follows that

$$\|Au\|^2 = M^2\|u\|^2 - \|Bu\|^2 - (Ru, u) \leqslant M^2\|u\|^2 + |(Ru, u)|$$

for $u \in C_0^\infty(\mathbb{R}^n)$. This means that everything reduces to the boundedness of the operator R, and this follows from the Young inequality because the kernel K_R of R is majorised by the kernel of a convolution operator with a rapidly decreasing function.

The same argument shows that if

$$\overline{\lim_{|x|+|\xi|\to\infty}} |a(x, \xi)| = 0 \tag{1.41}$$

under the hypotheses of Theorem 1.4, then the operator A is compact in $L_2(\mathbb{R}^n)$. Indeed, in this case, by performing truncation in a neighbourhood of infinity, we can replace A by a similar operator, with symbol $a(x, \xi)$ which is equal to zero for large $|x|$, that differs from A by an operator with an arbitrarily small norm. Assuming now that $a(x, \xi) = 0$ for large $|x|$, we repeat the above construction. We can then assume that $M > 0$ is as small as we please and that the symbol $r(x, \xi)$ vanishes for large $|x|$. The operator R will then be a Hilbert-Schmidt operator and hence a compact operator. Thus we finally find that

$$\|Au\|^2 \leqslant \varepsilon^2\|u\|^2 + |(R_\varepsilon u, u)|$$

for every $\varepsilon > 0$, where R_ε is a compact operator. This evidently implies that the operator A is itself compact. In particular, A is compact in $L_2(\mathbb{R}^n)$ if $A = a(x, D_x) \in L^m(\mathbb{R}^n)$, where $m < 0$ and $a(x, \xi) = 0$ for large $|x|$.

From Theorem 1.4 we easily have

Theorem 1.4′. *Let the operator* $A = a(x, D_x) \in L^m(\mathbb{R}^n)$ *have a symbol* $a(x, \xi)$ *which satisfies the estimates*

$$|\partial_\xi^\alpha \partial_x^\beta a(x, \xi)| \leqslant C_{\alpha\beta}(1 + |\xi|)^{m-|\alpha|}, \quad x, \xi \in \mathbb{R}^n. \tag{1.42}$$

Then A *can be extended to a continuous linear operator*

$$A\colon H^s(\mathbb{R}^n) \to H^{s-m}(\mathbb{R}^n)$$

for any $s \in \mathbb{R}$.

(Here $H^s(\mathbb{R}^n)$ denotes the standard Sobolev space in $\mathbb{R}^n$; see Egorov and Shubin [1988, § 3, Chap. 2].)

In fact, if we use the operator $(1 - \Delta)^{s/2}$ which is an isomorphism of $H^s(\mathbb{R}^n)$ onto $L_2(\mathbb{R}^n)$, then the proof of Theorem 1.4′ reduces to the boundedness in $L_2(\mathbb{R}^n)$ of the operator $(1 - \Delta)^{(s-m)/2} A (1 - \Delta)^{-s/2}$ which satisfies the hypotheses of Theorem 1.4.

Theorem 1.4′ evidently implies that if $A \in L^m(\Omega)$ then A defines a continuous linear operator

$$A\colon H^s_{\text{comp}}(\Omega) \to H^{s-m}_{\text{loc}}(\Omega).$$

This enables us to derive from the existence of a parametrix of an elliptic operator A of order m a precise regularity theorem for solutions of the corresponding elliptic equation $Au = f$. Thus, we have $u \in H^{s+m}_{\text{loc}}(\Omega)$ if $Au = f \in H^s_{\text{loc}}(\Omega)$. This follows because $u = Bf - Ru$, where B is a parametrix of A and $R \in L^{-\infty}(\Omega)$. Then, by Theorem 1.4′, $Ru \in C^\infty(\Omega)$ and $Bf \in H^{s+m}_{\text{loc}}(\Omega)$ since $B \in L^{-m}(\Omega)$ and B is properly supported. This also implies the local *a priori* estimates

$$\|u\|_{s,\Omega'} \leqslant C(\|Au\|_{s-m,\Omega} + \|u\|_{-N,\Omega}),$$

where Ω' is a subdomain of Ω such that $\bar{\Omega}'$ is a compact set contained in Ω and $\|\cdot\|_{s,\Omega}$ denotes the norm in $H^s(\Omega)$.

We also note that the operator A satisfying the conditions of Theorem 1.4 is bounded in $L_p(\mathbb{R}^n)$ for any $p \in (1, \infty)$ as well as in a Hölder space $C^\gamma(\mathbb{R}^n)$ for any non-integer $\gamma > 0$ (see Egorov and Shubin [1988, §2.13, Chap. 2]). This fact enables us to establish precise theorems on boundedness and regularity as well as *a priori* estimates, like those above, in the scales W^s_p and C^γ.

1.5. Change of Variables and Pseudodifferential Operators on a Manifold (Hörmander [1983, 1985], Kumano-go [1982], Shubin [1978], Taylor [1981], Trèves [1980]). Let us consider an operator $A\colon C_0^\infty(\Omega) \to C^\infty(\Omega)$ in a domain Ω. Let a diffeomorphism $\varkappa\colon \Omega \to \Omega_1$ be given. We introduce the induced map $\varkappa^*\colon C^\infty(\Omega_1) \to C^\infty(\Omega)$ of change of variables by the formula

$$(\varkappa^* f)(x) = (f \circ \varkappa)(x) = f(\varkappa(x));$$

$\varkappa^*$ also maps $C_0^\infty(\Omega_1)$ into $C_0^\infty(\Omega)$. On Ω_1 we define an operator A_1 by the commutative diagram

$$\begin{array}{ccc} C_0^\infty(\Omega) & \xrightarrow{\;A\;} & C^\infty(\Omega) \\ \uparrow{\scriptstyle \varkappa^*} & & \uparrow{\scriptstyle \varkappa^*} \\ C_0^\infty(\Omega_1) & \xrightarrow{\;A_1\;} & C^\infty(\Omega_1) \end{array}.$$

That is, $A_1 u = [A(u \circ \varkappa)] \circ \varkappa_1$, where $\varkappa_1 = \varkappa^{-1}$.

Theorem 1.5. *If $A \in L^m(\Omega)$, then $A_1 \in L^m(\Omega_1)$. Moreover, if $A = a(x, D_x) + R$, where $R \in L^{m-1}(\Omega)$, then $A_1 = a_1(y, D_y) + R_1$, where $R_1 \in L^{m-1}(\Omega_1)$, and the symbol a_1 is defined by the formula*

$$a_1(y, \eta) = a(\varkappa_1(y), ({}^t\varkappa_1'(y))^{-1}\eta). \tag{1.43}$$

Here $\varkappa_1'(y)$ *denotes the Jacobian matrix of the map* $\varkappa_1$ *at the point* y *and* ${}^t\varkappa_1'(y)$ *denotes the transpose of* $\varkappa_1'(y)$. *If* A *is a classical pseudodifferential operator of order* m, *then* A_1 *is also a classical pseudodifferential operator of order* m. *Furthermore, the principal symbol* a_{1m}^0 *of* A_1 *is given by the formula*

$$a_{1m}^0(y, \eta) = a_m^0(\varkappa_1(y), ({}^t\varkappa_1'(y))^{-1}\eta), \tag{1.44}$$

where a_m^0 *is the principal symbol of* A.

To establish this result, we express A in terms of the amplitude $a(x, y, \xi) \in S^m$ by the formula (1.22). This immediately yields

$$A_1 u(x) = (2\pi)^{-n} \iint e^{i(\varkappa_1(x)-y)\cdot\xi} a(\varkappa_1(x), y, \xi) u(\varkappa(y))\, dy\, d\xi.$$

Setting $y = \varkappa_1(z)$, we obtain

$$A_1 u(x) = (2\pi)^{-n} \iint e^{i(\varkappa_1(x)-\varkappa_1(z))\cdot\xi} a(\varkappa_1(x), \varkappa_1(z), \xi) \times |\det \varkappa_1'(z)| u(z)\, dz\, d\xi. \tag{1.45}$$

We now note that, to within an operator with a smooth kernel, we can assume that $a(x, y, \xi) = 0$ for $(x, y) \notin U$, where U is an arbitrarily small neighbourhood of the diagonal in $\Omega \times \Omega$. When x and z are close, we can transform the phase function in the exponent of the exponential function in (1.45) as follows:

$$(\varkappa_1(x) - \varkappa_1(z))\cdot\xi = [\psi(x, z)(x - z)]\cdot\xi = (x - z)[{}^t\psi(x, z)\xi],$$

where $\psi = \psi(x, z)$ is a matrix function which is defined and smooth for x and z close, and is such that $\psi(x, x) = \varkappa_1'(x)$. We now substitute ${}^t\psi(x, z)\xi = \eta$ in (1.45) and obtain

$$A_1 u(x) = (2\pi)^{-n} \iint e^{i(x-z)\cdot\eta} a(\varkappa_1(x), \varkappa_1(z), [{}^t\psi(x, z)]^{-1}\eta) |\det \varkappa_1'(z)| \times |\det[{}^t\psi(x, z)]|^{-1} u(z)\, dz\, d\eta.$$

This shows that $A_1 \in L^m(\Omega_1)$. The formulae (1.43) and (1.44) now follow easily from Theorem 1.1.

We can interpret the formula (1.44) in the following way. We identify Ω and Ω_1 by means of the diffeomorphism $\varkappa$. Then the operator A goes over to A_1. We also identify in a natural manner the tangent bundles $T\Omega$ and $T\Omega_1$, the fibre $T_x\Omega$ being identified with the fibre $T_{\varkappa(x)}\Omega_1$ by means of the linear isomorphism $\varkappa'(x)$. The contangent bundles $T^*\Omega$ and $T^*\Omega_1$ are similarly identified, the fibre $T_x^*\Omega$ being identified with the fibre $T_{\varkappa(x)}^*\Omega_1$ by means of the map ${}^t\varkappa'(x)$: $T_{\varkappa(x)}^*\Omega_1 \to T_x^*\Omega$ which is the dual of the map $\varkappa'(x)$: $T_x\Omega \to T_{\varkappa(x)}\Omega_1$. But then (1.44) implies that the principal symbols of the operators A and A_1 are identified if they are assumed to be defined on the cotangent bundles $T^*\Omega$ and $T^*\Omega_1$, that is, if the argument (x, ξ) in the principal symbol $a_m^0(x, \xi)$ is considered to be a point of $T^*\Omega \cong \Omega \times \mathbb{R}^n$ and, similarly, the argument (y, η) in $a_{1m}^0(y, \eta)$ is considered to be a point of $T^*\Omega_1 \cong \Omega_1 \times \mathbb{R}^n$. Thus we can say that the principal symbol of a classical pseudodifferential operator A is a well-defined function on the cotangent bundle space. Similarly, we can interpret the formula (1.43) by saying that the symbol $a = a(x, \xi) \in S^m$ of the operator $A \in L^m$ is well defined on the cotangent bundle space modulo symbols belonging to S^{m-1}.

Theorem 1.5 enables us to define a *pseudodifferential operator of class* S^m *as well as classical pseudodifferential operators on an arbitrary paracompact* C^∞-*manifold* M. To do this, let us consider an operator

$$A\colon C_0^\infty(M) \to C^\infty(M).$$

For any coordinate neighbourhood $\Omega \subset M$ (not necessarily connected) we define the restriction of A to Ω by the formula

$$A_\Omega = p_\Omega A i_\Omega\colon C_0^\infty(\Omega) \to C^\infty(\Omega),$$

where $i_\Omega\colon C_0^\infty(\Omega) \to C_0^\infty(M)$ is the natural embedding (that is, extension by zero beyond Ω) and $p_\Omega\colon C^\infty(M) \to C^\infty(\Omega)$ is the restriction operator which transforms f into f_Ω. We write $A \in L^m(M)$ or $A \in L^m_{cl}(M)$ if for any coordinate neighbourhood on Ω the restriction A_Ω belongs, respectively, to $L^m(\Omega)$ or $L^m_{cl}(\Omega)$ in the local coordinates on Ω. By Theorem 1.5, the membership of A_Ω to L^m or L^m_{cl} does not depend on the coordinates chosen in Ω. By the same Theorem 1.5, the principal symbol of A is a well-defined function on T^*M.

We note that since any two points x and y in M can be included in the same coordinate neighbourhood Ω (we did not require Ω to be connected), the kernel $K_A = K_A(x, y)$ of A [2] is of class C^∞ off the diagonal in $M \times M$. In other words, $K_A(x, y)$ is smooth for $x \neq y$. Thus the operator $A \in L^m(M)$ is pseudolocal.

The pseudodifferential operators of the classes L^m are defined in a similar fashion in the sections of vector bundles. To do this we have first to introduce matrix pseudodifferential operators of these classes on a domain $\Omega \in \mathbb{R}^n$. These operators are defined in exactly the same way as the usual scalar pseudodifferential operators, the only difference being that the symbol a of the operator A, and the principal symbol a_m of the operator $A \in L^m_{cl}$, must both be matrix functions, in general rectangular. Now $|a(x, \xi)|$ and $|\partial_\xi^\alpha \partial_x^\beta a(x, \xi)|$ denote the norms of the corresponding matrices. Suppose next that there are two smooth vector bundles E and F on the manifold M. Then the classes $L^m(M, E, F)$ and $L^m_{cl}(M, E, F)$ consist of the maps

$$A\colon C_0^\infty(M, E) \to C^\infty(M, F) \tag{1.46}$$

such that the restriction A_Ω of A to any coordinate neighbourhood Ω turns into a matrix pseudodifferential operator of the corresponding class for any choice of trivializations of E and F above Ω. We note that Theorem 1.5 and the composition theorem imply that this result is independent of the choice of the local coordinates and trivializations of the bundles E and F; here $C^\infty(M, F)$ is

---

[2] The kernel K_A can be defined, for example, by choosing a fixed positive smooth density $d\mu$ on M and writing A formally in the form $Au(x) = \int K_A(x, y)u(y)\, d\mu(y)$.

the space of smooth sections of F and $C_0^\infty(M, E)$ is the space of smooth sections of E having compact support. Now let x be the projection of the vector (x, ξ) onto M. For any non-zero $(x, \xi) \in T^*M$, the principal symbol $a_m^0 = a_m^0(x, \xi)$ of the operator $A \in L_{cl}^m(M, E, F)$ defines a linear map of fibres

$$a_m^0(x, \xi)\colon E_x \to F_x. \tag{1.47}$$

Thus altogether we have a bundle map

$$a_m^0\colon \pi_0^* E \to \pi_0^* F, \tag{1.48}$$

where $\pi_0\colon T^*M \backslash 0 \to M$ is the canonical projection of the cotangent bundle space without the zero section onto the base M; π_0^*E and π_0^*F are the induced bundles, with fibres E_x and F_x above each point $(x, \xi) \in T^*M\backslash 0$. An operator A of the form (1.46) is said to be *elliptic* if all its local representatives (obtained by all choices of the coordinate neighbourhood Ω, the coordinates on it and the trivializations $E|_\Omega$ and $F|_\Omega$) are elliptic. These representatives are matrix pseudodifferential operators and their ellipticity means that

$$|a^{-1}(x, \xi)| \leqslant C|\xi|^{-m}, \quad |\xi| \geqslant R, \quad x \in K, \tag{1.49}$$

where $C = C(K)$, $R = R(K)$ and K is an arbitrary compact set in Ω. We note that, in the scalar case, these estimates are equivalent to (1.35). For a classical pseudodifferential operator $A \in L_{cl}^m(M; E, F)$ the ellipticity means that all the maps (1.47) are invertible, that is, the map (1.48) is a bundle isomorphism.

Example 1.3 (*A singular integral operator on a smooth closed curve*). Let Γ be a smooth closed curve in the complex plane. Suppose that Γ is oriented, that is, a direction for going along the curve has been fixed. On Γ we consider an operator $A\colon C^\infty(\Gamma) \to C^\infty(\Gamma)$ defined by the formula

$$Au(z) = a(z)u(z) + \text{v.p.}\frac{1}{\pi i}\int_\Gamma \frac{L(z, w)}{z - w} u(w)\, dw,$$

where $a \in C^\infty(\Gamma)$, $L \in C^\infty(\Gamma \times \Gamma)$ and dw denotes the complex differential of the function $w\colon \Gamma \to \mathbb{C}$ defined by the embedding of Γ in $\mathbb{C}$; the principal value of the integral is understood in the same sense as in Example 1.1. By introducing local coordinates on Γ whose orientation is consistent with that of Γ, we easily find that in any local coordinates the operator A becomes the operator of Example 1.1. Therefore A is a classical pseudodifferential operator of order zero on Γ. We can assume that $u(z)$ is a vector function with N components, and that $a(z)$ and $L(z, w)$ and $N \times N$ matrix functions. Then A becomes a matrix classical pseudodifferential operator of order zero. Its principal symbol is a matrix function $\sigma = \sigma(z, \xi)$ on $T^*\Gamma\backslash 0 = \Gamma \times (\mathbb{R}\backslash 0)$ that is homogeneous in ξ of degree zero, and is given by

$$\sigma(z, \xi) = a(z) - b(z) \operatorname{sgn} \xi, \quad b(z) = L(z, z).$$

The ellipticity condition for A in the scalar case means that $a^2(z) - b^2(z) \neq 0$ for $z \in \Gamma$, while in the matrix case it means that the matrices $a(z) - b(z)$ and $a(z) + b(z)$ are invertible at all points $z \in \Gamma$.

Using a partition of unity, a pseudodifferential operator on a manifold M can be constructed by gluing. Namely, suppose that there is covering of M by the coordinate neighbourhoods; that is, $M = \bigcup_j \Omega_j$. Let $A_j \in L^m(\Omega_j)$ for any j. We construct a partition of unity $1 = \sum_j \varphi_j$ that is subordinate to the given covering, by which we mean that $\varphi_j \in C^\infty(M)$, the sum is locally finite and supp $\varphi_j \subset \Omega_j$. We choose functions $\psi_j \in C^\infty(M)$ such that supp $\psi_j \subset \Omega_j$ and $\psi_j \varphi_j \equiv \varphi_j$, and such that the sum $\sum_j \psi_j$ is also locally finite [3]. We denote by Φ_j and Ψ_j the operators of multiplication by φ_j and ψ_j respectively. Then we can examine the operator $A = \sum_j \Psi_j A_j \Phi_j$. It can be easily shown that $A \in L^m(M)$ and that $A \in L^m_{cl}(M)$ if $A_j \in L^m_{cl}(\Omega_j)$ for any j. Similarly, by using matrix pseudodifferential operators, we can glue a pseudodifferential operator in the bundles. By this procedure we can construct, for example, the parametrix $B \in L^{-m}$ of any elliptic operator of order m on M. This gives

$$B = \sum_j \Psi_j B_j \Phi_j, \tag{1.50}$$

where B_j denotes the parametrix of the operator A_{Ω_j}. Moreover, we have

$$B \circ A = I - R_1, \quad A \circ B = I - R_2, \quad R_j \in L^{-\infty}. \tag{1.51}$$

More precisely, if A is an elliptic operator on $C_0^\infty(M; E)$ into $C^\infty(M; F)$, then B is a properly supported pseudodifferential operator that maps $C_0^\infty(M; F)$ and $C^\infty(M; F)$ into $C_0^\infty(M; E)$ and $C^\infty(M; E)$ respectively, and

$$R_1 \in L^{-\infty}(M; E, E), \quad R_2 \in L^{-\infty}(M; F, F).$$

In the case of a compact manifold M, it is also convenient to introduce the Sobolev section spaces $H^s(M; E)$ that are defined as the spaces of sections belonging to H^s_{loc} in local coordinates on any coordinate neighbourhood $\Omega \subset M$ and for any choice of trivialization of E above Ω. If $R \in L^{-l}(M, E, E)$ and $l > 0$, it follows from the discussion of § 1.4 that R defines a compact linear operator in $H^s(M; E)$ for any $s \in \mathbb{R}$. By the well-known Riesz theorem, the operators $I - R_1$ and $I - R_2$ are Fredholm in the spaces $H^s(M; E)$ and $H^{s-m}(M; F)$ respectively. This result, together with (1.51), implies that the elliptic operator $A \in L^m(M; E, F)$ of order m defines a Fredholm operator

$$A\colon H^s(M; E) \to H^{s-m}(M; F) \tag{1.52}$$

for any $s \in \mathbb{R}$. The kernel Ker A of this operator belongs to $C^\infty(M; E)$ and is therefore independent of s. By using a formally adjoint operator A^*, constructed by means of any smooth density on M and smooth scalar products in fibres of the bundles E and F, we can easily show that the image of A in the space

---

[3] The sums $\sum_j \varphi_j$ and $\sum_j \psi_j$ will automatically be locally finite if the covering $M = \bigcup_j \Omega_j$ is itself locally finite, a property that can always be assumed to hold without loss of generality.

$H^{s-m}(M; F)$ can be described by means of orthogonality relations to a finite number of smooth sections. In this way we find that dim Coker A is also independent of s. Thus we have

Theorem 1.6. *If $A \in L^m(M; E, F)$ is an elliptic operator of order m on a compact manifold M, then A defines a Fredholm operator* (1.52) *for any $s \in \mathbb{R}$ such that both* dim Ker A *and* dim Coker A *are independent of s.*

In particular, the *index* defined by

$$\operatorname{ind} A = \dim \operatorname{Ker} A - \dim \operatorname{Coker} A \tag{1.53}$$

is also independent of s. We note that the index can be understood simply as the index of the operator $A: C^\infty(M; F) \to C^\infty(M; F)$.

We note further that if A is invertible under the hypotheses of Theorem 1.6 (either as an operator from $C^\infty(M; E)$ into $C^\infty(M; F)$ or as an operator (1.52) for any $s \in \mathbb{R}$), this being equivalent to the conditions that Ker $A = 0$ and Ker $A^* = 0$, then A^{-1} is again a pseudodifferential operator belonging to $L^{-m}(M; F, E)$, and it will be classical if the operator A is itself classical. Indeed, multiplying both sides of the second equation in (1.51) from the left by A^{-1}, we obtain

$$A^{-1} = B + A^{-1}R_2. \tag{1.54}$$

Now A^{-1} is a continuous map from $C^\infty(M)$ into $C^\infty(M)$, and therefore $A^{-1}R_2$ is an operator whose kernel $K_{A^{-1}R_2}(x, y) = [A^{-1}K_{R_2}(\cdot, y)](x)$ lies in $C^\infty(M \times M)$. Thus

$$A^{-1} - B \in L^{-\infty}(M), \tag{1.55}$$

which shows that A^{-1} coincides with B to within operators with smooth kernels. We see that the calculus of pseudodifferential operators enables us to describe the structure of the operator A^{-1} and even find it explicitly modulo $L^{-\infty}(M)$. In particular, if A is a classical operator, then the homogeneous components b^0_{-m-k} of the symbol of A^{-1} are given by (1.39).

1.6. Formulation of the Index Problem. The Simplest Index Formulae (Fedosov [1974a], Palais [1965]). According to well-known theorems of functional analysis, for any Fredholm operator $A: H_1 \to H_2$, where H_1 and H_2 are Hilbert spaces, there exists an $\varepsilon > 0$ such that the operator $A + B$ is Fredholm and

$$\operatorname{ind}(A + B) = \operatorname{ind} A$$

provided that the operator $B: H_1 \to H_2$ has the norm $\|B\| < \varepsilon$. In particular, ind A remains unchanged under any deformation of A which is continuous in the operator norm and does not take us out of the class of Fredholm operators. Furthermore, if $A: H_1 \to H_2$ is Fredholm and $T: H_1 \to H_2$ is a compact operator, then $A + T$ is also Fredholm and

$$\operatorname{ind}(A + T) = \operatorname{ind} A.$$

This result and Theorem 1.6 imply that the index of an elliptic operator on a compact manifold depends only on the principal symbol of this operator and remains unchanged under continuous deformations of this principal symbol. Thus the index is a homotopy invariant of the principal symbol, and therefore we can expect that the index can be expressed in terms of the homotopy invariants of the principal symbol. The problem of computing the index of an elliptic operator was formulated in 1960 by Gel'fand and it was solved in the general case in 1963 by Atiyah and Singer (see Palais [1965]). The Atiyah-Singer formula prescribes the construction of a certain differential form based on the symbol of the elliptic operator A, and integration of this form yields the index of A. Without writing down the general formula, we mention two of its special cases which were known before the publication of the Atiyah–Singer work.

A. *The Noether–Muskhelishvili formula.* This formula gives the index of the matrix elliptic singular integral operator of Example 1.3 on a closed oriented curve Γ which, for simplicity, we assume to be connected. The formula is of the form

$$\begin{aligned}\operatorname{ind} A &= \frac{1}{2\pi}\arg\det[\sigma(z,1)^{-1}\sigma(z,-1)]|_{\Gamma}\\ &= \frac{1}{2\pi}\arg\det[(a(z)-b(z))^{-1}(a(z)+b(z))]|_{\Gamma},\end{aligned}\tag{1.56}$$

where the notation of Example 1.3 has been used and $\arg f(z)|_{\Gamma}$ denotes the increment in the argument of $f(z)$ on going round Γ in the chosen direction.

B. *The Dynin–Fedosov formula.* This formula concerns a matrix elliptic operator $A = a(x, D_x)$ of order m in $\mathbb{R}^n$ that coincides in the neighbourhood of infinity with an operator $a_\infty(D_x)$ having a constant symbol $a_\infty(\xi)$ which is invertible for all $\xi \in \mathbb{R}^n$. Such an operator defines a linear Fredholm operator

$$A\colon H^s(\mathbb{R}^n) \to H^{s-m}(\mathbb{R}^n)$$

whose index is independent of $s \in \mathbb{R}$. The general problem of the index of matrix elliptic pseudodifferential operators on a sphere S^n can be easily reduced to the computation of the index of such operators A. The formula for the index is

$$\operatorname{ind} A = \frac{(-1)^{n-1}(n-1)!}{(2\pi i)^n(2n-1)!}\int_{\{(x,\xi):|\xi|=R\}} \operatorname{Tr}[(a^{-1}(x,\xi)\,da(x,\xi))^{2n-1}],\tag{1.57}$$

where $R > 0$ is sufficiently large. Here $a^{-1}\,da$ is understood as the matrix of differential 1-forms on $\mathbb{R}^n_x \times \mathbb{R}^n_\xi$, and $(a^{-1}\,da)^{2n-1}$ denotes the power of this matrix in the computation of which the elements of $a^{-1}\,da$ are multiplied by the wedge product. Thus $(a^{-1}\,da)^{2n-1}$ is the matrix of $(2n-1)$-forms and its trace is the usual $(2n-1)$-form which is further integrated over a $(2n-1)$-dimensional manifold $S_R = \{(x, \xi)\colon |\xi| = R\}$. This manifold is oriented as the boundary of the domain $\{(x, \xi)\colon |\xi| < R\}$ which is itself oriented with the aid of the $2n$-form $(dx_1 \wedge d\xi_1) \wedge \cdots \wedge (dx_n \wedge d\xi_n)$. We note that for $n \geqslant 2$ the matrix

$[a^{-1}(x, \xi)\, da(x, \xi)]^{2n-1}$ vanishes for large x. This is because the form $a^{-1}\, da$ is expressed only in terms of $d\xi_1, \ldots, d\xi_n$ if $a = a_\infty(\xi)$, and any exterior $(2n-1)$-form of n variables vanishes. Therefore, for $n \geqslant 2$, the integration in (1.57) is actually performed over a compact set and consequently makes sense. The integral also makes sense when $n = 1$, because if the 1-form contains only $d\xi$ and not dx, then its restriction to S_R is zero, that is, the integration is again performed over a compact set. For $n = 1$, the formula (1.57) is easily reduced to (1.56) in the preceding example (for the case of a classical pseudodifferential operator A on $\mathbb{R}^1$ which coincides with an invertible pseudodifferential operator $a_\infty(D)$ in a neighbourhood of infinity).

Let us mention two more simple particular cases of the Atiyah–Singer formula.

C. Suppose that A is a scalar elliptic pseudodifferential operator on a compact manifold M of dimension $n \geqslant 2$ where, by the term scalar, we mean that A acts on scalar functions rather than vector functions. Then ind $A = 0$. When M is a sphere, this result is a consequence of (1.57) because the wedge product of a scalar 1-form with itself is always equal to zero.

D. Let A be a differential (and not pseudodifferential !) operator which acts in the sections of vector bundles above an orientable compact manifold M whose dimension is an odd integer. Then ind $A = 0$. When $M = S^n$ (n is odd) and the bundles are trivial, this result can be easily derived from (1.57). We do this by reducing the problem to the case where the principal symbol of the operator (a polynomial matrix that is homogeneous in ξ) occurs in place of a; this is achieved by homotopy. Then we follow the action of the map $\xi \mapsto -\xi$ on the index and the integral.

1.7. Ellipticity with a Parameter. Resolvent and Complex Powers of Elliptic Operators (Agranovich and Vishik [1964], Seeley [1967, 1969a, 1969b], Shubin [1978]). We have already discussed (Egorov and Shubin [1988, § 2, Chap. 2]) for boundary-value problems the ellipticity condition with a parameter that guarantees the unique solvability of the boundary-value problem for large values of the parameter. Here we deal with the similar condition for pseudodifferential operators on a manifold M.

Let Λ be a closed angle with vertex at the origin of the complex plane $\mathbb{C}$, and we allow the possibility that Λ may degenerate into a ray. Let A be a classical pseudodifferential operator of order $m > 0$ on M. Here and in what follows we take A to be a scalar operator for simplicity, although all the results obtained here can be easily extended to the case of operators acting in the sections of the given vector bundle E. Suppose that $a_m^0 = a_m^0(x, \xi)$ is the principal symbol of A. We formulate the following basic condition.

(Ell_Λ) (*The ellipticity condition with a parameter* $\lambda \in \Lambda$)

$$a_m^0(x, \xi) - \lambda \neq 0 \quad \text{for all } (x, \xi) \in T^*M\backslash 0 \quad \text{and} \quad \lambda \in \Lambda.$$

If this condition is satisfied for a compact manifold M, we can construct a parametrix of the operator $A - \lambda I$, with a parameter λ, that is an approximation to the resolvent $(A - \lambda I)^{-1}$, which also exists for sufficiently large $|\lambda|$ when $\lambda \in \Lambda$ and the ellipticity condition (Ell_Λ) is satisfied. If we know the structure of the resolvent, we can construct the complex powers A^z of A and describe their structure. Seeley [1967] proposed a scheme for using pseudodifferential operators for describing the resolvent and the complex powers, and it plays an important role in the spectral theory which is discussed below.

Let us describe Seeley's procedure for constructing the parametrix for the operator $A - \lambda I$ for large values of λ, $\lambda \in \Lambda$. This is done locally in every coordinate neighbourhood and these local parametrices are then glued to obtain the global parametrix by means of the partition of unity as described in § 1.5.

We now proceed to construct the parametrix $B(\lambda)$ of the operator $A - \lambda I$ in a coordinate neighbourhood $\Omega \subset M$ by fixing local coordinates in it. In doing so, we shall identify $T^*\Omega$ and $\Omega \times \mathbb{R}^n$. We first note that the facts that M is compact and the function $a_m^0(x, \xi) - \lambda$ is homogeneous in $(\xi, \lambda^{1/m})$ imply that the condition (Ell_Λ) is equivalent to the condition that

$$|a_m^0(x, \xi) - \lambda| \geqslant \varepsilon > 0 \quad \text{for } \lambda \in \Lambda \quad \text{and} \quad |\xi|^m + |\lambda| = 1. \tag{1.58}$$

This condition (perhaps with a smaller ε) continues to be satisfied if the angle Λ is slightly enlarged. Therefore we can assume that the angle Λ does not degenerate into a ray.

Suppose that the asymptotic expansion of the symbol $a(x, \xi)$ of the operator A in Ω in terms of homogeneous functions is of the form (1.28).

It follows from (1.58) that it is possible to extend the function $a_m^0(x, \xi)$ from the set $\mathscr{L} = \{(x, \xi): x \in \Omega, |\xi| \geqslant 1\}$ to a smooth function $a_m(x, \xi)$ on $\Omega \times \mathbb{R}^n$ that satisfies

$$|a_m(x, \xi) - \lambda| \geqslant \varepsilon_1 > 0 \quad \text{for } (x, \xi) \in \Omega \times \mathbb{R}^n, \lambda \in \Lambda. \tag{1.59}$$

Also, in particular, $a_m \in S^m(\Omega \times \mathbb{R}^n)$. Let $a_{m-j} = a_{m-j}(x, \xi), j = 1, 2, \dots$, similarly denote any extension of the functions $a_{m-j}^0(x, \xi)$ from $\mathscr{L}$ to the functions $a_{m-j} \in C^\infty(\Omega \times \mathbb{R}^n)$. Then automatically $a_{m-j} \in S^{m-j}(\Omega \times \mathbb{R}^n)$.

The construction of the required parametrix proceeds on the same lines as if the function $a_m - \lambda$ is the principal symbol of $A - \lambda I$. We may then expect, on account of the composition theorem, that a good approximation to the resolvent in Ω will be the operator $B(\lambda)$ whose symbol is the sum

$$b(x, \xi, \lambda) = \sum_{k=0}^{K} b_{-m-k}(x, \xi, \lambda), \tag{1.60}$$

where $K > 0$ is sufficiently large and the components b_{-m-k} are found from the following equations:

$$\begin{gathered} b_{-m}(a_m - \lambda) = 1, \\ b_{-m-l}(a_m - \lambda) + \sum_{\substack{|\alpha|+j+k=l \\ k<l}} \frac{1}{\alpha!}[\partial_\xi^\alpha b_{-m-k}][D_x^\alpha a_{m-j}] = 0, \quad l > 0 \end{gathered} \tag{1.61}$$

(compare with (1.39)).

Suppose now that there is a finite covering $M = \bigcup_{j=1}^{N} \Omega_j$ of the manifold M by coordinate neighbourhoods Ω_j. Let $B_j(\lambda)$ be the parametrix of $A - \lambda I$ on Ω_j, constructed as above. We glue the operators $B_j = B_j(\lambda)$ in accordance with (1.50) to obtain $B = B(\lambda)$. This is the required parametrix of $A - \lambda I$. Let us describe its properties.

The fundamental fact that follows by analysing the composition $B(\lambda)(A - \lambda I)$ in the spirit of the proof of the composition theorem consists in the following. There exists an integer $N > 0$, depending on K in (1.60), such that (i) $N \to +\infty$ as $K \to +\infty$; (ii) if $K_R = K_R(x, y, \lambda)$ is the kernel of the operator $R(\lambda) = B(\lambda)(A - \lambda I) - I$, then $K_R \in C^N(M \times M)$ for each fixed λ; (iii) if L is a differential operator on $M \times M$ of order $\leqslant N$ with smooth coefficients, then

$$|LK_R(x, y)| \leqslant C_L|\lambda|^{-1}, \quad \lambda \in \Lambda, \quad |\lambda| \geqslant 1. \tag{1.62}$$

Let us now examine the obvious relation

$$B(\lambda)(A - \lambda I) = I + R(\lambda). \tag{1.63}$$

It follows from (1.62) that

$$\|R(\lambda)\|_{s,t} \leqslant C_N|\lambda|^{-1}, \quad \lambda \in \Lambda, \quad |\lambda| \geqslant 1; \quad s, t \in [-N, N], \tag{1.64}$$

where $\|R\|_{s,t}$ denotes the norm of R as an operator from $H^s(M)$ into $H^t(M)$. In particular, this result implies that the operator $I + R(\lambda)$ is invertible in $H^s(M)$ for $s \in [-N, N]$, $\lambda \in \Lambda$ and $|\lambda| \geqslant R_0$, provided that $R_0 > 0$ is sufficiently large. Moreover, $(I + R(\lambda))^{-1} = I + R_1(\lambda)$, where $R_1(\lambda)$ has the same norm estimates (1.64) as $R(\lambda)$ does. The formula (1.63) now shows that $B_1(\lambda)(A - \lambda I) = I$ if $B_1(\lambda) = (I + R_1(\lambda))B(\lambda)$, that is, $B_1(\lambda)$ is the left inverse of $A - \lambda I$ for $\lambda \in \Lambda$ and $|\lambda| \geqslant R_0$.

We note that the index of $A - \lambda I$ is zero. This is because the ellipticity condition with a parameter implies that the operator $A - \lambda I$ is homotopic (in the class of elliptic operators) to a self-adjoint operator. Therefore its left invertibility implies its invertibility, and

$$(A - \lambda I)^{-1} = (I + R_1(\lambda))B(\lambda) = B(\lambda) + R_1(\lambda)B(\lambda). \tag{1.65}$$

By analysing the proof of boundedness, it can be verified directly that

$$\|B(\lambda)\|_{s,s} \leqslant C_N|\lambda|^{-1}, \quad \lambda \in \Lambda, \quad |\lambda| \geqslant 1, \quad s \in \mathbb{R}. \tag{1.66}$$

Thus

$$\|R_1(\lambda)B(\lambda)\|_{s,t} \leqslant C'_N|\lambda|^{-2}, \quad \lambda \in \Lambda, \quad |\lambda| \geqslant 1; \quad s, t \in [-N, N]. \tag{1.67}$$

This result implies that the operator $T(\lambda) = R_1(\lambda)B(\lambda)$ has a kernel that is as smooth as desired (as $K \to \infty$). The derivatives of the kernel are estimated by $C|\lambda|^{-2}$, $\lambda \in \Lambda$ and $|\lambda| \geqslant R_0$ because the kernel $K_T = K_T(x, y)$ is given, just as in the case of the kernel of any operator with as mooth kernel, by the formula

$$K_T(x, y) = T[\delta(\cdot - y)](x),$$

where $\delta(\cdot - y)$ is a function of y of the class C^l with values in the space $H^{-l-n/2-\varepsilon}$ with $\varepsilon > 0$. Thus it follows from (1.65) and (1.66) that

$$(A - \lambda I)^{-1} = B(\lambda) + T(\lambda), \tag{1.68}$$

where the operator $T(\lambda)$ has a sufficiently smooth kernel that decays as $0(|\lambda|^{-2})$ for $\lambda \in \Lambda$, $|\lambda| \to \infty$, together with any large number of derivatives (depending on K). In this way we have established the desired result regarding the existence of the resolvent $(A - \lambda I)^{-1}$ and its approximate representation to within $0(|\lambda|^{-2})$.

From (1.66), (1.67) and (1.68), we have the following estimate for the norm of the resolvent:

$$\|(A - \lambda I)^{-1}\|_{s,s} \leqslant C_N |\lambda|^{-1}, \quad \lambda \in \Lambda, \quad |\lambda| \geqslant R_0; \quad s \in [-N, N]. \tag{1.69}$$

This is true for any $N > 0$ because we can take the parametrix $B(\lambda)$ to contain as many terms as we like.

It also follows from (1.68) that $(A - \lambda I)^{-1}$ is a classical pseudodifferential operator of order $-m$. This operator is compact in $L_2(M)$. This fact, together with well-known theorems of functional analysis (see Gokhberg and Krejn [1965]), implies that the spectrum $\sigma(A)$ of A in $L_2(M)$ is a discrete set of points with finite multiplicities. Here A should be regarded as an unbounded operator in $L_2(M)$ with domain of definition $H^m(M)$. Further, only a finite number of points of the spectrum can lie in Λ. Therefore by narrowing Λ we can arrange that $\sigma(A) \cap \Lambda$ contains no more than the single point 0. What is more, for the operator $A - \delta_0 I$, with any fixed $\delta_0 \in \mathbb{C} \setminus \sigma(A)$, we find that $0 \notin \sigma(A - \delta_0 I)$. On replacing A by $A - \delta_0 I$ if necessary, we can assume in the sequel that $0 \notin \sigma(A)$, that is, A is an invertible operator. Now, by narrowing Λ if necessary, we can arrange that $\sigma(A) \cap \Lambda = \varnothing$, and this will also be assumed to hold in the sequel.

Under these assumptions, we can construct the complex powers A^z of the operator A. To do this, we choose a ray $L = \{re^{i\varphi_0}: r \geqslant 0\}$ lying in Λ in the complex λ-plane, and construct a contour Γ in the following manner: $\Gamma = \Gamma_1 \cup \Gamma_2 \cup \Gamma_3$, where

$$\lambda = re^{i\varphi_0} \quad (r \text{ varies from } +\infty \text{ to } \rho > 0) \quad \text{on } \Gamma_1,$$

$$\lambda = \rho e^{i\varphi} \quad (\varphi \text{ varies from } \varphi_0 \text{ to } \varphi_0 - 2\pi) \quad \text{on } \Gamma_2$$

and

$$\lambda = re^{i(\varphi_0 - 2\pi)} \quad (r \text{ varies from } \rho \text{ to } +\infty) \quad \text{on } \Gamma_3.$$

The direction along Γ is provided by Γ_1, Γ_2 and Γ_3 in that order (see Fig. 1).

We must choose $\rho > 0$ so small that the disc $\{\lambda: |\lambda| \leqslant \rho\}$ does not intersect the spectrum of A. We now set

$$A_z = \frac{i}{2\pi} \int_\Gamma \lambda^z (A - \lambda I)^{-1} \, d\lambda, \tag{1.70}$$

where $z \in \mathbb{C}$, Re $z < 0$ and λ^z is defined as a holomorphic function of λ in $\mathbb{C} \setminus L$. Thus

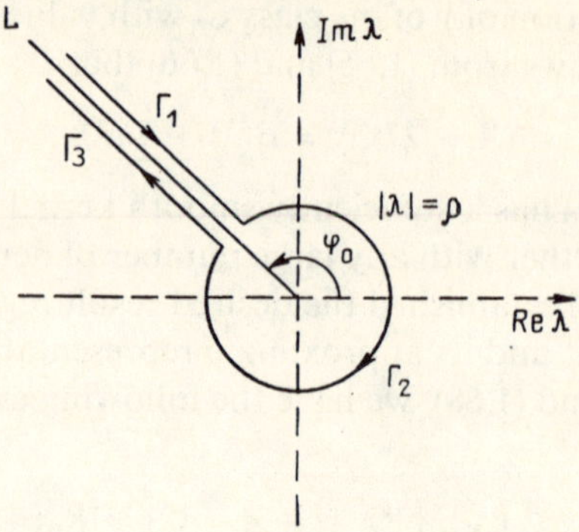

Fig. 1

$$\lambda^z = e^{z \ln \lambda} = e^{z \ln|\lambda| + iz \arg \lambda} = |\lambda|^z e^{iz \arg \lambda},$$

where $\arg \lambda$ is so chosen that $\varphi_0 - 2\pi \leqslant \arg \lambda \leqslant \varphi_0$ and, naturally, we take $\arg \lambda = \varphi_0$ on Γ_1 and $\arg \lambda = \varphi_0 - 2\pi$ on Γ_3.

In view of (1.69), when $\operatorname{Re} z < 0$, the integral in (1.70) converges in the operator norm in the space $H^s(M)$ for any $s \in \mathbb{R}$. We substitute the expression (1.68) for $(A - \lambda I)^{-1}$ into (1.70) and use the estimates (1.67) for the norms of $T(\lambda)$ to obtain

$$A_z = B_z + R_z, \quad B_z = \frac{i}{2\pi} \int_\Gamma \lambda^z B(\lambda)\, d\lambda, \tag{1.71}$$

where the operator R_z has a kernel that is as smooth as desired (depending on K) and depends holomorphically on z. In local coordinates the operator B_z is represented, to within an operator with a smooth kernel that depends analytically on z, in the form $B_z = b^{(z)}(x, D_x)$, where

$$b^{(z)}(x, \xi) = \sum_{k=0}^{K} \frac{i}{2\pi} \int_\Gamma \lambda^z b_{-m-k}(x, \xi, \lambda)\, d\lambda. \tag{1.72}$$

We note that each of the integrals in (1.72) is an integral of rational functions, and this integral can be expanded if we use the expressions for b_{-m-k} obtained from (1.61). For instance, by the Cauchy formula, the principal term of (1.72) has the form

$$\begin{aligned} b_0^{(z)}(x, \xi) &= \frac{i}{2\pi} \int_\Gamma \lambda^z b_{-m}(x, \xi, \lambda)\, d\lambda \\ &= \frac{i}{2\pi} \int_\Gamma \lambda^z (a_m(x, \xi) - \lambda)^{-1}\, d\lambda = a_m^z(x, \xi) \end{aligned}$$

where, in computing the powers a_m^z, we use the same values of the argument a_m as we used in computing λ^z in the integral (1.70). Likewise, the remaining integrals contain rational functions with the only pole $\lambda = a_m(x, \xi)$, and they turn out to be smooth functions of x and ξ and holomorphic functions of z. Furthermore, the fact that the functions $b_{-m-k}(x, \xi)$ are homogeneous of degree $-m-k$ in $(\xi, \lambda^{1/m})$, for $|\xi| \geqslant 1$, implies that

$$b_k^{(z)}(x, \xi) = \frac{i}{2\pi} \int_\Gamma \lambda^z b_{-m-k}(x, \xi, \lambda)\, d\lambda \tag{1.73}$$

is homogeneous in ξ of degree $mz - k$, for $|\xi| \geqslant 1$, and depends holomorphically on z. Therefore B_z is a classical pseudodifferential operator of order mz whose symbol depends holomorphically on z, and hence so is the operator A_z. We note that earlier we examined classical pseudodifferential operators of real order only, but classical pseudodifferential operators of any complex order are defined in an analogous way.

We note that with the aid of the Cauchy formula, we can easily deduce, first, the group property of the operators A_z, namely, that

$$A_z A_w = A_{z+w}, \quad \operatorname{Re} z < 0, \quad \operatorname{Re} w < 0, \tag{1.74}$$

and, secondly, that $A_{-1} = A^{-1}$. Hence also $A_{-k} = A^{-k}$ for any integer $k > 0$. Using these facts, we can correctly define the operators A^z for any $z \in \mathbb{C}$ by

$$A^z = A^k \circ A_{z-k}, \quad k \in \mathbb{Z}, \quad k > \operatorname{Re} z, \quad k \geqslant 0. \tag{1.75}$$

It follows easily from the above properties of A_z that A^z is a classical pseudodifferential operator of order mz whose principal symbol is $[a_m^0(x, \xi)]^z$. All the homogeneous components of the symbol of A^z (in local coordinates) depend holomorphically on z and, moreover,

$$A^z = \sum_{k=1}^{K} b_k^{(z)}(x, D_x) + T_K^{(z)},$$

where the operator $T_K^{(z)}$ has a kernel of class C^N for $\operatorname{Re} z < d_0$ and $N = N(K, d_0) \to +\infty$ as $K \to +\infty$ for any fixed d_0. The derivatives of this kernel up to order N are holomorphic in z in the half-plane $\{z\colon \operatorname{Re} z < d_0\}$. Thus the operators A^z depend, in a natural sense, holomorphically on z.

It is natural to refer to the operators A^z as the *complex powers of a pseudodifferential operator A*. In fact, (1.75) easily implies that for any integer z the operator A^z coincides with the usual integral power of A and, in particular, $A^0 = I$ and $A^1 = A$. Further, $A^z = A_z$ for $\operatorname{Re} z < 0$ and the group property is satisfied for all z:

$$A^z \circ A^w = A^{z+w}, \quad z, w \in \mathbb{C}. \tag{1.76}$$

Finally, suppose that A has an eigenfunction ψ: $A\psi = \lambda\psi$, this implying that $\psi \in C^\infty(M)$ since A is elliptic. Then, by the Cauchy formula, we immediately see that $A_z\psi = \lambda^z\psi$, that is, ψ is an eigenfunction of A^z corresponding to the eigenvalue λ^z. In particular, if the operator A is self-adjoint and positive (under our assumptions, positivity of A implies that we can take the ray $(-\infty, 0]$ for the ray L), then, by examining the values of A^z on a complete orthogonal system of eigenfunctions of A, we find that our definition of A^z coincides with the usual definition in spectral theory.

Thus the calculus of pseudodifferential operators enables us to describe the structure of such important objects of operator calculus as the resolvent and complex powers of elliptic operators on compact manifolds. We shall see later that it plays an equally important role in the theory of boundary-value problems for elliptic equations.

1.8. Pseudodifferential Operators in $\mathbb{R}^n$ (Hörmander [1983, 1985], Kumano-go [1982], Shubin [1978]). Although the Euclidean space $\mathbb{R}^n$ can be regarded as a particular case of a manifold, and thus one can speak about classes L^m of pseudodifferential operators on $\mathbb{R}^n$, there often arise other useful classes of pseudodifferential operators on $\mathbb{R}^n$ which are connected with the additional algebraic structures present on $\mathbb{R}^n$. As mentioned above, we can, for example, consider operators on $\mathbb{R}^n$ with uniform estimates (with respect to x) for symbols (see Theorem 1.4, the remark following the statement of this theorem and also Theorem 1.4′). To make this remark precise, we introduce a class of symbols $S_u^m(\mathbb{R}^n)$ consisting of functions $a = a(x, \xi) \in C^\infty(\mathbb{R}^n \times \mathbb{R}^n)$ which satisfy the uniform estimates (1.42). The corresponding operators $a(x, D_x)$ acting according to the usual formula (1.6) map the space $S(\mathbb{R}^n)$ and the space

$$C_b^\infty(\mathbb{R}^n) = \left\{u\colon u \in C^\infty(\mathbb{R}^n),\ \sup_{x \in \mathbb{R}^n} |\partial^\alpha u(x)| < \infty \quad \text{for all } \alpha\right\}$$

into themselves. In particular, we can form the composition of operators of this class without the requirement that one of the multiplying operators is properly supported. Furthermore, the composition theorem, Theorem 1.2, remains valid in this class as do the formulae (1.30) and (1.31) which define the symbols of the transpose and adjoint operators, the asymptotic expansions being understood up to symbols belonging to the classes $S_u^{-N}(\mathbb{R}^n)$, where $N \to \infty$. We can also use the amplitudes $a(x, y, \xi)$ which satisfy the estimates

$$|\partial_\xi^\alpha \partial_x^{\beta_1} \partial_y^{\beta_2} a(x, y, \xi)| \leqslant C_{\alpha\beta_1\beta_2}(1 + |\xi|)^{m-|\alpha|}, \quad x, y, \xi \in \mathbb{R}^n. \tag{1.77}$$

The operators with such amplitudes can also be defined by the symbol $\sigma_A \in S_u^m(\mathbb{R}^n)$, and Theorem 1.1 remains valid. If the operator $A = a(x, D_x)$ with symbol $a \in S_u^m(\mathbb{R}^n)$ is *uniformly elliptic*, that is, if there exist $\varepsilon > 0$ and $R > 0$ such that

$$|a(x, \xi)| \geqslant \varepsilon |\xi|^m, \quad x \in \mathbb{R}^n, \quad |\xi| \geqslant R, \tag{1.78}$$

then we can find the parametrix of A, this being an operator $B = b(x, D_x)$ with symbol $b \in S_u^{-m}(\mathbb{R}^n)$ such that $BA - I$ and $AB - I$ have symbols belonging to $S_u^{-\infty}(\mathbb{R}^n) = \bigcap_m S_u^m(\mathbb{R}^n)$. We note in passing that this statement does not imply in any way that the operator A is Fredholm. This is because the operators with symbols in $S_u^{-\infty}(\mathbb{R}^n)$ are not necessarily compact in $L_2(\mathbb{R}^n)$. For example, this set contains, among others, all those operators $a(D)$ with constant symbols $a =$

$a(\xi) \in S(\mathbb{R}^n)$ which, on applying the Fourier transformation, go over to the operators of multiplication by a function. A uniformly elliptic operator A in $\mathbb{R}^n$ can have an infinite-dimensional kernel. One such operator, for example, is the operator $a(D)$ with the elliptic symbol $a = a(\xi)$ that vanishes for $|\xi| \leqslant 1$.

A vital fact which simplifies working with pseudodifferential operators in $\mathbb{R}^n$ is that there is a one-one correspondence between the operators and the symbols: the symbol σ_A is given in terms of A by the formula

$$\sigma_A(x, \xi) = e^{-ix\cdot\xi} A(e^{ix\cdot\xi}), \tag{1.79}$$

where the operator A is applied with respect to x and ξ plays the role of a parameter. There are also other ways of establishing the one-one correspondence between the operators and the symbols in $\mathbb{R}^n$. One of these possibilities, already mentioned in § 1.3, involves using amplitudes of the form $a(y, \xi)$. This correspondence is dual to the one just mentioned, and has totally analogous properties that are established most easily by going over to the transpose or adjoint operators. It is more interesting to use the *Weyl symbols*, that is, amplitudes of the form $a\left(\frac{x+y}{2}, \xi\right)$. Thus, we denote by $a^w(x, D_x)$ an operator defined by the amplitude $a\left(\frac{x+y}{2}, \xi\right)$, that is,

$$a^w(x, D_x)u(x) = (2\pi)^{-n} \int e^{i(x-y)\cdot\xi} a\left(\frac{x+y}{2}, \xi\right) u(y)\, dy\, d\xi. \tag{1.80}$$

The function $a = a(x, \xi)$ is known as the *Weyl symbol* of the operator $a^w(x, D_x)$.

An important property of the Weyl symbol which distinguishes it from the usual symbol is that the transition to the adjoint operator is simple. Thus, if A has the Weyl symbol $a(x, \xi)$, then the formal adjoint operator A^* in $L_2(\mathbb{R}^n)$, is defined by the Weyl symbol $\overline{a(x, \xi)}$, the complex conjugate of a. In particular, if the symbol a is real valued, then the corresponding operator $A = a^w(x, D_x)$ is formally self-adjoint.

The composition formula for the Weyl symbols assumes the following form. If $A = a^w(x, D_x)$ and $B = b^w(x, D_x)$, where $a \in S_u^{m_1}(\mathbb{R}^n)$ and $b \in S_u^{m_2}(\mathbb{R}^n)$, then $A \circ B = C = c^w(x, D_x)$, where the function $c(x, \xi)$ has the asymptotic expansion

$$c(x, \xi) \sim \sum_{\alpha,\beta} \frac{(-1)^{|\beta|}}{\alpha!\,\beta!} 2^{-|\alpha+\beta|} (\partial_\xi^\alpha D_x^\beta a)(\partial_\xi^\beta D_x^\alpha b). \tag{1.81}$$

We note that the principal terms in this formula corresponding to the pairs (α, β), with $|\alpha + \beta| \leqslant 1$, are of the form

$$a\cdot b - \frac{i}{2}\{a, b\},$$

where

$$\{a, b\}(x, \xi) = \sum_{j=1}^n \left(\frac{\partial a}{\partial \xi_i}\frac{\partial b}{\partial x_j} - \frac{\partial a}{\partial x_j}\frac{\partial b}{\partial \xi_j}\right)$$

is the *Poisson bracket* of the functions a and b.

The connection between the Weyl symbol and the usual symbol of a given operator is easily established by Theorem 1.1. Thus, the usual symbol σ_A is expressed in terms of the Weyl symbol σ_A^w by the formula

$$\sigma_A(x, \xi) \sim \sum_\alpha \frac{1}{\alpha!} 2^{-|\alpha|} \partial_\xi^\alpha D_x^\alpha \sigma_A^w(x, \xi), \tag{1.82}$$

while the inverse correspondence following from this result is

$$\sigma_A^w(x, \xi) \sim \sum_\alpha \frac{1}{\alpha!} 2^{-|\alpha|} \partial_\xi^\alpha (-D_x)^\alpha \sigma_A(x, \xi). \tag{1.83}$$

The Weyl symbol is also often referred to as a *symmetric symbol*. The meaning of this terminology can be easily understood: the operator with the Weyl symbol $a = x_j \xi_j$ is $\frac{1}{2}(x_j D_j + D_j x_j)$ while the operator with the usual symbol $x_j \xi_j$ is $x_j D_j$ and the operator with the amplitude $y_j \xi_j$ is $D_j x_j$.

The problem of establishing a correspondence between functions on the phase space $\mathbb{R}_x^n \times \mathbb{R}_\xi^n$ and operators arises in a natural way in quantum mechanics, where such a correspondence is known as *quantization*. The presence of different kinds of symbols reflects the fact that quantization is, in principle, not unique. In particular, the correspondence $a \leftrightarrow a^w(x, D_x)$ between the operators and their Weyl symbols is often referred to as the *Weyl quantization*.

In examining the transition to classical mechanics from quantum mechanics in the problem of quantization, the presence of a small parameter h, known as the Planck constant, has to be taken into account. This parameter usually also appears in the quantization under which the momentum operator $\frac{h}{i}\frac{\partial}{\partial x_j}$ corresponds to the functions ξ_j. In view of this situation, one can associate with the function $a(x, \xi)$ the operator $a(x, hD_x)$ or the operator $a^w(x, hD_x)$ depending on the quantization chosen. It can easily be seen, for example, that

$$\begin{aligned} a^w(x, hD_x)u(x) &= (2\pi)^{-n} \int e^{i(x-y)\cdot\xi} a\left(\frac{x+y}{2}, h\xi\right) u(y)\, dy\, d\xi \\ &= (2\pi h)^{-n} \int e^{(i(x-y)\cdot\xi)/h} a\left(\frac{x+y}{2}, \xi\right) u(y)\, dy\, d\xi. \end{aligned} \tag{1.84}$$

The function $a(x, \xi)$ is known as the *Weyl h-symbol* of the operator $a^w(x, hD_x)$. The various formulae of operator calculus for h-symbols usually have the form of asymptotic expansions in powers of h. For instance, the composition theorem assumes the following form. If the operators A and B have Weyl h-symbols $a(x, \xi)$ and $b(x, \xi)$, then the operator $C = A \circ B$ has a Weyl h-symbol $c = c(h, x, \xi)$ (in general, depending explicitly on h) such that

$$c \sim \sum_{\alpha, \beta} \frac{(-1)^{|\beta|}}{\alpha! \beta!} \left(\frac{h}{2}\right)^{|\alpha+\beta|} (\partial_\xi^\alpha D_x^\beta a)(\partial_\xi^\beta D_x^\alpha b). \tag{1.85}$$

The precise meaning of this expansion is as follows. If $a \in S_u^{m_1}(\mathbb{R}^n)$ and $b \in S_u^{m_2}(\mathbb{R}^n)$, then

$$h^{-N}\left[c - \sum_{|\alpha+\beta|\leqslant N-1} \frac{(-1)^{|\beta|}}{\alpha!\,\beta!}\left(\frac{h}{2}\right)^{|\alpha+\beta|}(\partial_\xi^\alpha D_x^\beta a)(\partial_\xi^\beta D_x^\alpha b)\right] \in S_u^{m_1+m_2-N/2}$$

uniformly in h, where $0 < h \leqslant 1$.

The classes of symbols $S_u^m(\mathbb{R}^n)$ are described in such a way that the variables x and ξ do not enjoy equal status, and this situation is unnatural from the point of view of quantum mechanics. This deficiency can be overcome in a number of ways. The simplest of these is to consider for any ρ with $0 < \rho \leqslant 1$ the class G_ρ^m of symbols consisting of the functions $a = a(y) = a(x, \xi) \in C^\infty(\mathbb{R}^{2n})$ which satisfy the estimates

$$|\partial^\gamma a(y)| \leqslant C_\gamma(1 + |y|)^{m-\rho|\gamma|}, \tag{1.86}$$

where $C_\gamma = C_\gamma(a)$. This class (even with $\rho = 1$) contains all the polynomials in $y = (x, \xi)$. We can consider operators with the usual symbols or with Weyl symbols (or h-symbols) belonging to G_ρ^m. In particular, every differential operator with polynomial coefficients can be written as an operator with the usual symbol or Weyl symbol (or h-symbol) belonging to G_ρ^m. In the classes of operators just described the basic theorems of the calculus of pseudodifferential operators hold. For example, the composition theorem holds for the usual symbols or Weyl symbols (or h-symbols). However, we note that for $m < 0$ the operators with symbols belonging to G_ρ^m are compact in $L_2(\mathbb{R}^n)$, and this fact enables us to develop the theory in these classes in the same way as for a compact manifold. Further, as an analogue of $C^\infty(M)$ (with M a compact manifold) we have here the Schwartz space $S(\mathbb{R}^n)$. For instance, if $a \in G_\rho^m$ and $|a(y)| \geqslant \varepsilon|y|^m$ for $|y| \geqslant R$ (an analogue of the ellipticity!), then the inclusion $a(x, D_x)u = f \in S(\mathbb{R}^n)$ (or $a^w(x, D_x)u = f \in S(\mathbb{R}^n)$) and the *a priori* assumption $u \in S'(\mathbb{R}^n)$ imply that $u \in S(\mathbb{R}^n)$. Further, under the same condition, the operators $a(x, D_x)$ and $a^w(x, D_x)$ are Fredholm in $S(\mathbb{R}^n)$ and $S'(\mathbb{R}^n)$. The simplest example of an operator that satisfies the above ellipticity condition is the quantum mechanical energy operator of the harmonic oscillator $H = \frac{1}{2}[-\Delta + |x|^2]$ $\left(\text{or } H = -\dfrac{h^2}{2}\Delta + \frac{1}{2}|x|^2\right)$.

Thus the classes G_ρ^m enable us to take into account, apart from the smoothness, the behaviour of functions at infinity also. They prove to be useful in considering problems where the basic effects are localized. But in those problems where translation invariance is vital (for example, in studying operators with almost-periodic coefficients), the classes $S_u^m(\mathbb{R}^n)$ prove to be more convenient.

A detailed account of the above pseudodifferential operators in $\mathbb{R}^n$ and their wider classes can be found in the monographs by Hörmander [1983, 1985], Kumano-go [1982], and Shubin [1978].

§ 2. Singular Integral Operators and their Applications. Calderon's Theorem. Reduction of Boundary-value Problems for Elliptic Equations to Problems on the Boundary

2.1. Definition and Boundedness Theorems (Agranovich [1965], Bers, John, Schechter [1964], Calderon and Zygmund [1952], Mikhlin [1962], Mizohata [1965]). By a *singular integral operator* we mean an operator

$$A\colon u(x) \mapsto \int K(x, x-y)u(y)\,dy, \quad x \in \mathbb{R}^n, \quad y \in \mathbb{R}^n, \tag{2.1}$$

where the function $K(x, z)$ has a singularity for $z = 0$ only, and

$$K(x, tz) = t^{-n}K(x, z) \quad \text{for } t > 0,\, z \in \mathbb{R}^n\backslash 0, \tag{2.2}$$

$$\int_{|z|=1} K(x, z)\,dS = 0, \quad x \in \mathbb{R}^n. \tag{2.3}$$

In this case, the integral (2.1) can be defined in the sense of the principal value (v.p.), that is,

$$Au(x) = \lim_{\substack{\varepsilon \to +0 \\ R \to \infty}} \int_{\varepsilon < |x-y| < R} K(x, x-y)u(y)\,dy.$$

The existence of the integral can be established easily if, for example, $K(x, z)$ is bounded for $x \in \mathbb{R}^n$, $|z| = 1$ $u \in C_0(\mathbb{R}^n)$.

Example 2.1. The *Hilbert transformation*

$$Hu(x) = \frac{1}{\pi i}\,\text{v.p.} \int \frac{u(y)}{x-y}\,dy.$$

The boundedness of this operator in $L_2(\mathbb{R})$ is a consequence of the Parseval identity, because the transition to Fourier transforms yields $\widetilde{Hu}(\xi) = \operatorname{sgn} \xi \cdot \tilde{u}(\xi)$, and so $\|Hu\|_{L_2} = \|u\|_{L_2}$. Thus H is even an isometry. It can be shown that this operator is bounded in $L_p(\mathbb{R}^n)$ for all $p > 1$.

Theorem 2.1 (Calderon and Zygmund [1952]; see also Bers, John and Schechter [1964]). *The operator A defined by* (2.1) *is bounded in $L_p(\mathbb{R}^n)$ for $p > 1$ if*

$$\int_{|z|=1} |K(x, z)|^q\,dz < C \quad \textit{for } x \in \mathbb{R}^n,$$

where the number q is such that $p^{-1} + q^{-1} = 1$.

Theorem 2.2 (Hölder, Korn, Lichtenshtein and Giraud; see Bers, John and Schechter [1964]). *If $K \in C^1(\mathbb{R}^n \times S^{n-1})$, then the operator A defined by* (2.1) *is bounded in $C^\gamma(\mathbb{R}^n)$ for $\gamma \in (0, 1)$.*

2.2. Smoothness of Solutions of Second-order Elliptic Equations (Bers, John and Schechter [1964]). We first assume that Ω is a bounded domain in $\mathbb{R}^n$ with smooth boundary. Let $\Delta u = f$ in Ω. A well-known formula (see (2.9) in Egorov and Shubin [1988]; § 2, Chap. 2) enables us to express u in the form $u = u_1 + u_2$, where

$$u_1(x) = \int_\Omega E(x - y)f(y)\,dy,$$

and u_2 is expressed in terms of an integral over the boundary $\partial\Omega$. For $x \in \Omega$, the function u_2 is always infinitely differentiable, and so u has the same smoothness inside Ω as u_1. If $\alpha = (\alpha_1, \ldots, \alpha_n)$ and $|\alpha| = 2$, then $D^\alpha u_1$ is given by the singular integral

$$D^\alpha u_1(x) = \int_\Omega D^\alpha E(x - y)f(y)\,dy.$$

Applying Theorems 2.1 and 2.2, we find that the following statements hold.

1°. If $\Delta u \in L_p(\Omega)$ and $p \in (1, \infty)$, then $D^\alpha u \in L_p(\Omega')$, where $|\alpha| = 2$ and $\Omega' \Subset \Omega$.

2°. If $\Delta u \in C^\gamma(\Omega)$ and $\gamma \in (0, 1)$, then $D^\alpha u \in C^\gamma(\Omega')$, where $|\alpha| = 2$ and $\Omega' \Subset \Omega$.

These results remain valid, and are also proved by Theorems 2.1 and 2.2, for general elliptic operators with smooth coefficients. Thus, if $P(x, D)$ is an elliptic differential operator of order m with smooth coefficients, then the following assertions hold.

1°. If $P(x, D)u \in L_p(\Omega)$ and $p \in (1, \infty)$, then $D^\alpha u \in L_p(\Omega')$, where $|\alpha| = m$ and $\Omega' \Subset \Omega$.

2°. If $P(x, D)u \in C^\gamma(\Omega)$ and $\gamma \in (0, 1)$, then $D^\alpha u \in C^\gamma(\Omega')$, where $|\alpha| = m$ and $\Omega' \Subset \Omega$.

2.3. Connection with Pseudodifferential Operators (Agranovich [1965], Hörmander [1983, 1985], Taylor [1981]). A pseudodifferential operator $P(x, D)$, with symbol $p(x, \xi)$, in a domain $\Omega \subset \mathbb{R}^n$ can be expressed as an integral operator by the formula

$$P(x, D)u(x) = \int K(x, x - y)u(y)\,dy,$$

the kernel of which is the distribution

$$K(x, z) = (2\pi)^{-n} \int p(x, \xi)e^{i\xi z}\,d\xi.$$

When p is a positive homogeneous function in ξ of degree zero such that $p \in C^\infty(\Omega \times (\mathbb{R}^n\backslash 0))$, the kernel K satisfies the following conditions:

a) $K(x, z) \in C^\infty(\Omega \times (\mathbb{R}^n\backslash 0))$;

b) $K(x, tz) = t^{-n}K(x, z)$, $t > 0$, $x \in \Omega$, $z \in \mathbb{R}^n\backslash 0$

(more precisely, this relation is valid for all x and z if it is understood in the sense of distributions);

c) $\int_{|z|=1} K(x, z)\, dS_z = 0, x \in \Omega$.

It can be shown easily that these conditions lead to the relation

$$K(x, z) = K_0(x)\delta(z) + K_1(x, z),$$

where $\int_{|z|=1} K_1(x, z)\, dS_z = 0$. Thus $P(x, D)u(x) = K_0(x)u(x) + K_1 u(x)$, where K_1 is a singular integral operator.

Thus Theorems 2.1 and 2.2 imply that pseudodifferential operators of order zero are bounded in the spaces L_p $(1 < p < \infty)$ and C^γ $(0 < \gamma < 1)$. An easy consequence of this result is the following theorem.

Theorem 2.3. *Let $P(x, D)$ be a classical pseudodifferential operator of order m whose symbol satisfies the estimates* (1.7) *in which the constants $C_{\alpha,\beta}$ are independent of $x \in \mathbb{R}^n$. Then it is a bounded operator from the Sobolev space $W_p^k(\mathbb{R}^n)$ into $W_p^{k-m}(\mathbb{R}^n)$ for $p \in (1, \infty)$, $k \in \mathbb{Z}$, $k \geqslant m$, as well as an operator from the space $C^{k+\gamma}(\mathbb{R}^n)$ into $C^{k-m+\gamma}(\mathbb{R}^n)$ for $\gamma \in (0, 1)$, $k \in \mathbb{Z}$, $m \in \mathbb{Z}$ and $k \geqslant m \geqslant 0$.*

We note that in the general case, the kernel $K(x, z)$ of a pseudodifferential operator is a C^∞-function for $z \neq 0$, and for $z = 0$ it has a singularity which is not greater than that of the function $|z|^{-m-n-1}$, where m is the order of the operator in question. This is because the function $z^\alpha K(x, z)$ is continuous in x and z for $|\alpha| \geqslant m + n + 1$.

2.4. Diagonalization of Hyperbolic System of Equations (Mizohata [1965]). Let

$$L = \frac{\partial}{\partial t} I - \sum_{k=1}^{n} A_k(t, x) \frac{\partial}{\partial x_k}$$

be a strictly hyperbolic system of N equations with smooth coefficients. Thus the roots $\lambda_1, \ldots, \lambda_N$ of the characteristic equation

$$\det\left(\lambda I - \sum_{k=1}^{n} A_k(t, x)\xi_k\right) = 0$$

are real and distinct if $\xi \in \mathbb{R}^n \backslash 0$. We write

$$\sum_{k=1}^{n} A_k(t, x) \frac{\partial}{\partial x_k} = iH(t)\Lambda,$$

where $H(t)$ is a singular integral operator with symbol

$$\sigma(H(t)) = \sum_{k=1}^{n} A_k(t, x) \frac{\xi_k}{|\xi|},$$

and Λ is an operator with symbol $|\xi|$. The matrix $\sigma(H)$ has distinct real eigenvalues. Therefore the corresponding eigenvectors are real and linearly independent. Let $N_1(t)$ be an $N \times N$ square matrix whose rows are these eigenvectors normalized to have unit length. Then

$$N_1(t)\sigma(H) = D_1(t)N_1(t),$$

where $D_1(t)$ is the diagonal matrix with $\lambda_1, \ldots, \lambda_N$ along the diagonal. We denote by $N(t)$ and $D(t)$ the pseudodifferential operators with symbols $N_1(t)$ and $D_1(t)$ respectively. Changing the variables by formula $v = N(t)u$, we find that the system $Lu = f$ goes into a system of the form

$$\frac{\partial v}{\partial t} - iD(t)\Lambda v + Bv = Nf,$$

where B is a bounded operator in $\mathring{H}^s(\Omega)$ for each bounded domain $\Omega \subset \mathbb{R}^n$. The principal part of this system is in diagonal form, and this part decouples into separate equations. Such a diagonalization enables us, for example, to obtain energy estimates for the solution of the Cauchy problem, and to establish and uniqueness and existence of the solution of this problem. Further, the above diagonalization simplifies to a great extent the investigation of the well-posedness of boundary-value problems for the system $Lu = f$.

2.5. Calderon's Theorem (Calderon [1958], Hörmander [1983, 1985], Taylor [1981], Trèves [1980]). One of the most remarkable results of the general theory of partial differential equations is the proof of the uniqueness of solutions to the non-characteristic Cauchy problem for a general system of equations with smooth but not necessarily analytic coefficients. This result was obtained by Calderon with the aid of singular integral operators.

Let $P = p(t, x, D_t, D_x)$ be a differential operator of order m with principal symbol $p_0(t, x, \tau, \xi)$. Assume that $p_0(0, 0, 1, 0) \neq 0$ so that the plane $t = 0$ is non-characteristic at the origin.

Theorem 2.4. *Suppose that the equation $p_0(t, x, \tau, \xi) = 0$, with $|\xi| = 1$ and $\xi \in \mathbb{R}^n$ and $(t, x) \in \omega$, where ω is a neighbourhood of the origin in $\mathbb{R}^{n+1}$, has roots τ whose multiplicity does not exceed 2 and this multiplicity remains constant. Suppose that the condition $|\operatorname{Im} \tau| \geqslant c > 0$ holds for the roots of multiplicity 2 while for the simple rots τ let either $\operatorname{Im} \tau \geqslant 0$ or $\operatorname{Im} \tau \leqslant -c < 0$. Then the solution v of the equation $Pv = 0$ in ω, which equals zero if $t < 0$, vanishes everywhere in a neighbourhood of the origin in $\mathbb{R}^{n+1}$.*

The same theorem also holds for system of equations satisfying analogous conditions.

The main point of the proof lies in reducing the equation $Pv = 0$ to an equivalent system of first-order pseudodifferential equations and further diagonalization (more precisely, reduction to the Jordan normal form) of this system.

Let us demonstrate how this reduction can be performed. Let $Pu = f$. We set

$$u_j = \Lambda^{m-j} D_t^{j-1} u, \quad j = 1, \ldots, m,$$

where Λ is a pseudodifferential operator in the variables x with symbol $(1 + |\xi|^2)^{1/2}$, that is, $\Lambda = (I - \Delta_x)^{1/2}$. The Holmgren transformation $t \to t - \delta|x|^2$, where δ (>0) is a constant, maps the plane $t = 0$ onto the surface $t = \delta|x|^2$. We can therefore assume that $u(t, x) = 0$ when $t < \delta|x|^2$, and thus $u(t, x) = 0$ outside some ball for each small $t > 0$. Then

$$D_t u_j - \Lambda u_{j+1} = 0 \quad \text{for } j = 0, 1, \dots, m-1,$$

$$D_t u_m + \sum_{j=1}^{m} Q_j(t, x, D_x)\Lambda^{1-j} u_{m-j+1} = f,$$

where Q_j is a differential operator of order j. It is important that the characteristic equation of this system coincides with the equation $p_0(t, x, \tau, \xi) = 0$. The resulting system of first-order equations is reduced to the Jordan normal form in the principal part by the same method as the one used in § 2.4. After this reduction, it is a question of obtaining *a priori* estimates for the solution of the problem

$$D_t u - [A(t) + iB(t)]u = f, \quad u = 0 \quad \text{for } t \leqslant 0,$$

or of

$$D_t u - [A(t) + iB(t)]u = f, \quad u = 0 \quad \text{for } t < 0,$$

$$D_t v - [A(t) + iB(t)]v + \Lambda u = g, \quad v = 0 \quad \text{for } t < 0,$$

where $A(t)$ and $B(t)$ are first-order pseudodifferential operators with real symbols. The desired estimates are obtained quite simply by integration by parts (see Calderon [1958], Egorov [1984], Nirenberg [1973]).

2.6. Reduction of the Oblique Derivative Problem to a Problem on the Boundary (Egorov [1984], Hörmander [1983, 1985], Trèves [1980]). Using the technique of pseudodifferential operators, we can reduce the boundary-value problem for an elliptic system of differential equations to an equivalent system of pseudodifferential equations on a boundary manifold.

We start with a simple example. Let us consider the Laplace equation

$$\left(\frac{\partial^2}{\partial t^2} + \Delta_x\right) u = 0 \tag{2.4}$$

in the half-space $\mathbb{R}_+^{n+1} = \{(t, x) \in \mathbb{R}^{n+1}, t > 0\}$ with boundary conditions

$$a_0 \frac{\partial u}{\partial t} + \sum_{j=1}^{n} a_j \frac{\partial u}{\partial x_j} = f(x) \quad \text{for } t = 0. \tag{2.5}$$

We assume that $a_j \in C^\infty(\mathbb{R}^n)$ $(0 \leqslant j \leqslant n)$ and that $\sum_{j=0}^{n} |a_j| \neq 0$. Every solution of (2.4) belonging to the class $C(\mathbb{R}_+, L_2(\mathbb{R}^n))$ can be expressed as the Poisson integral

$$u(t, x) = (2\pi)^{-n} \iint v(y) e^{-t|\xi| + i(x-y)\cdot\xi}\, dy\, d\xi,$$

where $v(y) = u(0, y)$. This leads to

Theorem 2.5. *The problem* (2.4)–(2.5) *is equivalent to the pseudodifferential equation*

$$Av = f(x), \quad x \in \mathbb{R}^n, \tag{2.6}$$

where

$$A(x, D)v(x) = (2\pi)^{-n} \int \left[-a_0(x)|\xi| + i \sum_{j=1}^{n} a_j(x)\xi_j \right] \tilde{v}(\xi) e^{ix\cdot\xi}\, d\xi.$$

If $a_0(x) \neq 0$ in $\mathbb{R}^n$, then the symbol

$$a(x, \xi) = -a_0(x)|\xi| + i \sum_{j=1}^{n} a_j(x)\xi_j$$

of the operator A does not vaish for $|\xi| = 1$, and therefore the equation (2.6) is elliptic. For example, in this case the estimate

$$\|v\|_s \leqslant C(\|f\|_{s-1} + \|v\|_{s-1}), \quad v \in C_0^\infty(\mathbb{R}^n),$$

holds, and this immediately gives an estimate for the solution of the problem (2.4)–(2.5). Namely,

$$\|u\|_{s+1/2} \leqslant C(\|f\|_{s-1} + \|u\|_{s-1/2}), \quad u \in H^s(\mathbb{R}_+^{n+1}).$$

Here the norms of u denote the norms in the Sobolev spaces in $\mathbb{R}_+^{n+1}$ and for f the norm in $H^{s-1}(\mathbb{R}^n)$.

2.7. Reduction of the Boundary-value Problem for the Second-order Equation to a Problem on the Boundary (Egorov [1984], Hörmander [1983, 1985], Trèves [1980]). Let us show how to reduce a boundary-value problem for the general second-order elliptic equation

$$Lu = f$$

in a domain $\Omega \subset \mathbb{R}^n$ to a problem on the boundary, where

$$L = \sum_{j,k=1}^{n} a_{jk} D_j D_k + \sum_{j=1}^{n} b_j D_j + c$$

is an elliptic operator with real smooth coefficients. We fix a point P_0 on the boundary Γ of Ω, assuming that this boundary is infinitely differentiable. Using a smooth mapping, we transform the part of the boundary in the neighbourhood of P_0 into a part of the plane $x_n = 0$ in such a way that the image of Ω falls in the half-space $x_n > 0$. Suppose that $u \in C_0^\infty(\mathbb{R}_+^n)$ and that the intersection of supp u with the image of Γ lies in the plane $x_n = 0$. Let u^0 denote the function which coincides with u if $x_n \geqslant 0$ and is zero if $x_n < 0$. Then

$$D_n u^0 = (D_n u)^0 + \frac{1}{i}\delta(x_n) \otimes u_0(x'),$$

where u_0 is the trace of u when $x_n = 0$, and $(D_n u)^0 = D_n u$ if $x_n \geqslant 0$ and is zero if $x_n < 0$. Likewise,

$$D_n^2 u^0 = (D_n^2 u)^0 + \frac{1}{i}\delta(x_n) \otimes u_1(x') - \delta'(x_n) \otimes u_0(x'),$$

where u_1 is the trace of $D_n u$ when $x_n = 0$. Suppose that after the above transformation the operator L takes the form

$$L = \sum_{j,k=1}^{n} A_{jk} D_j D_k + \sum_{j=1}^{n} B_j D_j + C.$$

Then

$$L(u^0) = (Lu)^0 + A_{nn}\left(\frac{1}{i}\delta(x_n) \otimes u_1(x') - \delta'(x_n) \otimes u_0(x')\right)$$

$$+ 2 \sum_{j=|1|}^{n-1} A_{jn}\frac{1}{i}\delta(x_n) \otimes D_j u_0(x') + B_n \frac{1}{i}\delta(x_n) \otimes u_0(x').$$

Let us construct a left parametrix Q for L in $\mathbb{R}^n$ such that $QL = I + T$, where T is a smoothing operator. Then

$$u^0 + Tu^0 = Q(Lu)^0 + Q\left[\frac{1}{i}\delta(x_n) \otimes A_{nn} u_1(x') - \delta'(x_n) \otimes A_{nn} u_0(x')\right.$$

$$\left. + \frac{2}{i}\sum_{j=|1|}^{n-1} \delta(x_n) \otimes A_{jn} D_j u_0(x') + \frac{1}{i}\delta(x_n) \otimes B_n u_0(x')\right]. \tag{2.7}$$

In reducing the problem to one on the boundary an important role is played by the following result of Hörmander [1966].

Theorem 2.6. *Let Q be a pseudodifferential operator in a domain $\Omega \subset \mathbb{R}^n$. Let its symbol $q(x, \xi)$ be expanded as an asymptotic sum $\sum q^j(x, \xi)$ each term of which is a rational function of ξ. Let ω denote the intersection of Ω and the plane $x_n = 0$. Then for each function u in $C_0^\infty(\omega)$ the function $Q(\delta^{(\nu)}(x_n) \otimes u)$ and all its derivatives can be extended from $\Omega^+ = \{x: x \in \Omega,\ x_n > 0\}$ to $\Omega^+ \cup \omega$. The limiting values*

$$Q^{\mu\nu} u = \lim_{x_n \to +0} D_n^\mu Q(\delta^{(\nu)} \otimes u)$$

and defined by means of pseudodifferential operators $Q^{\mu\nu}$ with symbols

$$\sum_j \frac{1}{2\pi}\int_{\Gamma_{\xi'}} (D_n + \xi_n)^\mu q^j(x', 0, \xi)(i\xi_n)^\nu \, d\xi_n,$$

where for $\Gamma_{\xi'}$ we can take, for example, a disc in the half-plane $\operatorname{Im} \xi_n > 0$ that contains in its interior the poles of q^j lying in this half-plane.

The principal symbol of the parametrix Q equals $\left(\sum_{j,k=1}^{n} A_{jk}\xi_j\xi_k\right)^{-1}$. For each $\xi' \neq 0$, the equation

$$\sum_{j,k=1}^{n} A_{jk}\xi_j\xi_k = 0$$

in ξ_n has two roots α and β, with $\operatorname{Im} \alpha > 0$ and $\operatorname{Im} \beta < 0$. Therefore the contour $\Gamma_{\xi'}$ must contain the point $\xi_n = \alpha$ in its interior. It can be seen easily that

$$\frac{1}{2\pi}\int_{\Gamma_{\xi'}}(D_n+\xi_n)^\mu\frac{\xi_n^\nu\,d\xi_n}{\sum\limits_{j,k=1}^{n}A_{jk}\xi_j\xi_k}=\frac{1}{2\pi}\int_{\Gamma_{\xi'}}(D_n+\xi_n)^\mu\frac{\xi_n^\nu\,d\xi_n}{A_{nn}(\xi_n-\alpha)(\xi_n-\beta)}$$

$$=\frac{i}{A_{nn}}(D_n+\xi_n)^\mu\left(\frac{\xi_n^\nu}{\xi_n-\beta}\right)\bigg|_{\xi_n=\alpha}.$$

Thus after restriction to the plane $x_n = 0$, the equation (2.7) assumes the form

$$Au_0 = Bu_1 + g, \tag{2.8}$$

where $g = Q(Lu)^0|_{x_n=0}$, and A and B are pseudodifferential operators with principal symbols

$$A_0 = \frac{\sum\limits_{j=0}^{n-1} A_{jn}\xi_j + A_{nn}\beta}{A_{nn}(\beta-\alpha)}$$

$$B_0 = \frac{1}{\alpha-\beta}$$

respectively. It is easy to see that A and B are elliptic operators. Thus (2.8) can be written, for example, in the form

$$u_1 = C_0u_0 + C_1g, \tag{2.9}$$

where C_0 and C_1 are first-order operators and the principal symbol of C_0 is $-\sum\limits_{j=1}^{n-1}\frac{A_{jn}}{A_{nn}}\xi_j - \beta$, and thus its imaginary part is positive if $\xi \neq 0$.

If we consider the boundary-value problem with the condition

$$\sum_{j=1}^{n} a_j(x)D_ju + a_0(x)u = h(x) \quad \text{on } \Gamma,$$

then, by excluding the derivative of u along the normal by means of (2.9), we obtain a pseudodifferential equation for u_0 on Γ. On solving this equation, we can find the solution u of the boundary-value problem as the solution of the Dirichlet problem with the condition $u = u_0$ on Γ. We can also use (2.9) and (2.7) to express u in an "explicit form".

2.8. Reduction of the Boundary-value Problem for an Elliptic System to a Problem on the Boundary. We can reduce the elliptic boundary-value problem for equations or systems of higher orders to a problem on the boundary in the same way.

The following construction is due to Calderon [1963]. Let Ω be a domain, with smooth boundary Γ, in $\mathbb{R}^n$. Let $P(x, D_x)$ be a system of r differential equations of order m in a neighbourhood of Ω. To make the discussion simple, we assume that the coefficients of these operators are infinitely smooth.

We denote by I^s the operator with symbol $(1 + |\xi|^2)^{-s/2}$ acting on $\mathbb{R}^n$, and by J^s an analogous operator acting on Γ. Let D be a smooth vector field that does not vanish and is orthogonal to Γ at points of the boundary.

With a vector $u = (u_1, \ldots, u_r)$ of class $H^k(\Omega)$ $(k \geqslant 1)$ we associate the vector $Bu = (v_1, \ldots, v_m)$ on Γ by setting $v_j = J^{j-m}D^{j-1}u(y)$ for $y \in \Gamma$. We denote by K the class of functions v such that $v = Bu$, where $u \in H^m(\Omega)$ and $P(x, D_x)u = 0$ in Ω.

Theorem 2.7. *The space K coincides with the range of an $rm \times rm$ matrix C consisting of singular integral operators in Ω the rank of whose symbol is $rm/2$ and for which $C^2 = C$.*

The symbol $\sigma(C)$ of C can be written explicitly in terms of the symbol of P. In this way, the above construction enables us to reduce the question of solving a boundary-value problem for a system of equations $Pu = 0$ in Ω to that of solving a system of pseudodifferential equations on the boundary manifold Γ.

§ 3. Wave Front of a Distribution and Simplest Theorems on Propagation of Singularities

3.1. Definition and Examples (Egorov [1984, 1985], Guillemin and Sternberg [1977], Hörmander [1971, 1983, 1985], Taylor [1981], Trèves [1980]). It is well known that a distribution u with compact support is a smooth function if and only if its Fourier transform $\tilde{u}(\xi)$ decays rapidly at infinity. If, however, u is not smooth, then the directions along which $\tilde{u}(\xi)$ decays insufficiently rapidly may serve as a characteristic of singularities of u.

In the classical theory of wave propagation, developed by Huygens, the waves are propagated at every moment of time in a direction that is normal to the wave front. In analogy with this theory, for each distribution u we introduce its *wave front set* $\mathrm{WF}(u)$ as a subset in the cotangent bundle space. This subset consists of those points (x, ξ) for which the direction of the vector ξ is singular for u at the point x. The projection of this set onto the x-space coincides with $\operatorname{sing\,supp} u$. It turns out that the set $\mathrm{WF}(u)$ is independent of the coordinate system chosen and can be described locally.

The concept of a wave front was introduced in the works of Sato [1970] and Hörmander [1971]. This concept proved to be extremely useful for the development of the theory of partial differential equations.

Definition 3.1. Let $u \in \mathcal{D}'(\Omega)$, where Ω is a domain in $\mathbb{R}^n$. A point (x_0, ξ_0) from $T^*\Omega\backslash 0$ does not lie in $\mathrm{WF}(u)$ if there exist functions $\varphi \in C_0^\infty(\Omega)$ and $\psi \in C^\infty(\mathbb{R}^n\backslash 0)$ such that

$$\varphi(x_0) \neq 0, \quad \psi(\xi_0) \neq 0, \quad \psi(t\xi) = \psi(\xi) \quad \text{for } t > 0, \xi \neq 0$$

and

$$\psi(D)\varphi(x)u(x) \in C^\infty(\mathbb{R}^n).$$

Here $\psi(D)$ denotes, as usual, a pseudodifferential operator with symbol $\psi(\xi)$:

$$\psi(D)v(x) = (2\pi)^{-n} \int \psi(\xi)\tilde{v}(\xi)e^{ix\xi}\, d\xi.$$

Sometimes it is convenient to use another equivalent definition, provided by the next theorem.

Theorem 3.1. *A point* $(x_0, \xi_0) \in T^*\Omega\backslash 0$ *does not lie in* $\mathrm{WF}(u)$ *if and only if there exist a function* $\varphi \in C_0^\infty(\Omega)$ *and a cone* $\Gamma \subset \mathbb{R}^n$ *such that for every* $N > 0$

$$\varphi(x_0) \neq 0, \quad \xi_0 \in \Gamma, \quad |\widetilde{\varphi u}(\xi)| \leqslant C_N(1 + |\xi|)^{-N} \quad \text{for } \xi \in \Gamma.$$

A subset K of $T^*\Omega\backslash 0$ is called *conical* if it contains all the points $(x, t\xi)$ $(t > 0)$ whenever it contains the point (x, ξ).

Theorem 3.2. *The wave front set of each distribution* $u \in \mathscr{D}'(\Omega)$ *is a closed conical set in* $T^*\Omega\backslash 0$ *whose projection onto* Ω *coincides with* sing sup u. *If* $\varphi \in C_0^\infty(\Omega)$, *then*

$$\mathrm{WF}(\varphi u) \subset \mathrm{WF}(u).$$

Examples. 1°. $u = \delta(x)$; $\mathrm{WF}(u) = \{(0, \xi), \xi \in \mathbb{R}^n\backslash 0\}$.

2°. $u = \delta(x_1)$; $\mathrm{WF}(u) = \{(0, x', \xi_1, 0) \in T^*\mathbb{R}^n\backslash 0, x' \in \mathbb{R}^{n-1}, \xi_1 \in \mathbb{R}\backslash 0\}$.

3°. $u = \dfrac{1}{x_2 + ix_1^2}$; $\mathrm{WF}(u) = \{(0, 0, x'', \xi_1, 0, 0); x'' \in \mathbb{R}^{n-2}, \xi_1 \in \mathbb{R}\backslash 0\}$.

4°. Let Γ be any closed conical subset of $T^*\Omega\backslash 0$. Then the function

$$u(x) = \sum_{k=1}^{\infty} k^{-2}\varphi(k(x - x_k))e^{ik^3 x\theta_k}$$

is continuous in $\mathbb{R}^n$ and $\mathrm{WF}(u) = \Gamma$ if $\varphi \in C_0^\infty(\mathbb{R}^n)$, $\tilde{\varphi}(0) = 1$, $|\theta_k| = 1$, and the sequence of points (x_k, θ_k) is dense in the intersection of Γ with the set $|\xi| = 1$ (see Hörmander [1983, 1985]).

3.2. Properties of the Wave Front Set. Let $X \subset \mathbb{R}^m$ and $Y \subset \mathbb{R}^n$ be open sets and let $f: X \to Y$ be a smooth map. If $\varphi \in C_0^\infty(Y)$ then the function $f^*\varphi(x) = \varphi(f(x))$ is smooth in X. An important question arises: when can the map f^* be extended by continuity to distributions? The following result answers this question.

Theorem 3.3. (Hörmander [1983, 1985]). *The distribution* $f^*u \in \mathscr{D}'(X)$ *is defined for each* $u \in \mathscr{D}'(Y)$ *for which* $N_f \cap \mathrm{WF}(u) = \varnothing$, *where*

$$N_f = \{(f(x), \eta) \in T^*(Y)\backslash 0: {}^t f'(x)\eta = 0\}.^1 \tag{3.1}$$

Furthermore,

$$\mathrm{WF}(f^*u) \subset f^*\mathrm{WF}(u) = \{(x, {}^t f'(x)\eta) \in T^*X\backslash 0: (f(x), \eta) \in \mathrm{WF}(u)\}.$$

[1] Here and in a similar definition further on the extension of the operation to distributions with a wave front in a fixed close cone is done by continuity with respect to a natural topology in the space of such distributions. This topology is defined by constants in estimates like estimates in Theorem 3.1.

Clearly, $N_f = \emptyset$ if $m = n$ and $\det f'(x) \neq 0$, and then Theorem 3.3 is applicable for any u.

In particular, this theorem enables us to define $\mathrm{WF}(u)$ for $u \in \mathscr{D}'(X)$, where X is a smooth manifold. Moreover, $\mathrm{WF}(u) \subset T^*X\backslash 0$, and is independent of the local coordinates used.

Let X be a smooth manifold and $Y \subset X$ a smooth submanifold of smaller dimension. Let $i: Y \hookrightarrow X$ be the embedding of Y into X. The set N_i, introduced in Theorem 3.3, coincides with the set

$$N(Y) = \{(y, \eta): y \in Y \text{ and } \eta \text{ is normal to } Y\}.$$

Thus we have

Theorem 3.4. *If $u \in \mathscr{D}'(X)$ and $\mathrm{WF}(u) \cap N(Y) = \emptyset$, then the restriction $u|_Y \in \mathscr{D}'(Y)$ is defined. Furthermore,*

$$\mathrm{WF}(u|_Y) \subset \{(y, \eta) \in T^*Y\backslash 0: \exists \xi \in N(Y), (y, \eta + \xi) \in \mathrm{WF}(u)\}.$$

An important question from the point of view of applications is that of defining the product uv of two distributions. It is known that this product is always defined if $u \in C^\infty(X)$. In the general case, it is important that the wave fronts $\mathrm{WF}(u)$ and $\mathrm{WF}(v)$ are compatible in the sense that, at those points which are singular for u the function v must be smooth and vice-versa (see, for example, Hörmander [1971, 1983, 1985], Shubin [1978]).

Let $u, v \in \mathscr{D}'(X)$, where X is a smooth manifold, and $\Delta: X \to X \times X$ be the diagonal map such that $\Delta(x) = (x, x)$. The set N_Δ given by (3.1) assumes the form

$$N_\Delta = \{(x, x, \xi, \eta) \in T^*(X \times X)\backslash 0: \xi + \eta = 0\}.$$

Theorem 3.5. *If $\mathrm{WF}(u) + \mathrm{WF}(v) \subset T^*X\backslash 0$, then the product $uv \in \mathscr{D}'(X)$ is defined, and*

$$\mathrm{WF}(uv) \subset \{(x, \xi + \eta) \in T^*X\backslash 0: (x, \xi) \in \mathrm{WF}(u) \text{ or } \xi = 0,$$
$$(x, \eta) \in \mathrm{WF}(v) \text{ or } \eta = 0\}.$$

This result is obtained from Theorem 3.3 by applying Δ^* to the distribution $u(x) \otimes v(y) \in \mathscr{D}'(X \times X)$.

Theorems 3.3 and 3.5 enable us to examine the question of applying to the distribution $u \in \mathscr{D}'(Y)$ the integral operator of the form

$$Au(x) = \int K(x, y)u(y)\,dy$$

with kernel K lying in $\mathscr{D}'(X \times Y)$.

In the particular case where $u \in C_0^\infty(Y)$, we have $Au \in \mathscr{D}'(X)$ and

$$\mathrm{WF}(Au) \subset \{(x, \xi) \in T^*X\backslash 0: (x, y, \xi, 0) \in \mathrm{WF}(K) \text{ for some } y \in \operatorname{supp} u\}.$$

This result follows immediately from Theorem 3.5. To tackle the general case where u lies in $\mathscr{E}'(Y)$, we introduce the following sets:

$$\mathrm{WF}'(K) = \{(x, y, \xi, \eta) \in T^*(X \times Y)\backslash 0 \colon (x, y, \xi, -\eta) \in \mathrm{WF}(K)\},$$

$$\mathrm{WF}(K)_X = \{(x, \xi) \in T^*(X)\backslash 0 \colon \exists y \in Y, (x, y, \xi, 0) \in \mathrm{WF}(K)\},$$

$$\mathrm{WF}'(K)_Y = \{(y, \eta) \in T^*(Y)\backslash 0 \colon \exists x \in X, (x, y, 0, -\eta) \in \mathrm{WF}(K)\}.$$

Theorem 3.6. *Let* $u \in \mathscr{E}'(Y)$ *and let* $\mathrm{WF}(u) \cap \mathrm{WF}'(K)_Y = \varnothing$. *Then the distribution*

$$Au(x) = \int K(x, y)u(y)\,dy \in \mathscr{D}'(X)$$

is defined, and

$$\mathrm{WF}(Au) \subset \mathrm{WF}(K)_X \cup \{(x, \xi) \in T^*(X)\backslash 0 \colon \exists (y, \eta) \in \mathrm{WF}(u), (x, y, \xi, \eta) \in \mathrm{W}'(K)\}.$$

In the important particular case where A is a pseudodifferential operator, we have

$$\mathrm{WF}(K) \subset \{(x, x, \xi, \xi) \in T^*(X \times X)\backslash 0\},$$

and hence $\mathrm{WF}(K)_X = \varnothing$ and $\mathrm{WF}'(K)_Y = \varnothing$, giving

$$\mathrm{WF}(Au) \subset \mathrm{WF}(u).$$

Thus we find that a pseudodifferential operator is pseudolocal in terms of wave front sets also (see § 1.1).

3.3. Applications to Differential Equations. Let $P = P(x, D)$ be a differential or a pseudodifferential operator of order m with principal symbol $p_0(x, \xi)$. We set

$$\mathrm{Char}\, P = \{(x, \xi) \in T^*X\backslash 0 \colon p_0(x, \xi) = 0\}.$$

Theorem 3.7 (Hörmander [1983, 1985]). *For each* $u \in \mathscr{D}'(X)$,

$$\mathrm{WF}(u) \subset \mathrm{Char}\, P \cup \mathrm{WF}(Pu).$$

Corollary 3.1. *If P is an elliptic operator* (*that is,* Char $P = \varnothing$), *then* $\mathrm{WF}(u) = \mathrm{WF}(Pu)$ *for all* $u \in \mathscr{D}'(X)$.

The proof of the theorem is based on the construction of the microlocal parametrix. If the point (x_0, ξ_0) lies outside Char P, then we can construct an operator $Q(x, D)$ such that the symbol of $QP - I$ will be zero in a conical neighbourhood of the point (x_0, ξ_0). This symbol can be constructed in the same way as in Theorem 1.3. Now if $(x_0, \xi_0) \notin \mathrm{WF}(Pu)$, then $(x_0, \xi_0) \notin \mathrm{WF}(QPu)$, and consequently $(x_0, \xi_0) \notin \mathrm{WF}(u)$.

Example 3.1. Let $u \in \mathscr{D}'(X)$, $X \subset \mathbb{R}^n$ and $D_1 u \in C^\infty(X)$. If $(x_0, \xi_0) \in T^*(X)\backslash 0$ and $(x_0, \xi_0) \notin \mathrm{WF}(u)$, then each point $(x^{(1)}, \xi_0)$ of $T^*(X)\backslash 0$ lies outside $\mathrm{WF}(u)$ provided that the point $x^{(1)}$ can be connected with x_0 by a line segment parallel to the x_1-axis and lying in X.

This simple example describes the actual situation that is typical for operators with simple characteristics.

Theorem 3.8 (Hörmander [1983, 1985]). *Let $X \subset \mathbb{R}^n$. Let P be a pseudodifferential operator in X with real principal symbol p_0. Let γ be a segment, lying in $T^*X\backslash 0$, of the integral curve of the Hamiltonian system of equations*

$$\dot{x}(t) = \partial p_0(x(t), \xi(t))/\partial \xi, \quad \dot{\xi}(t) = -\partial p_0(x(t), \xi(t))/\partial x.$$

If $\gamma \cap \mathrm{WF}(Pu) = \varnothing$, then either $\gamma \subset \mathrm{WF}(u)$ or $\gamma \cap \mathrm{WF}(u) = \varnothing$.

3.4. Some Generalizations. In the theory of differential equations some generalizations of the concept of the wave front set prove to be useful, such as the analytic wave front, the Gevrey wave front, the oscillation front, etc. (See Hörmander [1983, 1985].)

We mention one such generalization connected with the Sobolev spaces H^s ($s \in \mathbb{R}$).

Definition 3.2. Let $u \in \mathscr{D}'(X)$ and $(x_0, \xi_0) \in T^*(X)\backslash 0$. We shall write $u \in H^s(x_0, \xi_0)$ if $u = u_1 + u_2$, where $u_1 \in \overset{\circ}{H}{}^s(X)$ and $(x_0, \xi_0) \notin \mathrm{WF}(u_2)$.

Definition 3.3. By the *wave front set* $\mathrm{WF}_s(u)$ of a distribution $u \in \mathscr{D}'(X)$ we mean the complement in $T^*(X)\backslash 0$ of the set of points (x, ξ) for which $u \in H^s(x, \xi)$.

Theorem 3.9 (Hörmander [1983, 1985]). *For each $u \in \mathscr{D}'(X)$*

$$\mathrm{WF}_s(u) \subset \mathrm{Char}\, P \cup \mathrm{WF}_{s-m}(Pu).$$

Theorem 3.10 (Hörmander [1983, 1985]). *Suppose that the hypotheses of Theorem 3.8 hold. If $\gamma \cap \mathrm{WF}_{s-m}(Pu) = \varnothing$, then either $\gamma \subset \mathrm{WF}_s(u)$ or $\gamma \cap \mathrm{WF}_s(u) = \varnothing$.*

§4. Fourier Integral Operators

4.1. Definition and Examples. The theory of pseudodifferential operators, discussed in § 1, is well suited for investigating various problems connected with elliptic differential equations. However, this theory fails to be adequate for studying equations of hyperbolic type, and one is then forced to examine a wider class of operators, the so-called Fourier integral operators (Egorov [1975], Hörmander [1968, 1971, 1983, 1985], Kumano-go [1982], Shubin [1978], Taylor [1981], Trèves [1980]).

Definition 4.1. By a *Fourier integral operator* we mean an operator of the form

$$\Phi: u \mapsto \iint a(x, y, \theta)u(y)e^{iS(x, y, \theta)}\, dy\, d\theta. \tag{4.1}$$

Here the function a is called the *amplitude* and $a \in C^\infty(X \times Y \times \mathbb{R}^N)$, where $X \subset \mathbb{R}^{n_1}$ and $Y \subset \mathbb{R}^{n_2}$, and the estimates

$$|D_\theta^\alpha D_x^\beta D_y^\gamma a(x, y, \theta)| \leqslant C_{\alpha,\beta,\gamma}(1 + |\theta|)^{m-|\alpha|}, \quad x \in X, \quad y \in Y, \quad \theta \in \mathbb{R}^N, \tag{4.2}$$

hold for any multi-indices α, β and γ. The function S is called the *phase function* and $S \in C^\infty(X \times Y \times (\mathbb{R}^N \setminus 0))$; S is real valued, positive and homogeneous in θ of degree 1.

Usually the phase function S is subjected to some non-degeneracy conditions too; these will be mentioned below.

Example 4.1. If $n_1 = n_2 = N$, $S(x, y, \theta) = (x - y)\cdot\theta$, then (4.1) becomes a pseudodifferential operator (see § 1).

Example 4.2. The solution to the Cauchy problem

$$\frac{\partial^2 u}{\partial t^2} = \Delta u \quad \text{for } t > 0,\ u = 0, \quad \frac{\partial u}{\partial t} = \varphi \quad \text{for } t = 0,$$

is given by the formula

$$u(t, x) = (2\pi)^{-n} \iint e^{i(x-y)\xi + it|\xi|} \frac{\varphi(y)}{2i|\xi|}\, dy\, d\xi$$

$$- (2\pi)^{-n} \iint e^{i(x-y)\xi - it|\xi|} \frac{\varphi(y)}{2i|\xi|}\, dy\, d\xi.$$

This formula contains two operators which have the form of Fourier integral operators with phase functions $(x - y)\xi + t|\xi|$, but with amplitudes of the form $(2i|\xi|)^{-1}$ which have a singularity at $\xi = 0$. By cutting off this singularity, that is, by replacing $(2i|\xi|)^{-1}$ with the amplitude $(1 - \chi(\xi))(2i|\xi|)^{-1}$, where $\chi \in C_0^\infty(\mathbb{R}^n)$ and $\chi(\xi) = 1$ in a neighbourhood of the origin, we add an operator with a smooth kernel to the right-hand side. Thus $u(t, x)$ is expressed in terms of $\varphi(x)$ by means of two Fourier integral operators and an operator with a smooth kernel.

Example 4.3. Let us consider the Cauchy problem

$$\frac{\partial u}{\partial t} = p(t, x, D_x)u \quad \text{for } t > 0,\ u(0, x) = \varphi(x),$$

where $p(t, x, D_x)$ is a pseudodifferential operator of order 1 in x that depends smoothly on t and whose symbol $p(t, x, \xi)$ is positive and is a real-valued homogeneous function of $\xi \in \mathbb{R}^n$ of degree 1. The parametrix of this problem has the form

$$T\varphi(x) = (2\pi)^{-n} \iint a(t, x, \xi)\varphi(y)e^{i(x-y)\xi + iS(t, x, \xi)}\, dy\, d\xi,$$

where the phase function S is determined as a solution of the non-linear equation

$$\frac{\partial S}{\partial t} = p\left(t, x, \xi + \frac{\partial S}{\partial x}\right)$$

satisfying the initial condition

$$S(0, x, \xi) = 0.$$

The techniques for finding the amplitude $a(t, x, \xi)$ are discussed below in §4.3.

4.2. Some Properties of Fourier Integral Operators. Let us first examine the question of convergence of the integral (4.1). It is clear that the integral cannot converge when $m > -n_2$ even if $u \in C_0^\infty(Y)$. We try to define this integral as the limit

$$\lim_{\varepsilon \to 0} \iint \chi(\varepsilon\theta) a(x, y, \theta) u(y) e^{iS(x, y, \theta)}\, dy\, d\theta,$$

where $\chi \in C_0^\infty(\mathbb{R}^n)$ and $\chi(\theta) = 1$ in a neighbourhood of $0 \in \mathbb{R}^n$. Assume that

$$d_{y,\theta} S(x, y, \theta) \neq 0 \quad \text{for } \theta \in \mathbb{R}^N \backslash 0,\ x \in X,\ y \in Y. \tag{4.3}$$

Setting

$$L = \frac{1}{i} \sum_{k=1}^{n_2} \frac{\partial S}{\partial y_k} \frac{\partial}{\partial y_k} + \frac{1}{i} \sum_{k=1}^{N} |\theta|^2 \frac{\partial S}{\partial \theta_k} \frac{\partial}{\partial \theta_k},$$

we have

$$\begin{aligned} L e^{iS(x, y, \theta)} &= \left[\sum_{k=1}^{n_2} \left(\frac{\partial S}{\partial y_k} \right)^2 + |\theta|^2 \sum_{j=1}^{N} \left(\frac{\partial S}{\partial \theta_j} \right)^2 \right] e^{iS(x, y, \theta)} \\ &= \psi(x, y, \theta) e^{iS(x, y, \theta)}, \end{aligned}$$

where $\psi(x, y, \theta) \neq 0$ and $\psi(x, y, t\theta) = t^2 \psi(x, y, \theta)$ for $t > 0$. We choose $h \in C_0^\infty(\mathbb{R}^n)$ such that $h(\theta) = 1$ if $|\theta| < 1/2$ and $h(\theta) = 0$ if $|\theta| > 1$, and set

$$L_1 = [1 - h(\theta)] \psi^{-1}(x, y, \theta) L + h(\theta).$$

Then $L_1 e^{iS} = e^{iS}$, and

$$L_1 = \sum_{k=1}^{n_2} a_k(x, y, \theta) \frac{\partial}{\partial y_k} + \sum_{j=1}^{N} b_j(x, y, \theta) \frac{\partial}{\partial \theta_j} + c(x, y, \theta).$$

The functions a_k, b_j and c satisfy estimates of the form (4.2) for $m = -1, 0$ and -1 respectively. Hence

$$\iint \chi(\varepsilon\theta)a(x, y, \theta)u(y)e^{iS(x,y,\theta)}\, dy\, d\theta$$

$$= \iint \chi(\varepsilon\theta)a(x, y, \theta)u(y)L_1^k e^{iS(x,y,\theta)}\, dy\, d\theta$$

$$= \iint ({}^tL_1)^k(\chi(\varepsilon\theta)a(x, y, \theta)u(y))e^{iS(x,y,\theta)}\, dy\, d\theta,$$

where

$${}^tL_1 v = -\sum_{k=1}^{n_2} \frac{\partial}{\partial y_k}(a_k v) - \sum_{j=1}^{N} \frac{\partial}{\partial \theta_j}(b_j v) + cv.$$

If k is chosen so large that $k > n_2 + m$, then the resulting integral converges absolutely and uniformly in ε for $\varepsilon \in [0, 1]$ when $u \in C_0^\infty(Y)$, and it enables us to take the limit as $\varepsilon \to 0$ under the integral sign.

Since the same construction can be carried out for the integrals obtained from (4.1) by differentiation with respect to x, we have

Theorem 4.1. *If* (4.3) *holds, then Φ is a bounded operator from $C_0^\infty(Y)$ into $C^\infty(X)$.*

Since the transpose of Φ is of the form

$${}^t\Phi v(y) = \iint e^{iS(x,y,\theta)}a(x, y, \theta)v(x)\, dx\, d\theta,$$

duality yields the following

Theorem 4.2. *If the condition*

$$d_{x,\theta}S(x, y, \theta) \neq 0 \quad \text{for } \theta \neq 0, x \in X, y \in Y \tag{4.4}$$

holds, then the map Φ defined by the integral (4.1) *can be extended to a continuous map*

$$\Phi: \mathscr{E}'(Y) \to \mathscr{D}'(X).$$

The Schwartz kernel of Φ is a distribution $K(x, y) \in \mathscr{D}'(X \times Y)$ defined by the integral

$$\langle K, w\rangle = \iint e^{iS(x,y,\theta)}a(x, y, \theta)w(x, y)\, dx\, dy\, d\theta, \quad w \in C_0^\infty(X \times Y).$$

Theorem 3.6 and arguments analogous to those used earlier enable us to establish the following result.

Theorem 4.3. *Let $u \in \mathscr{E}'(\Omega)$ and let Φ be an operator of the form* (4.1). *Assume that the following conditions hold.*

1°. *If $d_x S(x, y, \theta) = 0$ and $d_\theta S(x, y, \theta) = 0$, then* $\left(y, -\dfrac{\partial S}{\partial y}\right) \notin \mathrm{WF}(u)$.

2°. *If* $d_\theta S(x, y, \theta) = 0$ *and* $\theta \neq 0$, *then* $d_y S(x, y, \theta) \neq 0$.
Then the distribution $\Phi u \in \mathscr{D}'(X)$ *is defined, and*

$$\mathrm{WF}(\Phi u) \subset \left\{(x, \xi)\colon \xi = \frac{\partial S(x, y, \theta)}{\partial x}, \right.$$

$$\left. \eta = -\frac{\partial S(x, y, \theta)}{\partial y}, \frac{\partial S(x, y, \theta)}{\partial \theta} = 0, \quad (y, \eta) \in \mathrm{WF}(u)\right\}.$$

4.3. Composition of Fourier Integral Operators with Pseudodifferential Operators. Let A be a pseudodifferential operator of order m in a domain $X \subset \mathbb{R}^{n_1}$ and let $a(x, \xi)$ be its symbol. We first examine the behaviour of the function $A(fe^{i\lambda\psi(x)})$ as $\lambda \to +\infty$, assuming A to be properly supported, $f \in C^\infty(X)$ and $\psi(x) \in C^\infty(X)$ with $\psi'(x) \neq 0$ for $x \in X$.

Theorem 4.4. *For any integer* $N \geqslant 0$ *and* $\lambda \geqslant 1$, *we have*

$$e^{-i\lambda\psi} A(fe^{i\lambda\psi}) = \sum_{|\alpha| < N} \frac{1}{\alpha!} \partial_\xi^\alpha a(x, \lambda\psi'(x)) [D_y^\alpha(f(y) e^{i\lambda\varphi_x(y)})]_{y=x}$$
$$+ \lambda^{m-N/2} R_N(\lambda, x), \tag{4.5}$$

where $\varphi_x(y) = \psi(y) - \psi(x) - \psi'(x)(y - x)$ *and* R_N *is such that*

$$|\partial_x^\beta R_N(x, \lambda)| \leqslant C_{\beta, N, K}, \quad x \in K,$$

in which K is a compact subset of X and the constants $C_{\beta, N, K}$ *are independent of* λ.

We note that the summand corresponding to α in (4.5) is estimated by $C\lambda^{m-|\alpha|/2}$, because $\dfrac{\partial}{\partial y_j}\varphi_x(y) = 0$ when $y = x$, and hence

$$|[D_y^\alpha(f(y) e^{i\lambda\varphi_x(y)})]_{y=x}| = O(\lambda^{|\alpha|/2}).$$

Theorem 4.4 enables us to define the operator $A \circ \Phi$, where Φ is an operator of order m' of the form (4.1) with amplitude $b(x, y, \theta)$, that is,

$$\Phi u(x) = \iint b(x, y, \theta) u(y) e^{iS(x, y, \theta)}\, dy\, d\theta.$$

Theorem 4.5. *The composition* $A \circ \Phi$ *is an operator of the form* (4.1) *with the same phase function S and amplitude*

$$c(x, y, \theta) = e^{-iS(x, y, \theta)} a(x, D_x) [b(x, y, \theta) e^{iS(x, y, \theta)}].$$

The amplitude can be expanded in an asymptotic series

$$c(x, y, \theta) \sim \sum_\alpha \frac{1}{\alpha!} \partial_\xi^\alpha a\left(x, \frac{\partial S(x, y, \theta)}{\partial x}\right) D_z^\alpha (b(z, y, \theta) e^{i\varphi_x(z, y, \theta)})|_{z=x}$$

as $|\theta| \to \infty$, *where* $\varphi_x(z, y, \theta) = S(z, y, \theta) - S(x, y, \theta) - (z - x)\dfrac{\partial S(x, y, \theta)}{\partial x}$.

A similar result also holds for the composition $\Phi \circ A$, and this follows easily from Theorem 4.5 by expressing $\Phi \circ A$ as the transpose of the operator ${}^tA \circ {}^t\Phi$.

Theorem 4.6. *The operator $\Phi \circ A$ can be represented in the form* (4.1) *with the same phase function S and with an amplitude that is asymptotically equal to*

$$\sum_{\alpha,\beta} \frac{(-1)^{|\alpha|}}{\alpha!\beta!}[\partial_\xi^{\alpha+\beta} D_y^\alpha a(y, \xi)]_{\xi=-\partial S(x,y,\theta)/\partial y}[D_z^\beta(b(x, z, \theta)e^{i\varphi_y(x,z,\theta)})]_{z=y},$$

where $\varphi_y(x, z, \theta) = S(x, z, \theta) - S(x, y, \theta) - (z - y)\dfrac{\partial S(x, y, \theta)}{\partial y}$.

We now return to Example 4.3, examined in §4.1. Let us apply Theorem 4.5 to compute the symbol of the operator $P \circ T$. In view of this theorem, the symbol $a(t, x, \xi)$ must satisfy the asymptotic equation

$$\frac{\partial a}{\partial t} = \sum_\alpha \frac{1}{\alpha!}\partial_\xi^\alpha p\left(t, x, \xi + \frac{\partial S}{\partial x}\right)[D_z^\alpha(a(t, z, \xi)e^{i\varphi_x(t,z,\xi)})]_{z=x},$$

where $\varphi_x(t, z, \xi) = S(t, z, \xi) - S(t, x, \xi) - (z - x)\dfrac{\partial S(t, x, \xi)}{\partial x}$. Moreover, $a(0, x, \xi) = 1$. If we represent a as an asymptotic sum,

$$a \sim \sum_{j=0}^\infty a_j(t, x, \xi), \quad a_j(t, x, \lambda\xi) = \lambda^{-j}a_j(t, x, \xi) \quad \text{for } \lambda > 0,$$

we find that

$$\frac{\partial a_0}{\partial t} = \sum_{|\alpha|=1} \partial_\xi^\alpha p\left(t, x, \xi + \frac{\partial S}{\partial x}\right) D_x^\alpha a_0(t, x, \xi), \quad a_0(0, x, \xi) = 1;$$

$$\frac{\partial a_j}{\partial t} = \sum_{|\alpha|=1} \partial_\xi^\alpha p\left(t, x, \xi + \frac{\partial S}{\partial x}\right) D_x^\alpha a_j(t, x, \xi) + F_j(t, x, \xi),$$

$$a_j(0, x, \xi) = 0, \quad j = 1, 2, \ldots$$

Here the functions F_j can be computed if the functions $a_0, a_1, \ldots, a_{j-1}$ are known. Moreover, $F_j(t, x, \lambda\xi) = \lambda^{-j}F_j(t, x, \xi)$ for $\lambda > 0$. These equations are referred to as the *transport equations*.

The parametrix T thus constructed enables us to establish the existence of a unique solution to the Cauchy problem.

4.4. Canonical Transformations. Let us consider a Fourier integral operator of the special form

$$\Phi u(x) = (2\pi)^{-n} \int a(x, \xi)\tilde{u}(\xi)e^{iS(x,\xi)}\, d\xi. \tag{4.6}$$

Suppose that $\det\left(\dfrac{\partial^2 S(x, \xi)}{\partial x \partial \xi}\right) \neq 0$. The operator (4.1) takes the form (4.6) if a and

S are independent of y and $N = n_1 = n_2 = n$. The Schwartz kernel of this operator has the form

$$K(x, y) = (2\pi)^{-n} \int e^{iS(x,\xi) - iy\xi} a(x, \xi)\, d\xi.$$

The method of stationary phase (see, for example, Fedoryuk [1971]) enables us to show that

$$\mathrm{WF}(K) \subset \left\{(x, y, \xi, -\eta) \colon y = \frac{\partial S(x, \eta)}{\partial \eta}, \xi = \frac{\partial S(x, \eta)}{\partial x}\right\}.$$

Therefore Theorem 3.6 implies that

$$\mathrm{WF}(\Phi u) \subset \left\{(x, \xi) \colon \exists \eta \in \mathbb{R}^n, \xi = \frac{\partial S(x, \eta)}{\partial x}, \left(\frac{\partial S(x, \eta)}{\partial \eta}, \eta\right) \in \mathrm{WF}(u)\right\}.$$

The equations

$$y = \frac{\partial S(x, \eta)}{\partial \eta}, \quad \xi = \frac{\partial S(x, \eta)}{\partial x}$$

define a transformation

$$T \colon (x, \xi) \mapsto (y, \eta)$$

which, in classical mechanics, is known as the *canonical transformation* (see Arnol'd [1979]).

Let us recall some of the notions connected with this transformation.

Let M be a smooth manifold whose dimension is even, $2n$ say. A *symplectic structure* on M is a symplectic form on M, i.e. an exterior smooth differential 2-form

$$\omega = \frac{1}{2} \sum \omega_{ij}\, dx_i \wedge dx_j, \quad \omega_{ij} = -\omega_{ji},$$

which is *closed*, that is, $d\omega = 0$, and is *non-degenerate*, that is, $\det \|\omega_{ij}\| \neq 0$. A manifold on which a symplectic form ω is given, is called *symplectic*.

For example, if Ω is a domain of $\mathbb{R}^n$, then the manifold $T^*\Omega$ has the symplectic structure, given by the form

$$\omega = dx \wedge d\xi = \sum_{j=1}^{n} dx_j \wedge d\xi_j,$$

where $x_1, \ldots, x_n$ are local coordinates in Ω and $\xi_1, \ldots, \xi_n$ are corresponding coordinates in fibres of the cotangent bundle. This form is independent of the coordinates $x_1, \ldots, x_n$ chosen in Ω.

The transformation $T \colon T^*\Omega \to T^*\Omega$ is called *canonical* if it is smooth and preserves the symplectic structure.

Examples. 1°. The change of coordinates $y = F(x)$ generates a canonical transformation T given by

$$y = F(x), \quad \eta = {}^t F'(x)^{-1} \xi.$$

2°. A real-valued smooth function $H(x, \xi)$ defines a canonical transformation

$$(x, \xi) \mapsto (x(t), \xi(t))$$

for each $t \in \mathbb{R}$, where $x(t)$, $\xi(t)$ is the solution of the Hamiltonian system of equations

$$\dot{x}(t) = \frac{\partial H(x(t), \xi(t))}{\partial \xi}, \quad \dot{\xi}(t) = -\frac{\partial H(x(t), \xi(t))}{\partial x}$$

with initial conditions

$$x(0) = x, \quad \xi(0) = \xi,$$

where we assume that a solution exists for each $t \in \mathbb{R}$.

A transformation T: $T^*\Omega \to T^*\Omega$ is canonical if and only if, under the transformation induced on the functions, the *Poisson bracket*

$$\{f, g\} = \sum_{j=1}^{n} \left(\frac{\partial f}{\partial x_j} \frac{\partial g}{\partial \xi_j} - \frac{\partial f}{\partial \xi_j} \frac{\partial g}{\partial x_j} \right)$$

is preserved for any two functions $f, g \in C^\infty(T^*\Omega)$.

Each smooth function $S(x_1, \ldots, x_n, \eta_1, \ldots, \eta_k, y_{k+1}, \ldots, y_n)$ $(0 \leqslant k \leqslant n)$ satisfying the condition

$$\det \left\| \frac{\partial^2 S}{\partial x \partial \eta}, \frac{\partial^2 S}{\partial x \partial y} \right\| \neq 0$$

defines a canonical transformation T by the formulae

$$\xi_j = \frac{\partial S}{\partial x_j}, \quad j = 1, \ldots, n;$$

$$y_j = \frac{\partial S}{\partial \eta_j}, \quad j = 1, \ldots, k; \quad \eta_j = -\frac{\partial S}{\partial y_j}, \quad j = k+1, \ldots, n.$$

The function S is called the *generating function* of T.

4.5. Connection Between Canonical Transformations and Fourier Integral Operators. As we noted above, the wave front set of a distribution is transformed under the action of a Fourier integral operator of the form (4.6) in accordance with the canonical transformation whose generating function is the phase function of the Fourier integral operator.

It turns out that the principal symbol of each pseudodifferential operator changes in the same manner.

Theorem 4.7 (Egorov [1984]). *Let P be a properly supported pseudodifferential operator in $\Omega \subset \mathbb{R}^n$ with the symbol $p \in S^m(\Omega \times \mathbb{R}^n)$ and let Φ be the Fourier integral operator* (4.6) *with phase function S. Then there exists a properly supported pseudodifferential operator Q with symbol $q \in S^m(\Omega' \times \mathbb{R}^n)$, where Ω'*

is the image of Ω *under the map defined by* S, *such that the operator* $T = P\Phi - \Phi Q$ *is smoothing and* $q - q_0 \in S^{m-1}(\Omega' \times \mathbb{R}^n)$, *where* $q_0\left(\dfrac{\partial S(x,\eta)}{\partial \eta}, \eta\right) = p\left(x, \dfrac{\partial S(x,\eta)}{\partial x}\right)$.

In the investigation of various *a priori* estimates the following related result proves to be useful.

Theorem 4.8 (Egorov [1984]). *Let* P *be a properly supported pseudodifferential operator with symbol* $p \in S^m(\Omega \times \mathbb{R}^n)$ *and let* Φ *be the Fourier integral operator of the form* (4.6) *with amplitude* $a \in S^0(\Omega \times \mathbb{R}^n)$ *and phase function* S. *Then the operator* $Q = \Phi^* P \Phi$ *is pseudodifferential with symbol* $q \in S^m(\Omega' \times \mathbb{R}^n)$, *and*

$$q(y,\eta) - p(x,\xi)|a(x,\xi)|^2 |\det S_{x\xi}|^{-1} \in S^{m-1},$$

where (y, η) *is the image of* (x, ξ) *under the canonical transformation defined by the generating function* S.

Corollary 4.1. *Let* Φ *be a Fourier integral operator of the form* (4.6) *with amplitude* $a \in S^0(\Omega \times \mathbb{R}^n)$ *and phase function* S. *Then for any subdomain* $\Omega' \Subset \Omega$ *there exists a constant* C *such that*

$$\|\Phi u\|_0 \leqslant C\|u\|_0, \quad u \in C_0^\infty(\Omega'),$$

and if $|a(x,\xi)| \geqslant c_0 > 0$ *for large* $|\xi|$, *then*

$$\|u\|_0 \leqslant c(\|\Phi u\|_0 + \|u\|_{-1}), \quad u \in C_0^\infty(\Omega').$$

Here $\|\cdot\|_s$ *denotes, as usual, the norm in* $H^s(\mathbb{R}^n)$.

This statement follows immediately from Theorem 4.8 with $P = I$, since

$$\|\Phi u\|_0^2 = (\Phi^*\Phi u, u)_0 = (Qu, u)_0 = (Au, u)_0 + (Bu, u)_0,$$

where A is a pseudodifferential operator with symbol $|a(x,\xi)|^2 |\det S_{x\xi}(x,\xi)|^{-1}$ and B is a pseudodifferential operator of order -1.

Corollary 4.2. *If the amplitude* a *of the operator* (4.6) *lies in* $S^m(\Omega \times \mathbb{R}^n)$, *then, for any* $\Omega' \Subset \Omega$ *and any* $s \in \mathbb{R}$, *there exists a constant* $C = C(s, \Omega')$ *such that*

$$\|\Phi u\|_s \leqslant C\|u\|_{s+m}, \quad u \in C_0^\infty(\Omega').$$

Definition 4.2. A Fourier integral operator of the form (4.6) is called *elliptic* if its phase function S is non-degenerate, that is, $\det\|\partial^2 S/\partial x_i \partial \xi_j\| \neq 0$ and its amplitude a satisfies the inequality

$$|a(x,\xi)| \geqslant C|\xi|^m - C_1, \quad C > 0, \quad (x,\xi) \in T^*\Omega \backslash 0.$$

Theorem 4.9 (Egorov [1984]). *Let* Φ *be an elliptic Fourier integral operator of the form* (4.6). *Then there exists a Fourier integral operator* Ψ *of the form* (4.6) *with the same phase function* $S(x,\xi)$ *and with amplitude* $b(x,\xi) \in S^{-m}$ *such that* $\Psi^*\Phi - I$ *is a pseudodifferential operator of order* $-\infty$.

4.6. Lagrangian Manifolds and Phase Functions. Let M be a symplectic manifold of dimension $2n$ and let ω be its symplectic form. A submanifold $\Lambda \subset M$ is called a *Lagrangian manifold* if, first $\omega_x(\xi, \eta) = 0$ for any $x \in \Lambda$ and any vectors ξ and η in $T_x\Lambda$ and, secondly, $\dim \Lambda = n$.

In the investigation of Fourier integral operators of the form (4.1) an important role is played by the set

$$C_S = \{(x, y, \theta) \in X \times Y \times (\mathbb{R}^N \backslash 0): d_\theta S(x, y, \theta) = 0\}.$$

We assume that the differentials of the functions $\partial S/\partial\theta_j$ ($j = 1, \ldots, N$) with respect to (x, y, θ) are linearly independent at each point of C_S. Then C_S is a smooth submanifold in $X \times Y \times (\mathbb{R}^N \backslash 0)$ of codimension N.

We further assume that $n_1 = n_2 = n$, $X = Y = \Omega$ and

$$D(S) = \det \begin{pmatrix} \dfrac{\partial^2 S}{\partial x \partial y} & \dfrac{\partial^2 S}{\partial x \partial \theta} \\ \dfrac{\partial^2 S}{\partial \theta \partial y} & \dfrac{\partial^2 S}{\partial \theta^2} \end{pmatrix} \neq 0. \tag{4.7}$$

If these requirements are fulfilled, the phase function S is called *non-degenerate*.

Consider the set

$$\Lambda_S = \{(x, \xi, y, \eta) \in \Omega \times (\mathbb{R}^n \backslash 0) \times \Omega \times (\mathbb{R}^n \backslash 0):$$
$$\exists\theta, \xi = \partial S(x, y, \theta)/\partial x, \eta = -\partial S(x, y, \theta)/\partial y, (x, y, \theta) \in C_S\}.$$

The condition (4.7) implies that for the local coordinates on C_S we can use either (x, ξ) or (y, η), where

$$\xi = \partial S(x, y, \theta)/\partial x, \quad \eta = -\partial S(x, y, \theta)/\partial y.$$

Indeed,

$$\frac{\partial\left(x, \dfrac{\partial S}{\partial x}, \dfrac{\partial S}{\partial \theta}\right)}{\partial(x, y, \theta)} = D(S) \neq 0, \quad \frac{\partial\left(y, -\dfrac{\partial S}{\partial y}, \dfrac{\partial S}{\partial \theta}\right)}{\partial(x, y, \theta)} = D(S) \neq 0,$$

so that the maps $(x, y, \theta) \mapsto \left(x, \dfrac{\partial S}{\partial x}, \dfrac{\partial S}{\partial \theta}\right)$ and $(x, y, \theta) \mapsto \left(y, -\dfrac{\partial S}{\partial y}, \dfrac{\partial S}{\partial \theta}\right)$ define local coordinates in a neighbourhood of the surface C_S. Thus if we consider the projections

$$p: C_S \to \Omega_x \times (\mathbb{R}^n \backslash 0), \quad q: C_S \to \Omega_y \times (\mathbb{R}^n \backslash 0),$$

we find that these projections are local diffeomorphisms under the condition (4.7). The map $r = p \circ q^{-1}$ is also a local diffeomorphism and can be defined by the equations

$$x = x(y, \eta), \quad \xi = \xi(y, \eta),$$

where x and ξ are positive and homogeneous in η of degree 0 and 1 respectively. The set Λ_S is conical and coincides with the graph of r. It is important to note that

$$\xi\, dx - \eta\, dy = \frac{\partial S}{\partial x} dx + \frac{\partial S}{\partial y} dy = dS(x, y, \theta) = 0$$

on C_S because $S = \theta \cdot \dfrac{\partial S}{\partial \theta} = 0$ on C_S. Therefore under r the form $\xi\, dx$ goes over to the form $\eta\, dy$, that is, the form $\xi\, dx - \eta\, dy$ is identically equal to zero on vectors which are tangent to Λ_S. Accordingly, the symplectic form

$$d\xi \wedge dx - d\eta \wedge dy,$$

defined on $\Omega \times (\mathbb{R}^n \setminus 0) \times \Omega \times (\mathbb{R}^n \setminus 0)$, vanishes identically on Λ_S, that is, Λ_S is a Lagrangian manifold.

We now assume that X is a smooth manifold of dimension n. Let L be a conical Lagrangian manifold in $T^*X \setminus 0$. Let us see how to associate with L *local phase functions*, these being real-valued smooth functions $S(x, \theta)$ which are positive and homogeneous in θ of degree 1, and defined in an open subset $\Gamma \subset \mathbb{R}^n \times (\mathbb{R}^N \setminus 0)$ with $d_{x,\theta} S \neq 0$ on Γ. Let S be such a function, and let the set

$$C_S = \left\{ (x, \theta) \in \Gamma \colon \frac{\partial S(x, \theta)}{\partial \theta} = 0 \right\}$$

be a smooth manifold. Let the rank of the differential of the map

$$\Gamma \ni (x, \theta) \mapsto \left(x, \frac{\partial S(x, \theta)}{\partial x} \right) \in T^*X \setminus 0 \tag{4.8}$$

be n at each point of C_S. We fix a point $(x_0, \theta_0) \in \Gamma$ and let $\xi_0 = \dfrac{\partial S}{\partial x}(x_0, \theta_0)$.

Definition 4.3. A phase function S is said to be *associated with a Lagrangian manifold* L in a neighbourhood of the point $(x_0, \theta_0) \in L$ if there exist conical neighbourhoods Γ^0 and Γ^1 of the points (x_0, θ_0) and (x_0, ξ_0) respectively, such that $L \cap \Gamma^1$ is the image of $C_S \cap \Gamma^0$ under the map (4.8).

The existence of such local phase functions is guaranteed by the following

Proposition 4.1. *Let L be a conical Lagrangian manifold in $T^*X \setminus 0$, and let $(x^0, \xi^0) \in L$. The coordinates $(x_1, \ldots, x_n)$ may be chosen in a neighbourhood of $x^0 \in X$ so that in the neighbourhood of (x^0, ξ^0) the manifold L is defined by the equations $x_j = \partial H(\xi)/\partial \xi_j$ $(j = 1, \ldots, n)$ where $H \in C^\infty$ in a conical neighbourhood of ξ^0 and $H(t\xi) = tH(\xi)$ for $t > 0$. Thus the phase function S associated with L in a neighbourhood of (x^0, ξ^0) can be chosen in the form*

$$S = x\xi - H(\xi).$$

Definition 4.4. The phase functions S and $\tilde{S}$ are said to be *locally eqivalent* if $S(x, \theta) = \tilde{S}(x, \tilde{\theta}(x, \theta))$ and the map $(x, \theta) \mapsto (x, \tilde{\theta}(x, \theta))$ is a local diffeomorphism, and $\tilde{\theta}(x, t\theta) = t\tilde{\theta}(x, \theta)$ for $t > 0$.

Theorem 4.10 (Hörmander [1971, 1983, 1985]). *The non-degenerate phase functions $\tilde{S}(x, \theta)$ and $\tilde{S}(x, \tilde{\theta})$, $\theta \in \mathbb{R}^{N_1}$, $\tilde{\theta} \in \mathbb{R}^{N_2}$, are locally equivalent if and only if*

1) $N_1 = \tilde{N}_2$;
2) *S and $\tilde{S}$ are associated with the same Lagrangian manifold L;*
3) *The matrices $\frac{\partial^2 S}{\partial \theta^2}$ and $\frac{\partial^2 \tilde{S}}{\partial \tilde{\theta}^2}$ have the same rank and the same number of positive and negative eigenvalues.*

In the investigation of Fourier integral operators of the form (4.1) an important role is played by a conical Lagrangian manifold Λ_S in $(T^*X\backslash 0) \times (T^*Y\backslash 0)$. In this case, Proposition 4.1 asserts that Λ_S can be defined locally by means of the phase function

$$S = x\xi - y\eta - H(\xi, \eta).$$

It is easy to see that this phase function is non-degenerate.

4.7. Lagrangian Manifolds and Fourier Distributions. Consider the integral

$$I_a(x) = \int e^{iS(x,\theta)} a(x, \theta)\, d\theta, \tag{4.9}$$

where S is a non-degenerate phase function and $a \in S^m(X \times \mathbb{R}^N)$. Repeating the arguments of § 4.2, we can give a meaning to the integral

$$\iint e^{iS(x,\theta)} a(x, \theta) u(x)\, dx\, d\theta, \quad u \in C_0^\infty(X). \tag{4.10}$$

This integral depends continuously on u, and thus $I_a \in \mathscr{D}'(X)$.

By the stationary phase method we can easily check that

$$\mathrm{WF}(I_a) \subset \left\{ \left(x, \frac{\partial S(x, \theta)}{\partial x} \right) \in T^*X\backslash 0\colon \frac{\partial S(x, \theta)}{\partial \theta} = 0 \right\} = L.$$

Definition 4.5. A *Fourier distribution* belonging to the class $I^m(X, L)$ is an integral of the form (4.9), where $a \in S^{m+(n-2N)/4}(X \times \mathbb{R}^N)$ and S is the phase function associated with the Lagrangian manifold L.

It can be shown that the class $I^m(X, L)$ actually depends only on L and not on the phase function S associated with L. If L can be defined only locally by the phase functions, then $I^m(X, L)$ is defined as the class consisting of finite sums of distributions (4.9) with phase functions S associated with L in neighbourhoods of points of L.

As an example of a Fourier distribution we mention the Schwartz kernel of a Fourier integral operator.

4.8. Global Definition of a Fourier Integral Operator. Let X and Y be two smooth manifolds of the same dimension n. A distribution $A \in \mathscr{D}'(X \times Y)$ defines a continuous bilinear form on $C_0^\infty(X) \times C_0^\infty(Y)$, and hence a continuous map $C_0^\infty(Y) \to \mathscr{D}'(X)$.

If Λ is a closed, conical Lagrangian manifold in $T^*(X \times Y)\backslash 0$, then the space of Fourier distributions $I^m(X \times Y, \Lambda)$ can be regarded as a space of continuous linear maps from $C_0^\infty(Y)$ into $\mathcal{D}'(X)$.

Definition 4.6. A closed conical submanifold C in $T^*(X \times Y)\backslash 0$ is called a *homogeneous canonical relation* from T^*Y into T^*X if C is contained in $(T^*X\backslash 0) \times (T^*Y\backslash 0)$ and is Lagrangian with respect to the form $\sigma_X - \sigma_Y$, where σ_X and σ_Y are the canonical symplectic forms on T^*X and T^*Y respectively.

As in § 4.6, we can show that a canonical homogeneous relation C can be defined locally by means of a non-degenerate phase function $S(x, y, \theta)$. Thus, with

$$C_S = \left\{(x, y, \theta): \frac{\partial S(x, y, \theta)}{\partial \theta} = 0\right\},$$

the map

$$(x, y, \theta) \mapsto \left(x, \frac{\partial S}{\partial x}, y, -\frac{\partial S}{\partial y}\right)$$

is a local homogeneous diffeomorphism of C_S onto C.

Since C is closed, there is a local finite covering of $T^*(X \times Y)\backslash 0$ by open conical sets Γ_j $(j = 1, 2, \dots)$ which are coordinate neighbourhoods in $T^*(X \times Y)\backslash 0$ and have the following property: for each j the set $C \cap \Gamma_j$ is defined by a non-degenerate phase function $S_j(x, y, \theta)$, $\theta \in \mathbb{R}^{N_j}$. Let $\{g_j\}$ $(j = 1, 2, \dots)$ be a smooth partition of unity in $T^*(X \times Y)\backslash 0$ subordinated to this covering, where the functions g_j are positively homogeneous in θ of degree 0.

The operator

$$\Phi u(x) = \sum_j \iint g_j(x, y, \theta) e^{iS_j(x, y, \theta)} a_j(x, y, \theta) u(y)\, dy\, d\theta$$

is called a *global Fourier integral operator* if $a_j \in S^{m+(n-2N_j)/4}$ $(j = 1, 2, \dots)$.

Example 4.4. If $X = Y$ and $C = \Delta_x^*$ is the diagonal in $(T^*X\backslash 0) \times (T^*Y\backslash 0)$, then we can take $(x - y)\theta$ as the phase function, where $\theta \in \mathbb{R}^n\backslash 0$. In this case the operator Φ is a pseudodifferential operator.

§ 5. Pseudodifferential Operators of Principal Type

5.1. Definition and Examples. Let $P(D)$ be a differential operator of order m with constant coefficients, and let $p_0(\xi)$ be its principal symbol.

Definition 5.1. The operator $P(D)$ is called an *operator of principal type* if $d_\xi p_0(\xi) \neq 0$ for $\xi \neq 0$, $\xi \in \mathbb{R}^n$.

The terminology is explained by the following

Theorem 5.1 (see Hörmander [1963]). *If $P(D)$ is an operator of principal type, then for each operator $Q(D)$ of order $m - 1$ with constant coefficients, there exists a constant $C > 0$ such that*

$$\|[P(D) + Q(D)]u\| \leqslant C(\|P(D)u\| + \|u\|), \quad u \in C_0^\infty(\mathbb{R}^n),$$

where $\|\cdot\|$ denotes the $L_2(\mathbb{R}^n)$-norm.

This theorem shows that many properties of P are determined by the principal symbol and they do not depend on the lower terms.

Example 5.1. The operators Δ and $\square$ are operators of principal type but the operators $\frac{\partial}{\partial t} \pm \Delta$ and $\frac{\partial}{\partial t} \pm i\Delta$ are not.

Definition 5.1 can be extended to general pseudodifferential operators in the following way.

Definition 5.2. A pseudodifferential operator $P(x, D)$, with principal symbol $p_0(x, \xi)$, is called an *operator of principal type* in a domain $\Omega \subset \mathbb{R}^n$ if, for all $(x, \xi) \in T^*\Omega \backslash 0$, the form $d_{x,\xi}p_0(x, \xi)$ is not proportional to the form $\xi\, dx$.

We observe that if the form $d_{x,\xi}p_0(x, \xi)$ of an operator $P(x, D)$ is proportional to the form $\xi\, dx$ at some point $(x, \xi) \in T^*\Omega \backslash 0$, then $d_\xi p_0(x, \xi) = 0$, and the Euler identity implies that $p_0(x, \xi) = 0$. Among non-elliptic operators the simplest operator of principal type is D_1. As we shall see below, every operator of principal type with real-valued principal symbol p_0 is microlocally equivalent to this simplest operator.

5.2. Operators with Real Principal Symbol. Let $P(x, D)$ be an operator of principal type of order m, with real principal symbol $p_0(x, \xi)$, and let $p_0(x_0, \xi^0) = 0$, where $x_0 \in \Omega$ and $\xi^0 \in \mathbb{R}^n \backslash 0$. Since elliptic operators are microlocally invertible in the algebra of pseudodifferential operators, the equation $P(x, D)u = f$ can be microlocally replaced by the equivalent equation $QPu = Qf$, where Q is an elliptic operator of order $1 - m$. Thus, without loss of generality, we can assume that $m = 1$.

Theorem 5.2 (Egorov [1984]). *Let $P(x, D)$ be a first-order pseudodifferential operator of principal type with real principal symbol $p_0(x, \xi)$. Let $(x^0, \xi^0) \in T^*\Omega \backslash 0$ and $p_0(x^0, \xi^0) = 0$. Then there exists an elliptic Fourier integral operator Φ of the form* (4.6) *such that one of the operators $P\Phi - \Phi D_1$ or $\Phi P - D_1\Phi$ has amplitude equal to zero in a conical neighbourhood of the point (x^0, ξ^0).*

We first examine the case when $d_\xi p_0(x^0, \xi^0) \neq 0$. To be definite, assume that $\partial_{\xi_1} p_0(x^0, \xi^0) \neq 0$. Consider the following auxiliary Cauchy problem

$$p_0\left(x, \frac{\partial S(x, \xi)}{\partial x}\right) = \xi_1, \quad S = \sum_{j=2}^{n} x_j \xi_j \quad \text{for } x_1 = x_1^0. \tag{5.1}$$

This problem has a solution in a conical neighbourhood of the point (x^0, ξ^0), since the plane $x_1 = x_1^0$ is non-characteristic in a neighbourhood of this point.

It can be seen easily that, in ω,

$$S(x, t\xi) = tS(x, \xi) \quad \text{for } t > 0 \quad \text{and} \quad \det\|\partial^2 S/\partial x_i \partial \xi_j\| \neq 0.$$

This last inequality is a consequence of the fact that $\det\|\partial^2 S(x^0, \xi^0)/\partial x_i \partial \xi_j\| = \partial^2 S(x^0, \xi^0)/\partial x_1 \partial \xi_1$, and

$$\sum_{j=1}^{n} \frac{\partial p_0(x, \partial S/\partial x)}{\partial \xi_j} \frac{\partial^2 S}{\partial x_j \partial \xi_1} = 1,$$

by (5.1). But in view of the initial conditions, $\partial^2 S/\partial x_j \partial \xi_1 = 0$ $(j = 2, \ldots, n)$ when $x_1 = x_1^0$. Hence $\partial^2 S(x^0, \xi^0)/\partial x_1 \partial \xi_1 \neq 0$. Consequently, S is a generating function for some canonical transformation. We set

$$\Phi u(x) = (2\pi)^{-n} \int \tilde{u}(\xi) e^{iS(x, \xi)}\, d\xi, \tag{5.2}$$

where S is defined arbitrarily outside ω but in such a way that $S(x, t\xi) = tS(x, \xi)$ for $t > 0$ and $\det\|\partial^2 S/\partial x_i \partial \xi_j\| \neq 0$. Then, by Theorem 4.7, we find that the operator

$$P\Phi - \Phi D_1$$

is of order zero in ω. Thus P is equivalent to an operator of the form $D_1 + A$ in ω, where A is a pseudodifferential operator of order zero. A further transformation can be carried out with the aid of the following

Lemma 5.1 (Egorov [1984]). *Let $P(x, D) = D_1 + A(x, D)$, where A is a pseudodifferential operator of order zero. Then there exists an elliptic pseudodifferential operator T of order zero such that*

$$SPT = D_1 + R,$$

where R is an infinitely smoothing operator and S is a parametrix of T.

Thus the problem of solving the equation $Pu = f$, where P is a pseudodifferential operator of principal type with real principal symbol, leads to that of solving the equation $D_1 v = g + Rv$, where R is of order $-\infty$ in ω.

If $d_\xi p_0(x^0, \xi^0) = 0$ but the form $d_x p_0(x^0, \xi^0)$ is not proportional to the form $\xi^0\, dx$, then the same construction is applicable but the equation (5.1) has to be replaced by another equation. With the aid of rotation in the x-space we can always arrange that $\operatorname{grad}_x p_0(x^0, \xi^0)$ has the form $(a, 0, \ldots, 0)$, where $a \neq 0$ and the vector $\xi^0 \neq 0$ is such that $\xi_1^0 = 0$.

Let us seek the generating function S in the form $S = S(y, \xi)$ such that S is a solution of the Cauchy problem

$$p_0\left(\frac{\partial S}{\partial \xi}, \xi\right) = \frac{\partial S}{\partial y_1}, \quad S = \sum_{j=2}^{n} y_j \xi_j + \frac{\xi_1^2}{2|\xi|} \quad \text{for } y_1 = 0.$$

Clearly, this problem has a solution in a conical neighbourhood ω of the point (y^0, ξ^0). Also, $y_1^0 = 0$, $S(y, t\xi) = tS(y, \xi)$ $(t > 0)$ and

$$\det \left\| \frac{\partial^2 S(y, \xi)}{\partial y_i \partial \xi_j} \right\| \neq 0 \quad \text{in } \omega.$$

On extending S to $T^*\Omega$, we find that the operator

$$\Phi P - D_1 \Phi$$

has order zero if Φ is the operator defined by (5.2).

Thus in this case too the equation $Pu = f$ leads to the equation $D_1 v = g + Rv$ in a conical neighbourhood of the point (x^0, ξ^0), where the symbol of R is zero in this neighbourhood.

5.3. Solvability of Equations of Principle Type with Real Principal Symbol. Theorems 3.7–3.10 on singularities of solutions of the equation $Pu = f$, together with Theorem 5.2, enable us to tackle the question of local solvability of this equation.

Definition 5.4. A pseudodifferential operator $P(x, D)$ is said to be *solvable at a point* $x^0 \in \Omega$ if there exist neighbourhoods U and V of this point in Ω such that $U \Subset V$ and if for every function $f \in C^\infty(V)$ there exists a distribution $u \in \mathscr{E}'(V)$ such that $Pu = f$ in U.

Definition 5.5. A pseudodifferential operator $P(x, D)$ is said to be *semi-globally solvable* in a domain $\Omega \subset \mathbb{R}^n$ if for each compact set K in Ω and for each function f in the subspace S with a finite codimension in $C^\infty(\Omega)$ there exists a distribution $u \in \mathscr{E}'(\Omega)$ such that $Pu = f$ in K.

The solvability conditions for P are determined by the behaviour of its bicharacteristics.

Example 5.2. The operator $x_2 D_1 - x_1 D_2$ is insolvable at the origin although it is of principal type.

The following condition on P in Ω depends on the choice of the compact set $K \Subset \Omega$.

Condition A. Let $K \Subset \Omega$. For each point $(x^0, \xi^0) \in T^*\Omega \backslash 0$ such that $p_0(x^0, \xi^0) = 0$, the *bicharacteristic*, that is, the integral curve of the Hamiltonian system

$$\dot{x}(t) = \partial p_0(x(t), \xi(t))/\partial \xi, \quad \dot{\xi}(t) = -\partial p_0(x(t), \xi(t))/\partial x$$

passing through the point (x^0, ξ^0), contains the points $(x, \xi) \in T^*\Omega \backslash 0$ whose projections onto Ω lie outside K.

Theorem 5.3 (Duistermaat and Hörmander [1972]). *The operator $P(x, D)$ is solvable at a point x^0 if there exists a neighbourhood K of this point where condition A holds. In particular, P is solvable at x^0 if $d_\xi p_0(x^0, \xi) \neq 0$ for all $\xi \in \mathbb{R}^n \backslash 0$.*

Theorem 5.4 (Duistermaat and Hörmander [1972]). *The operator $P(x, D)$ is semi-globally solvable in a domain $\Omega \subset \mathbb{R}^n$ if condition A holds for each compact* *set $K \Subset \Omega$. Moreover, if $f \in H^s(\Omega) \cap S$, then there exists a function $u \in \mathring{H}^{s+m-1}(\Omega)$ satisfying the equation $Pu = f$ in K and*

$$\|u\|_{s+m-1} \leqslant C\|f\|_s,$$

where the constant $C = C(s, K)$ is independent of f.

5.4. Solvability of Operators of Principal Type with Complex-valued Principal Symbol. If the pseudodifferential operator $P(x, D)$ of principal type has complex-valued symbol $p_0 = a + ib$, then the equation $Pu = f$, where $f \in C^\infty(\Omega)$, can generally be solved only when f belongs to a subspace having infinite codimension.

Example 5.3. From the historical point of view, H. Lewy gave the first example

$$\frac{\partial u}{\partial x_1} + i\frac{\partial u}{\partial x_2} + 2i(x_1 + ix_2)\frac{\partial u}{\partial x_3} = f(x_1, x_2, x_3)$$

with $f \in C^\infty(\mathbb{R}^3)$, which has no solution in any open set $\omega \subset \mathbb{R}^3$ (see Hörmander [1963]).

Example 5.4. An even simpler example is furnished by the equation

$$\frac{\partial u}{\partial x_1} + ix_1\frac{\partial u}{\partial x_2} = f(x_1, x_2).$$

There exists a second category set of functions f in $C^\infty(\mathbb{R}^n)$ for which this equation is insolvable in any open set on the plane which contains points of the $0x_2$-axis (see Hörmander [1963]).

Example 5.5. The oblique derivative problem for a second-order elliptic equation leads, after reduction to a problem on the boundary, to an insolvable pseudodifferential equation if there are "repelling" submanifolds of dimension $n - 2$. For example, in the case of the Laplace equation in the half-space $x_n > 0$, the problem with the condition

$$\frac{\partial u}{\partial x_1} + ax_1^k\frac{\partial u}{\partial x_n} = f \quad \text{for } x_n = 0$$

can be transformed to the equation

$$\frac{\partial v}{\partial x_1} - ax_1^k \Lambda v = f,$$

where v is the trace of the function u on the boundary and Λ is an operator with symbol $|\xi|$. If k is odd and $a > 0$, then the boundary-value problem has no solution u in any neighbourhood of the point lying on the manifold $x_n = 0$, $x_1 = 0$.

At present, the solvability theory for equations of principal type has advanced considerably. Let us list some of the results.

Theorem 5.5 (Hörmander [1963]). *If* $p_0(x^0, \xi^0) = 0$ *for some* $(x^0, \xi^0) \in T^*\Omega \backslash 0$, *and* $c_1^0(x^0, \xi^0) > 0$, *where* $c_1^0(x, \xi) = \operatorname{Im} \sum_{j=1}^{n} \frac{\partial p_0(x, \xi)}{\partial \xi_j} \frac{\partial p_0(x, \xi)}{\partial x_j}$, *then in every neighbourhood* ω *of* x_0 *there exists a function* $f \in C_0^\infty(\omega)$ *such that the equation* $Pu = f$ *has no solution belonging to* $\mathscr{D}'(\omega)$.

Theorem 5.6 (Beals and Fefferman [1973]). *Let P be an operator of principal type satisfying*

Condition ($\mathscr{P}$). *On each bicharacteristic of the function* $\operatorname{Re} p_0$ *along which* $\operatorname{Re} p_0 \equiv 0$, *the function* $\operatorname{Im} p_0$ *does not change sign.*

Then for each point $x^0 \in \Omega$ *there exists a neighbourhood* ω *such that for any function* f *in* $H^s(\omega)$ *there exists a function* $u \in H^{s+m-1}(\Omega)$ *satisfying the equation* $Pu = f$ *in* Ω. *Moreover,* $\|u\|_{s+m-1} \leqslant C\|f\|_s$, *where C is independent of f.*

Theorem 5.7 (Egorov [1975]). *Let P be an operator of principal type that satisfies the following two conditions.*

Condition (Ψ). *On each bicharacteristic of the function* $\operatorname{Re} p_0$ *along which* $\operatorname{Re} p_0 \equiv 0$, *the function* $\operatorname{Im} p_0$ *does not change sign from* $-$ *to* $+$ *in moving in the positive direction.*

Condition (B). *For each point* $(x^0, \xi^0) \in T^*\Omega \backslash 0$ *where* $p_0(x^0, \xi^0) = 0$, *there exists a function obtained from* p_0 *and* $\bar{p}_0$ *by means of the operation of forming the Poisson brackets, that is,* $\{\ldots\{\{p_0, \bar{p}_0\}, p_1\}\ldots p_j\}$ *in which each of the functions* $p_1, \ldots, p_j$ *is either* p_0 *or* $\bar{p}_0$ *and* $j \leqslant k - 1$, *and whose value at* (x^0, ξ^0) *is not zero.*

Then for each point $x^0 \in \Omega$ *a neighbourhood* ω *can be found such that for any function* $f \in H^s(\omega)$ *there exists a function in* $H^{s+m-\delta}(\Omega)$, *where* $\delta = k(k+1)^{-1}$, *which satisfies the equation* $Pu = f$ *in* ω. *Moreover,*

$$\|u\|_{s+m-\delta} \leqslant C\|f\|_s,$$

where the constant C is independent of f.

The operators P which are adjoint to those satisfying conditions (Ψ) and (B) are called *subelliptic*. For example, such operators arise in the study of the oblique derivative problem for second-order elliptic equations if the order of tangency of the field, given on the boundary of the domain, with this boundary does not exceed k.

§6. Mixed Problems for Hyperbolic Equations

6.1. Formulation of the Problem. In investigating vibrations of bounded bodies, it is necessary to take into account the interaction between the bodies and the external medium. This interaction is usually expressed by the boundary conditions which are imposed on the boundary of the body. We have (Egorov and Shubin [1988, §§ 4.13, 4.14, Chap. 2]) dealt with the results of the classical theory for a second-order hyperbolic equation with boundary conditions of one of the three kinds mentioned there. However, even for an outwardly similar mixed problem with the boundary condition $\frac{\partial u}{\partial \alpha} = g$, where α is a smooth vector field which does not coincide identically with the field of conormals, the classical theory does not furnish any methods to solve this problem. The main results in the general theory of boundary-value problems for hyperbolic equations were obtained in the 1970s. These results make significant use of such achievements of the general theory as the techniques of Fourier integral operators and the propagation of singularities.

6.2. The Hersh-Kreiss Condition. Let us consider a first-order system of equations of hyperbolic type with constant coefficients in a quadrant. Thus we consider the system

$$\frac{\partial u}{\partial t} + A\frac{\partial u}{\partial x} + \sum_{j=1}^{n} B_j \frac{\partial u}{\partial y_j} = 0, \quad t > 0, \quad x > 0,$$

with the conditions

$$u = 0 \quad \text{for } t = 0, x \geqslant 0,$$

$$Mu = g \quad \text{for } x = 0, t \geqslant 0,$$

where $u = (u_1, \ldots, u_N)$, A and B_j are $N \times N$ matrices and M is a constant $k \times N$ matrix. We assume that the eigenvalues of A are real and distinct and the plane $x = 0$ is non-characteristic, that is, the eigenvalues of A are not zero.

On applying a Fourier transformation with respect to y, we obtain a boundary-value problem in the quadrant of the (t, x)-plane:

$$\frac{\partial v}{\partial t} + A\frac{\partial v}{\partial x} + Cv = 0, \quad C = \sum_{j=1}^{n} i\eta_j B_j, \quad t > 0, \quad x > 0,$$

$$v = 0 \quad \text{for } t = 0, x \geqslant 0,$$

$$Mv = \tilde{g} \quad \text{for } x = 0, t \geqslant 0.$$

The substitution $v = Tw$, where T is a matrix whose columns consist of eigenvectors of A, transforms the differential part of the system in question to the diagonal form, and the whole system thus assumes the form

$$\frac{\partial w_j}{\partial t} + \lambda_j \frac{\partial w_j}{\partial x} + \sum_{k=1}^{N} \alpha_{jk} w_k = 0, \quad j = 1, \ldots, N,$$

where the λ_j are eigenvalues of A.

Let us renumber the coordinates $(w_1, \ldots, w_N)$ so that the first μ eigenvalues $\lambda_1, \ldots, \lambda_\mu$ are positive and the remaining ones negative. The resulting system is equivalent to the system

$$w_j(t, x) + \int_{l_j} \sum_{k=1}^{N} \alpha_{jk} w_k \, ds = w_{j0}, \quad j = 1, \ldots, N, \quad t > 0, \quad x > 0$$

of integral equations, where l_j is the segment of the line $x' = \lambda_j t' + c$ which ends at the point (t, x), and starts at the point P_j which lies on either the plane $x' = 0$ or the plane $t' = 0$, and w_{j0} is the value of w_j at this point P_j (see Egorov and Shubin [1988, §4.7, Chap. 2]). When the w_{j0} are known, this system can be solved easily (by the method of successive approximations, for example).

Since $x > 0$, only those lines for which $j = 1, \ldots, \mu$ can intersect the plane $x = 0$. Therefore the boundary-value problem in question is well-posed if the following two conditions hold:

1°. $\mu = k$;

2°. The first k columns of the matrix MT are linearly independent.

These conditions can also be expressed in the following equivalent form. Let E be the space spanned by the eigenvectors of A corresponding to the positive eigenvalues. Let K be the space spanned by the rows of M. Then the orthogonal projection of K onto E is of dimension k and $L \cap E = \{0\}$, where L is the orthogonal complement of K in $\mathbb{R}^N$.

The conditions just mentioned were obtained by Hersh [1963]. Balaban [1969] had earlier used these conditions under additional assumptions. Kreiss [1970] had formulated an analogous result for systems of equations with variable coefficients and remarked that the proof can be extended to this case. The complete proofs were published by Agranovich [1971] and Rauch [1972]. Let us cite the appropriate result.

Consider the system of equations

$$Lu \equiv \frac{\partial u}{\partial t} + \sum_{j=1}^{n} A_j(t, x) \frac{\partial u}{\partial x_j} + B(t, x)u = f \quad \text{in } [0, T] \times \Omega, \tag{6.1}$$

where Ω is a domain of $\mathbb{R}^n$ with smooth boundary Γ. Let the boundary conditions be of the form

$$M(t, x)u = g \quad \text{in } [0, T] \times \Gamma, \tag{6.2}$$

$$u = h \quad \text{in } \Omega \quad \text{for } t = 0. \tag{6.3}$$

Here $M(t, x)$ is a smooth $k \times N$ matrix with rank k, where k is the number of positive eigenvalues of the matrix $\sum_{j=1}^{n} \nu_j A_j$ and the vector $\nu = (\nu_1, \ldots, \nu_n)$ is normal to Γ. We assume that L is a strictly hyperbolic operator, and thus the eigenvalues of the matrix $\sum_{j=1}^{n} \xi_j A_j$ are real and distinct for all real $\xi \in \mathbb{R}^n \backslash 0$. We also assume that the boundary Γ does not contain characteristic points, that is, the matrix $\sum_{j=1}^{n} \nu_j A_j$ is non-degenerate throughout Γ.

Let us fix a point $x^0 \in \Gamma$. A smooth transformation in the neighbourhood of this point transforms Γ into a part of the plane $x_n = 0$ so that the image of Ω lies in $x_n > 0$. Having performed this transformation, we fix the values of the arguments of L and the matrix M by setting $t = t_0 > 0$ and $x = x^0$, and set $B = 0$. This results in a boundary-value problem in the segment $t > 0$, $x_n > 0$ of the space.

Theorem 6.1. *The problem* (6.1)–(6.3) *has a solution* $u \in C^1([0, T] \times \Omega)$ *for any* $f \in C^1([0, T] \times \bar{\Omega})$, $g \in C^1([0, T] \times \Gamma)$ *and* $h \in C^1(\bar{\Omega})$ *if and only if the above boundary-value problem with constant coefficients in the segment of the space satisfies conditions* 1° *and* 2° *for any point* (t_0, x^0). *Moreover,*

$$\|u\| \leqslant C(\|f\| + \|g\| + \|h\|),$$

where the norms are taken in the spaces L_2 *over suitable sets, namely, for* u *and* f *over* $\Omega \times [0, T]$, *for* g *over* $\Gamma \times [0, T]$ *and for* h *over* Ω.

Remark 6.1. If $f = 0$, $g = 0$ and $h \in L_2$, then the problem (6.1)–(6.3) has a unique solution $u(t, x)$ which lies in L_2 for each $t > 0$.

6.3. The Sakamoto Conditions. Let $P(D_t, D_x, D_y)$ be a homogeneous hyperbolic operator of order m with constant coefficients. Consider the boundary-value problem

$$Pu = 0 \quad \text{for } t > 0,\ x > 0,\ y \in \mathbb{R}^n,$$

$$u = 0 \quad \text{for } t \leqslant 0,\ x > 0;$$

$$B_j u = g_j \quad \text{for } t > 0,\ x = 0,\ j = 1, \ldots, k,$$

where $B_j = B_j(D_t, D_x, D_y)$ are homogeneous differential operators of order m_j.

As in the case of systems, the Fourier transformation with respect to y reduces this problem to a boundary-value problem on the plane (t, x). In view of hyperbolicity, the roots $\lambda_1, \ldots, \lambda_m$ of the characteristic equation

$$P(\lambda, \xi, \eta) = 0$$

are real for real $(\xi, \eta) \neq 0$. We assume that the plane $x = 0$ is non-characteristic with respect to P and consider the characteristic equation as an equation in ξ with $\operatorname{Im} \lambda < 0$. By hypothesis, this equation cannot have real roots. Let $\xi_1, \ldots, \xi_\mu$ be the roots of this equation for which $\operatorname{Im} \xi_j > 0$ and $\operatorname{Im} \xi_l < 0$ for $l = \mu + 1, \ldots, m$. We set $A_+(\xi) = \prod_{j=1}^{\mu} (\xi - \xi_j)$.

Applying again the techniques described in Egorov and Shubin [1988, § 1.7, Chap. 2], we find that for the mixed problem in question to be well posed it is necessary and sufficient that the following two conditions are satisfied:

1°. $\mu = k$;

2°. The polynomials $\xi \mapsto B_j(\tau, \xi, \eta)$ $(j = 1, \ldots, k)$ are linearly independent $\operatorname{mod}(A_+(\xi))$ for $(\tau, \eta) \neq 0$ and $\operatorname{Im} \tau \leqslant 0$.

Let

$$B_j^+(\tau, \xi, \eta) = \sum_{l=0}^{k-1} B_{jl}(\tau, \eta)\xi^l$$

be the remainder when B_j, as a polynomial in ξ, is divided by $A_+(\xi)$ for $j = 1, \dots, k$. The $k \times k$ matrix $\|B_{jl}(\tau, \eta)\|$ is known as the *Lopatinskij matrix*. Its determinant $R(\tau, \eta)$ is referred to as the *Lopatinskij determinant* corresponding to the system $(A_+, B_j, j = 1, \dots, k)$.

Sakamoto has shown that the conditions 1° and 2°, also known as the *uniform Lopatinskij condition*, are necessary and sufficient for the well-posedness (in the sense made precise in Theorem 6.2 below) of the boundary-value problem

$$P(t, x, y, D_t, D_x, D_y)u = f \quad \text{for } t > 0, x > 0,$$

$$u = 0 \quad \text{for } t \leqslant 0, x > 0,$$

$$B_j(t, y, D_t, D_x, D_y)u = g_j, \quad j = 1, \dots, k \quad \text{for } x = 0, t > 0.$$

We assume that conditions 1° and 2° are satisfied for each boundary-value problem that is obtained from the given one by dropping the lower members of the operators P and B_j and by fixing the values of the coefficients at each point $(t, 0, y)$ for $t \geqslant 0$.

In the investigation of the given boundary-value problem it proves to be more convenient to use the function spaces $H_{p,\gamma}(\mathbb{R}_+^{n+2})$ in which the norm is defined by the formula

$$\|v\|_{p,\gamma}^2 = \sum_{i+j+l+|\alpha|=p} \int_{\substack{x>0\\t>0}} |e^{-\gamma t}\gamma^i D_t^j D_x^l D_y^\alpha v(t, x, y)|^2 \, dt \, dx \, dy.$$

We also introduce the space $H_{p,\gamma}(\mathbb{R}_+^{n+1})$ with the norm defined by the formula

$$\langle v \rangle_{p,\gamma}^2 = \sum_{i+j+|\alpha|=p} \int_0^\infty dt \int_{\mathbb{R}^n} |e^{-\gamma t}\gamma^i D_t^j D_y^\alpha v(t, y)|^2 \, dy.$$

Theorem 6.2 (Sakamoto [1970]). *If the uniform Lopatinskij condition holds, then there exist positive numbers c and γ_0 such that the inequality*

$$\gamma \|u\|_{m-1,\gamma}^2 + \sum_{j=0}^{m-1} \langle D_x^j u \rangle_{m-1-j,\gamma}^2 \leqslant c \left\{ \frac{1}{\gamma} \|Au\|_{0,\gamma}^2 + \sum_{j=1}^{k} \langle B_j u \rangle_{m-1-m_j,\gamma}^2 \right\}$$

holds for $u \in H_{m,\gamma}(\mathbb{R}_+^{n+2})$ with $\gamma \geqslant \gamma_0$. When $\gamma \geqslant \gamma_0$, the boundary-value problem has a unique solution in the class $H_{m-1,\gamma}(\mathbb{R}_+^{n+2})$ for any functions $f \in H_{0,\gamma}(\mathbb{R}_+^{n+2})$ and $g_j \in H_{m-1-m_j,\gamma}(\mathbb{R}_+^{n+2})$.

We observe that in this case too an assertion similar to Remark 6.1 is valid. Agranovich [1971] has shown that the above boundary-value problem for a higher-order equation can be reduced to an equivalent problem of the form (6.1)–(6.3), and thus the Sakamoto conditions are indeed equivalent to the Kreiss conditions.

6.4. Reflection of Singularities on the Boundary. A new stage in the development of the theory of mixed problems for hyperbolic equations is concerned with the application of the technique of Fourier integral operators and the investigation of singularities of the solution. For interior points the singularities of the solution are propagated along the bicharacteristics (see Theorems 3.8 and 3.10). When the bicharacteristic meets the boundary of the domain, reflection occurs in accordance with the law of geometrical optics, and if the bicharacteristic is transverse to the boundary, then on reflection the singularity of the solution is propagated inside the domain along the reflected bicharacteristic.

This fact was first noted by Povzner and Sukharevskij [1960]. They studied the equation

$$\frac{\partial^2 u}{\partial t^2} = \frac{\partial^2 u}{\partial x^2} + \frac{\partial^2 u}{\partial y^2}$$

with the conditions that $u = 0$ on $\partial\Omega$, and $u = 0$ and $\dfrac{\partial u}{\partial t} = \delta(x - x_0, y - y_0)$ for $t = 0$ and $(x_0, y_0) \in \Omega$. It was assumed that Ω is a convex domain. In this case the rays, that is, the projections of the zero bicharacteristics onto the (t, x, y)-space, are straight lines making an angle of 45° with the t-axis. The rays passing through the point $(0, x_0, y_0)$ constitute a cone. In intersecting $[0, T] \times \partial\Omega$, every generator of this cone goes over to a ray that is symmetrical to the initial ray with respect to the normal to $[0, T] \times \partial\Omega$ and lying in the plane generated by this normal and the initial ray. If $T = \infty$, then the process of reflection of each ray is repeated infinitely many times, because the convexity condition gurantees that each such ray is always transverse to the boundary. Beyond the set of points (t, x, y) lying on these rays, the solution u is an infinitely differentiable function.

This result was generalized by Lax and Nirenberg (see Nirenberg [1973]). Let P be a differential operator of order m with real principal symbol p that has simple characteristic roots. Let Ω be a domain of $\mathbb{R}^n$ with smooth boundary S which is defined locally by the equation $\varphi(x) = 0$, and $\varphi > 0$ in Ω. It is assumed that the normal to S is not in the characteristic direction, that is, $p(x, \operatorname{grad} \varphi(x)) \neq 0$.

Let $x^0 \in S$ and $p(x^0, \xi^0) = 0$, $\xi^0 \neq 0$. Consider the equation

$$p(x^0, \xi^0 + \tau \operatorname{grad} \varphi(x_0)) = 0.$$

Suppose that this equation has k real roots $\tau_1, \ldots, \tau_k$ (one of which, by assumption, is zero) and that these roots are simple, that is,

$$\frac{\partial}{\partial \tau} p(x^0, \xi^0 + \tau_j \operatorname{grad} \varphi(x^0)) = \sum_{k=1}^{n} \frac{\partial p}{\partial \xi_k} \frac{\partial \varphi}{\partial x_k} \neq 0.$$

Let Γ_j $(j = 1, \ldots, k)$ be the part of the bicharacteristic p that lies in Ω and passes through the point $(x^0, \xi^0 + \tau_j \operatorname{grad} \varphi(x^0))$. Thus Γ_j is part of the integral curve of the system of equations

$$\dot{x}(s) = \frac{\partial p(x(s), \xi(s))}{\partial \xi}, \quad \dot{\xi}(s) = -\frac{\partial p(x(s), \xi(s))}{\partial x}$$

with initial conditions $x(0) = x^0$, $\xi(0) = \xi^0 + \tau_j \operatorname{grad} \varphi(x^0)$. These curves are not tangent to the surface S because

$$\frac{d}{ds}\varphi(x(s)) = \sum_{k=1}^{n} \frac{\partial \varphi}{\partial x_k}\dot{x}_k = \sum_{k=1}^{n} \frac{\partial \varphi}{\partial x_k}\frac{\partial p}{\partial \xi_k} \neq 0.$$

The embedding $i: S \to \mathbb{R}^n$ induces the map $i^*: T^*(\mathbb{R}^n) \to T^*S$ under which all the points of the form $(x^0, \xi + \lambda \operatorname{grad} \varphi(x^0))$, with $x^0 \in S$, are mapped into $i^*(x^0, \xi)$.

The curves $\Gamma_1, \ldots, \Gamma_k$ constructed above are known as the *reflected family of semi-bicharacteristics* corresponding to the point $i^*(x^0, \xi^0) \in T^*S$.

Theorem 6.3 (Nirenberg [1973]). *Assume that* $\Gamma_1, \ldots, \Gamma_{k_0}$ *do not contain points of* $\mathrm{WF}(u)$, *where* $0 \leqslant k_0 \leqslant k$. *Suppose that* $(x^0, \xi^0) \notin \mathrm{WF}\left(\left.\frac{\partial^j u}{\partial \nu^j}\right|_{\varphi=0}\right)$, *where* $0 \leqslant j \leqslant \frac{m+k}{2} - k_0 - 1$ *and* ν *is the normal to* S. *Then* $\Gamma_1, \ldots, \Gamma_k$ *do not contain points of* $\mathrm{WF}(u)$ and $(x^0, \xi^0) \notin \mathrm{WF}\left(\left.\frac{\partial^j u}{\partial \nu^j}\right|_{\varphi=0}\right)$ *for any* $j \geqslant 0$.

Examples 1. If $k = m$ and $k_0 = 0$, we obtain a theorem on the propagation of singularities for the solution of the Cauchy for a hyperbolic equation.

2. If $k = 0$, we have a theorem on the smoothness of the solution of the Dirichlet problem for an elliptic equation.

3. When $k = m > k_0 > 0$, we obtain a theorem on the smoothness of the solution of the boundary-value problem for a hyperbolic equation.

6.5. Friedlander's Example. The investigation of boundary-value problems in which the rays are allowed to be tangent to the boundary started with Friedlander's example [1976].

Consider the boundary-value problem

$$Pu \equiv (1 + x)\frac{\partial^2 u}{\partial y_n^2} - \frac{\partial^2 u}{\partial y_1^2} - \cdots - \frac{\partial^2 u}{\partial y_{n-1}^2} - \frac{\partial^2 u}{\partial x^2} = 0 \quad \text{for } y_n > 0,\ x > 0,$$

$$u = f(y) \quad \text{for } x = 0; \quad u = 0 \quad \text{for } y_n \leqslant 0,$$

with $f \in \mathscr{E}'(\mathbb{R}^n)$ and $\operatorname{supp} f \subset \{y, y_n > 0\}$. It can be seen easily that y_n increases with x along the bicharacteristics that are either tangent to the boundary $x = 0$ or have *hyperbolic direction* $(\xi, \eta_1, \ldots, \eta_n)$ at points of the boundary, that is, $\eta_n^2 > |\eta_1|^2$. It turns out that the singularities of the solution in this case are concentrated on those bicharacteristics that pass through the points belonging to $\mathrm{WF}(f)$.

Clearly, the bicharacteristics are defined by the equations

$$\dot{x} = -2\xi, \quad \dot{y}_j = -2\eta_j, \quad j = 1, \ldots, n-1; \quad \dot{y}_n = 2(1 + x)\eta_n,$$
$$\dot{\xi} = -\eta_n^2, \quad \dot{\eta}_j = 0, \quad j = 1, \ldots, n.$$

Integration of these equations yields

$$x = a_n^2 t^2 + 2a_0 t + b_0, \quad y_j = 2a_j t + b_j, \quad j = 1, \ldots, n-1,$$

$$y_n = \frac{2}{3} a_n^3 t^3 + 2a_0 a_n t^2 + 2a_n(1 + b_0)t + b_n; \quad a_j, b_j = \text{const}.$$

The Fourier transform $v(x, \eta)$ of the function u with respect to y reduces the equation to the form

$$\frac{d^2 v}{dx^2} + [(1 + x)\eta_n^2 - \eta_1^2 - \cdots - \eta_{n-1}^2]v = 0.$$

Setting $z = \eta_n^{-4/3}[\eta_1^2 + \cdots + \eta_{n-1}^2 - (1 + x)\eta_n^2]$, we obtain the equation

$$\frac{d^2 v}{dz^2} = zv,$$

which is known as the *Airy equation*. Its solutions are functions of the form

$$\operatorname{Ai}(z) = \int e^{i(\theta z - \theta^3/3)}\, d\theta,$$

as well as $\operatorname{Ai}(ze^{\pm 2\pi i/3})$. The function $\operatorname{Ai}(z)$ is bounded for positive real z and $\operatorname{Ai}(0) = 1$.

Thus the investigation of the boundary-value problem leads to the solution of the Airy equation with the boundary condition

$$v = \tilde{f}(\eta) \quad \text{for } z = \eta_n^{-4/3}(\eta_1^2 + \cdots + \eta_{n-1}^2 - \eta_n^2)$$

and to the inverse Fourier transform. It can be shown that the solution u of the boundary-value problem in question can be expressed as

$$u(x, y) = \frac{1}{(2\pi)^n} \iint e^{i[S(x,y,\xi) - S(0,z,\xi)]} f(z)\sigma(\zeta(0, z, \xi))$$

$$\times \frac{a(x, y, z, \xi)\operatorname{Ai}(\zeta(x, y, \xi)) + b(x, y, z, \xi)\operatorname{Ai}'(\zeta(x, y, \xi))}{\operatorname{Ai}(\zeta(0, z, \xi))}\, d\xi\, dz,$$

where S is a non-degenerate phase function that is homogeneous in ξ of degree 1, ζ is a non-degenerate phase function that is homogeneous in ξ of degree 2/3 and where σ is a smooth function such that $\sigma(s) = 0$ for $s \geqslant s_0 > 0$, and a and b are the standard symbols.

It can be shown that the following more general problem also leads to the computation of integrals of the same type. Let Ω be a bounded convex domain in $\mathbb{R}^n$ with smooth boundary Γ at each point of which the curvature is positive. Consider the following boundary-value problem:

$$\frac{\partial^2 u}{\partial t^2} = \Delta u \quad \text{for } t \geqslant 0, x \in \mathbb{R}^n \setminus \Omega,$$

$$Bu = f \quad \text{for } t > 0, x \in \Gamma; \quad u = 0 \quad \text{for } t \leqslant 0, x \in \mathbb{R}^n \setminus \Omega.$$

$$\text{Here} \quad Bu = u \quad \text{or} \quad Bu = \sum_{j=1}^{n} a_j \frac{\partial u}{\partial x_j} + bu.$$

Melrose [1978] has constructed the calculus of Fourier–Airy integral operators of the kind just mentioned.

6.6. Application of Canonical Transformations. Melrose later succeeded in simplifying the above class of "exterior" problems by means of the following result which was stated by Sato as a conjecture.

Let X be a symplectic manifold, and let F and G be two hypersurfaces in X given by the equations $f = 0$ and $g = 0$ respectively. Let $p \in F \cap G$. Then F and G are called *glancing surfaces* at the point p if

(1) df_p and dg_p are linearly independent;

(2) $\{f, g\}(p) = 0$;

(3) $\{\{f, g\}, f\}(p) \neq 0$, $\{\{f, g\}, g\}(p) \neq 0$.

The Sato Conjecture. If F and G are glancing surfaces at a point $p \in X$ and F' and G' are glancing surfaces at a point $p' \in X'$, where $\dim X = \dim X'$, then there exists a germ of a symplectic diffeomorphism

$$\Phi: (X, p) \to (X', p')$$

under which the pair (F, G) is mapped into the pair (F', G').

Oshima [1978] showed that the conjecture is false for real analytic surfaces if an analytic map Φ is sought. Melrose established that the conjecture is true in the case of infinitely smooth surfaces.

Theorem 6.4 (Melrose [1976]). *Every pair of glancing surfaces can be put in the form*

$$f = x_1, \quad g = \xi_1^2 - x_1 - \xi_n$$

by a symplectic transformation, or the form

$$f = x_1, \quad g = \xi_1^2 - x_1\xi_{n-1}^2 - \xi_n\xi_{n-1}$$

if only transformations that are homogeneous in ξ are considered.

Let Ω be a strictly convex domain in $\mathbb{R}^n$, and let $g(x, D)$ be a hyperbolic operator of the second order in $\mathbb{R}^n$ with symbol $g(x, \xi)$. Assume that the bicharacteristic passing through the point (x, ξ), where $x \in \Omega$, is *glancing*, that is, there is a point on it where the field $\sum_{j=1}^{n} \frac{\partial g}{\partial \xi_j}(x, \xi) \frac{\partial}{\partial x_j}$ is tangent to Γ. If Γ is described by the equation $f(x) = 0$, then at this point $\{f, g\}(x, \xi) = 0$. By an application of Melrose's theorem, we can "straighten" the boundary locally, that is, reduce it to the form $x_1 = 0$ so that the principal part of the operator is of the form $g(x, D) = D_1^2 - x_1 D_{n-1}^2 - D_n D_{n-1}$ in a conical neighbourhood of the point (x, ξ). The resulting boundary-value problem can be solved by means of the Fourier–Airy integral operators.

6.7. Classification of Boundary Points (Hörmander [1983, 1985]). Let us take up the classification of boundary points from the point of view of boundary-value problems for second-order hyperbolic equations.

Let P be a second-order differential operator, with principal symbol p, in a domain Ω having smooth boundary $\partial\Omega$. The function p is a constant along the integral curves of the Hamiltonian vector field H_p in $T^*\Omega$. By *null bicharacteristics* we mean those integral curves of H_p along which $p = 0$. Let $i: \partial\Omega \to \overline{\Omega}$ be the embedding and $i^*: T^*\Omega|_{\partial\Omega} \to T^*(\partial\Omega)$ the induced surjective map. The points of $T^*(\partial\Omega)\backslash 0$ are classified in terms of the behaviour of the bicharacteristics passing through the points of $T^*\Omega|_{\partial\Omega}$ which are projected onto them under i^*.

Suppose that $z \in T^*(\partial\Omega)\backslash 0$. If $p \neq 0$ on $i^{*-1}(z)$, then z is known as an *elliptic boundary point* for P. If the field H_p is transverse to $\partial T^*\Omega = T^*\Omega|_{\partial\Omega}$ and the set $i^{*-1}(z) \cap p^{-1}(0)$ consists of two distinct points, then z is known as a *hyperbolic point of the reflection*. It can be seen easily that in this case the field H_p is directed into Ω at one of these characteristic points and outwards at the other point.

For elliptic points the microlocal parametrix can be constructed by the theory of elliptic boundary-value problems, whereas the local theory of Fourier integral operators is adequate for dealing with the hyperbolic points.

The sets of elliptic and hyperbolic points are open. Let G denote the complement of the union of these two sets. Then G is a set of *glancing points* and is a closed conical set.

If $z \in G$, then $i^{*-1}(z) \cap p^{-1}(0)$ consists of one point, and if $v \in T^*\Omega|_{\partial\Omega}\backslash 0$ and $p(v) = 0$, then $i^*v \in G$ if and only if $\nu(H_p) = 0$ at the point v; here $\nu \in T_v^*(T^*\Omega)$ is the conormal to the boundary $\partial T^*\Omega$.

Let ν be a smooth section in $T^*(T^*\Omega)$ such that $\nu|_{\partial T^*\Omega}$ is a conormal covector to $\partial T^*\Omega$ directed into Ω in a neighbourhood of v. Clearly, $\nu(H_p) = 0$ at v, and hence $H_p(\nu(H_p))$ is independent of the choice of ν at the point v. If $H_p(\nu(H_p)) > 0$, then v is called a *diffractive point*. If $H_p(\nu(H_p)) < 0$, then v is called a *gliding point* and, if $H_p(\nu(H_p)) = 0$, then v is called a *point of contact of higher order*.

A point $z \in G$ is called *non-degenerate* if the derivative of the restriction of p to the fibre $T^*\Omega$, which contains $i^{*-1}(z)$, does not vanish at v. In a neighbourhood of a non-degenerate point in $T^*(\partial\Omega)\backslash 0$ there is a conical hypersurface which consists of diffractive points or gliding points. We note that if $v \in \partial T^*\Omega\backslash 0$, $p(v) = 0$ and $\nu(H_p)(v) = 0$, then the point i^*v is a diffractive point if and only if the bicharacteristic in $T^*\Omega$ is tangent to $\partial T^*\Omega$ at v and this contact is of the first order. This point is non-degenerate if the projection of the above bicharacteristic onto Ω does not have a singularity at the point πv (where π: $T^*\Omega \to \Omega$ is the canonical projection) and is tangent to the boundary $\partial\Omega$ there.

6.8. Taylor's Example. The following example of Taylor's [1981] shows that the singularities of the solution of a mixed problem can be propagated along the boundary without going inside the domain.

Let Ω be a strictly convex bounded domain in $\mathbb{R}^n$ with smooth boundary Γ. Let u be a solution of the wave equation

$$\frac{\partial^2 u}{\partial t^2} = \Delta u$$

in $\mathbb{R} \times \Omega$ such that $u = 0$ on $\mathbb{R} \times \Gamma$. We show that there exists a solution which is smooth in $\mathbb{R} \times \Omega$ but fails to be smooth in $\mathbb{R} \times \bar{\Omega}$.

Let $p_0 \in \Gamma$, $p_j \in \Omega$, and let $p_j \to p_0$ as $j \to \infty$. Let γ_j be the rays that pass through p_j, when $t = 0$, in the direction of a tangent to Γ at p_0 and continue in accordance with the rules of geometric optics after intersecting $\mathbb{R} \times \Gamma$. Let Ω_j be a sequence of domains such that $\Omega_j \Subset \Omega_{j+1} \Subset \Omega$ for all j, with $\bigcup_j \Omega_j = \Omega$, and such that the ray $\gamma_j(t)$ lies in $\Omega \setminus \Omega_j$ for $|t| \leqslant 1$.

Now let $\varphi_j \in L_{2,\mathrm{comp}}(\Omega)$, and let $\mathrm{WF}(\varphi_j)$ consist of a single bicharacteristic which lies above the ray γ_j such that $\varphi_j \in C^\infty(\Omega \setminus \{\gamma_j\})$ and $\varphi_j \in H^1(\Omega)$. Let v_j be a solution of the problem

$$\frac{\partial^2 v_j}{\partial t^2} = \Delta v_j \quad \text{in } \mathbb{R} \times \Omega, \quad v_j = 0 \quad \text{on } \mathbb{R} \times \Gamma,$$

$$v_j = \varphi_j, \quad \frac{\partial v_j}{\partial t} = 0 \quad \text{for } t = 0, x \in \Omega.$$

This solution has a singularity only on the ray γ_j inside Ω in view of the Lax-Nirenberg theorem quoted earlier.

Let $\operatorname{supp} \varphi_j \subset V_j$, where $V_j \Subset \Omega$ and V_j is a small neighbourhood of the point p_j. We set $u_j = c_j \int \rho_j(\tau) v_j(t - \tau, x)\, d\tau$, where $\rho_j \in C_0^\infty(\mathbb{R})$ is a function with small support and c_j are constants such that

$$\sum_{|\alpha| \leqslant j} \int_0^1 \int_{\Omega_j} |D^\alpha u_j|^2 \, dx\, dt \leqslant 2^{-j},$$

$$\int_{V_j} |\nabla u_j(0, x)|^2 \, dx \geqslant 10^j \left(1 + \sum_{k<j} \int_\Omega |\nabla u_k(0, x)|^2 \, dx\right),$$

$$\|u_j(0, x)\|_{L_2(\Omega)} \leqslant 2^{-j}.$$

Finally, let $u = \sum_{j=1}^\infty u_j$. Then $u \in C^\infty(\mathbb{R} \times \Omega)$, and u is a solution of the wave equation with $u = 0$ on $\mathbb{R} \times \Gamma$. But $u(0, x) \notin H^1(\Omega)$ and so u cannot lie in $C^\infty(\mathbb{R} \times \bar{\Omega})$. Thus the solution obtained has singularities only at points of $\mathbb{R} \times \Gamma$.

6.9. Oblique Derivative Problem. Since a separate paper will be devoted to hyperbolic equations, we do not give an exhaustive account of the results obtained during the past few years but confine our attention to the following typical example which required the powerful modern tools of the theory of differential equations for its investigation.

Let $A(x, D)$ be a second-order hyperbolic operator in $\mathbb{R}^{n+1}$ and let $x = (x_0, x_1, \ldots, x_n) \in \mathbb{R}^{n+1}$ where x_0 plays the role of time. Let G be a domain in $\mathbb{R}^n$ with smooth compact boundary ∂G. Let $\Omega = \mathbb{R} \times G$ and $\Gamma = \mathbb{R} \times \partial\Omega$. We examine the problem

$$A(x, D)u = 0 \quad \text{in } \Omega,$$

$$u = 0 \quad \text{for } x_0 \leqslant 0, x \in \Omega,$$

$$B(x, D)u = h(x) \quad \text{for } x \in \Gamma,$$

where B is a first-order differential operator.

When the domain G is unbounded, we assume that the coefficients of A are independent of x for large $|x|$. We also assume that the coefficients of A and B are smooth functions and that the boundary Γ does not contain characteristic points of A. In a neighbourhood of Γ we introduce the coordinates (x', x_n) such that x' is the coordinate on Γ while x_n denotes the distance from Γ. Let (ξ', ξ_n) be the dual coordinates in $T^*\Omega$ so that $\xi' \in T^*\Gamma$ and $\xi_n \in \mathbb{R}$. In this coordinate system, the principal symbol of $A(x, D)$ has the form $A_0(x, \xi) = (\xi_n - \lambda(x, \xi'))^2 - \mu(x, \xi')$ up to a factor $k(x)$ which does not vanish anywhere on Γ. In accordance with §6.7, we set

$$N_0 = \{(x', \xi') \in T^*(\Gamma)\backslash 0, \mu(x', 0, \xi') = 0\},$$

$$N_+ = \{(x', \xi') \in T^*(\Gamma)\backslash 0, \mu(x', 0, \xi') > 0\},$$

$$N_- = \{(x', \xi') \in T^*(\Gamma)\backslash 0, \mu(x', 0, \xi') < 0\}.$$

We put $b(x', \zeta') = B_0(x', 0, \zeta', \lambda(x', 0, \zeta') - \sqrt{\mu(x', 0, \zeta')})$, where $B_0(x', x_n, \xi', \xi_n)$ is the principal symbol of $B(x, D)$, $\zeta' = (\xi_0 + i\tau, \xi_1, \ldots, \xi_{n-1})$, $\tau > 0$, and the values of $\sqrt{\mu}$ are chosen so that $\operatorname{Im} \sqrt{\mu(x', 0, \zeta')} > 0$ when $(x', \xi') \in T^*(\Gamma)\backslash 0$ and $\tau > 0$.

It has been established that for the boundary-value problem to be well posed it is necessary that

(I) $b(x', \xi') \neq 0$ for all $(x', \xi') \in T^*(\Gamma)\backslash 0$, $\tau > 0$,

(see Kajitani [1974]). It can be shown further that the solution of the boundary-value problem on $N_+ \cup N_-$ can be microlocally reduced to the solution of the pseudodifferential equation

$$b(x', D')u_0 = h, \quad \text{where } u_0 = u|_\Gamma.$$

Note that the function $b(x', \xi')$ is continuous on $T^*\Gamma\backslash 0$ and is smooth outside N_0. We assume that the following conditions are satisfied:

(II) $\dfrac{\partial b(\bar{x}, \bar{\xi})}{\partial \xi_0} \neq 0$, if $b(\bar{x}, \bar{\xi}) = 0$, $(\bar{x}, \bar{\xi}) \in N_+ \cup N_-$,

(III) $\dfrac{\partial B_0(\bar{x}, 0, \bar{\xi}, \lambda(\bar{x}, 0, \bar{\xi}))}{\partial \xi_n} \neq 0$, if $b(\bar{x}, \bar{\xi}) = 0$, $(\bar{x}, \bar{\xi}) \in N_0$.

Finally, we also assume that

(IV) $\{\xi_n - \lambda(x', x_n, \xi'), \mu(x', x_n, \xi')\} > 0$ for $x_n = 0$, if $\mu(x', 0, \xi') = 0$; or

(IVa) $\{\xi_n - \lambda(x', x_n, \xi'), \mu(x', x_n, \xi')\} < 0$ for $x_n = 0$, if $\mu(x', 0, \xi') = 0$.

Theorem 6.5. *If the conditions* (I)–(IV) *hold, then for each function* h *in* $H^{s,\tau}(\Gamma)$ *which vanishes for* $x_0 < 0$ *there exists a unique solution* $u \in H^{s,\tau}(\Omega)$ *provided that* $\tau > 0$ *is sufficiently large.*

Here $H^{s,\tau}(\Omega)$ denotes the space with the norm $\|u\|_{s,\tau} = \|e^{-x_0\tau}u\|_s$.

Example 6.1. The hypotheses of the theorem are satisfied for the wave equation

$$\frac{\partial^2 u}{\partial x_0^2} - \sum_{k=1}^{n} \frac{\partial^2 u}{\partial x_k^2} = 0 \quad \text{in } \Omega = \mathbb{R} \times G,$$

where G is the exterior of a strictly convex compact domain, which the boundary conditions

$$\sum_{k=0}^{n} a_k(x') \frac{\partial u}{\partial x_k} = h(x'), \quad x' \in \Gamma,$$

$$u = 0 \quad \text{for } x_0 \leqslant 0,$$

if

1) the vector $(a_0, \ldots, a_n)$ is not tangent to the boundary Γ at any point;

2) $a_0(x') < \sum_{k=1}^{n} a_k(x') \nu_k(x')$, where $(\nu_1, \ldots, \nu_n)$ is the unit normal to Γ which is directed into G.

The boundary-value problem under condition (IVa), instead of (IV), has been found to be more difficult to handle. It follows from condition (III) that, in a neighbourhood of each point $(\bar{x}, \bar{\xi})$ belonging to N_0 and such that $b(\bar{x}, \bar{\xi}) = 0$, the following relation holds:

$$b(x', \zeta') = b_1(x', \zeta', \sqrt{\mu(x', 0, \zeta')})(\sqrt{\mu(x', 0, \zeta')} + \lambda_1(x', \zeta')),$$

where $b_1(x', \zeta', \sqrt{\mu(x', 0, \zeta')}) \neq 0$.

Assume that, for each point $(\bar{x}, \xi) \in (T^*(\Gamma)\backslash 0) \cap N_0$ such that $b(\bar{x}, \bar{\xi}) = 0$, there exists a neighbourhood $U_0 \subset N_0$ where

(V) $-\operatorname{Re} \lambda_1(x', \xi') \leqslant C_0 \operatorname{Im} \lambda_1(x', \xi')^2 |\xi'|^{-1}$ for $(x', \xi') \in U_0$.

Theorem 6.6. *Let the conditions* (I)–(III), (IVa) *and* (V) *be satisfied. Then the boundary-value problem has a unique solution* $u \in H^{s,\tau}(\Omega)$ *for each function* $h \in H^{s,\tau}(\Gamma)$ *which vanishes for* $x_0 < 0$.

When the function λ_1 is real valued, the condition (V) means that $\lambda_1(x', \xi') \geqslant 0$ on N_0.

Theorem 6.7. *Let the function* λ_1 *be real valued and let there be a point* $(\bar{x}, \bar{\xi}) \in N_0$ *such that* $\lambda_1(\bar{x}, \bar{\xi}) = 0$ *and condition* (V) *holds at this point. Assume that there is a sequence of points* $(x^{(k)}, \xi^{(k)}) \in N_0$ *which converges to* $(\bar{x}, \bar{\xi})$ *and is such that* $\lambda_1(x^{(k)}, \xi^{(k)}) < 0$. *Finally, let*

$$\{\lambda_1(x', \xi'), \mu(x', 0, \xi')\} = 0 \quad \text{if } (x', \xi') = (\bar{x}, \bar{\xi}).$$

Then the boundary-value problem is ill posed. More precisely, there is a function $h \in C_0^\infty(\Gamma)$ *such that* $h = 0$ *when* $x_0 < 0$ *but there is no distribution* $u \in \mathcal{D}'(\Omega)$ *that would satisfy the equation* $Au = 0$ *and the boundary conditions for* $-\infty < x_0 < \bar{x}_0 + \varepsilon$, *no matter how small* $\varepsilon > 0$. *Here* $\bar{x}_0$ *denotes the first coordinate of the point* $\bar{x}$.

The above Theorems 6.5–6.7 are due to Eskin [1984].

Examples 1. Let A be the wave operator and let

$$B(x, D) = \frac{\partial u}{\partial \nu} + a(x_0)\frac{\partial u}{\partial x_0}$$

where a is a real-valued function and ν denotes the direction of the normal.

In the exterior of a convex domain the only condition that is necessary and sufficient for the well posedness is of the form

$$a(x_0) > -1.$$

If, however, the domain is convex and the problem is to be solved inside it, then, for example, for $a = (x_0 - \bar{x}_0)^3$ the problem is ill posed. This statement remains valid, in general, in each case when there is a sequence of points $x_0^{(k)}$ which converges to $\bar{x}_0$ and is such that $a(x_0^{(k)}) < 0$, with $a(\bar{x}_0) = 0$ and $a'(\bar{x}_0) = 0$.

2. Let A be the wave operator with $n = 2$, and let $B = \frac{\partial}{\partial \nu} + b(s)\frac{\partial}{\partial s}$, where $\frac{\partial}{\partial s}$ denotes the derivative along the boundary.

In the exterior of a convex domain the boundary-value problem is always well posed. The interior boundary-value problem is well posed, however, if and only if all the zeros of b are simple.

§7. Method of Stationary Phase and Short-wave Asymptotics

The method of stationary phase is one of the simplest methods for finding asymptotics of integrals, and yet it leads already to important results in the theory of hyperbolic equations and in various asymptotic problems for elliptic equations. With this method is closely associated the classical WKB method, so called in honour of the physicists Wentzel, Kramers and Brillouin who were the first to apply this method to the problems of quantum mechanics. The development of the WKB method leads to significant analytic and geometrical constructions, and one of the most important among these is the Maslov canonical operator which enables us to solve a large class of asymptotic problems.

7.1. Method of Stationary Phase (Fedoryuk [1971, 1977], Guillemin and Sternberg [1977], Hörmander [1983, 1985], Vajnberg [1982]). The *method of stationary phase* is a method for finding asymptotics of integrals of the form

$$I(a, \lambda) = \int e^{i\lambda f(x,a)} g(x, a, \lambda)\, dx, \tag{7.1}$$

where $\lambda \to +\infty$, $x \in \mathbb{R}^n$, the parameter $a \in \mathbb{R}^p$, and where f and g are smooth functions. The main idea of this method consists of the observation that if the phase function f is real valued, then the exponential function $e^{i\lambda f(x,a)}$ oscillates rapidly at those points x where $f_x'(x, a) = \partial f(x, a)/\partial x \neq 0$. Due to interference, the effect of neighbourhoods of such points on the integral is small. Therefore the main contribution comes from the *points of stationary phase*; that is, from the critical points of f with respect to x, being the points x where $f_x'(x, a) = 0$.

Let us formulate the precise results. We assume that the parameter a runs through a compact set $A \subset \mathbb{R}^p$ and the amplitude g has, as a function of x, a compact support uniformly with respect to $a \in A$. Thus, there is a compact set $K \subset \mathbb{R}^n$ such that $g(x, a, \lambda) = 0$ when $x \notin K$, $a \in A$ and $\lambda \geqslant 1$. To simplify the discussion, we assume further that $f \in C^\infty(\mathbb{R}^n \times \mathbb{R}^p)$ and $g(\cdot, \cdot, \lambda) \in C^\infty(\mathbb{R}^n \times V)$ for each fixed λ, where V is a fixed neighbourhood of the compact set A in $\mathbb{R}^p$. Suppose that

$$|\partial_x^\alpha g(x, a, \lambda)| \leqslant c_\alpha \lambda^{m+\delta|\alpha|}, \quad x \in K, \quad a \in A, \quad \lambda \geqslant 1, \tag{7.2}$$

where α is any multi-index and $\delta < 1$ is fixed. We put

$$\textstyle\sum_f = \{x\colon x \in K, \partial_x f(x, a) = 0 \text{ for some } a \in A\}.$$

The set $\sum_f$ is the one that must play the main role in the asymptotics of the integral (7.1). It can be seen easily that, for example, if there exists a neighbourhood U of $\sum_f$ such that $g(x, a, \lambda) = 0$ for $x \in U$, $a \in A$ and $\lambda \geqslant 1$ or, more generally, if $g(x, a, \lambda) = O(\lambda^{-N})$ $(\lambda \to +\infty)$ uniformly in $(x, a) \subset U \times A$ for any fixed N, then $I(a, \lambda) = O(\lambda^{-N})$ $(\lambda \to +\infty)$ also uniformly in $a \in A$ for any fixed N. To establish this assertion, we perform the integration by parts after using a partition of unity and an identity of the form $L^N(e^{i\lambda f}) = \lambda^N e^{i\lambda f}$ for $x \in U$, where $L = \sum_{j=1}^n c_j(x, a) D_j$, $c_j = \left(\sum_{k=1}^n \left|\frac{\partial f}{\partial x_k}\right|^2\right)^{-1} \frac{\partial f}{\partial x_j}$ $(j = 1, \ldots, n)$. Then, if we assume that $g(x, a, \lambda) = 0$ for $x \in U$, $a \in A$ and $\lambda \geqslant 1$, we obtain

$$I(a, \lambda) = \lambda^{-N} \int e^{i\lambda f(x,a)} ({}^tL)^N g(x, a, \lambda)\, dx,$$

where the transpose ${}^tL$ of L is a differential operator with respect to x (also of the first order). This immediately implies that $I(a, \lambda) = O(\lambda^{-(1-\delta)N})$ uniformly in $a \in A$, and this is enough to prove our assertion since N can be chosen as large as we please.

Let us now describe the contribution of all those points of stationary phase that are *non-degenerate critical points*, that is, points x for which $f_x'(x, a) = 0$ and the matrix $f''_{xx}(x, a)$ consisting of the second-order derivatives is non-degenerate. Let us fix one such point x_0 and the corresponding value a_0 of the parameter. By the implicit function theorem, x_0 is an isolated point in the set of all points of stationary phase (for the given a_0), and if the parameter a is varied, then for a close to a_0 there is a smooth vector function $x = x(a)$ such that $x(a_0) = x_0$ and $f'_x(x(a), a) \equiv 0$. That is, $x(a)$ is a critical point which depends smoothly on the parameter, and such a vector function $x(a)$ is locally unique with respect to a. For the given a_0, we confine our attention to a small neighbourhood of a_0. By using a partition of unity with respect to x, we can reduce the problem to the case when the amplitude g, as a function of x, is concentrated in a small neighbourhood of the point x_0. By the Morse Lemma, we can choose the local coordinates $y = y(x, a)$ in this neighbourhood, depending smoothly on a, such that $y(x, a) = x - x(a) + O(|x - x(a)|^2)$ as $x - x(a) \to 0$ and, in the new coordinates, f takes the form $f(x, a) = f(x(a), a) + \frac{1}{2}\langle Q(a)y, y\rangle$, where $Q(a) = f''_{xx}(x(a), a)$. This implies that in the new coordinates the critical point is the origin, and up to an additive term, which is independent of y, $f(x, a)$ becomes a non-degenerate quadratic form with matrix $Q(a) = f''_{xx}(x(a), a)$ which is equal to the matrix of the second differential of f in the original coordinates. Thus in the new coordinates we have

$$I(a, \lambda) = e^{i\lambda f(x(a), a)} \int e^{i\lambda \langle Q(a)y, y\rangle} g_1(y, a, \lambda)\, dy, \tag{7.3}$$

where g_1 is the new amplitude of the form

$$g_1(y, a, \lambda) = g(x(y, a), a, \lambda) \left| \det \frac{\partial x(y, a)}{\partial y} \right|.$$

We now assume that $\delta < 1/2$, and we note that $\delta = 0$ in the most important particular case when $g(x, a, \lambda) = g(x, a)$ is independent of λ. Then the following result is true.

Theorem 7.1. *The asymptotics of the integral* (7.3) *have the form*

$$I(a, \lambda) \sim (2\pi)^{n/2} e^{i\lambda f(x(a), a)} \lambda^{-n/2} |\det Q(a)|^{-1/2} \exp\left[\frac{\pi i}{4} \operatorname{sgn} Q(a)\right]$$

$$\times \sum_{k=0}^{\infty} \lambda^{-k} [R^k g_1(0, a, \lambda)]/k!. \tag{7.4}$$

Here sgn $Q(a)$ *denotes the signature of a non-degenerate symmetric matrix* $Q(a)$ *and is equal to the signature of the corresponding quadratic form. Thus* sgn $Q(a)$ *is equal to the difference between the number of positive eigenvalues of the matrix* $Q(a)$ *and the number of negative eigenvalues, with all multiplicities counted* (sgn $Q(a)$ *is independent of* a *for* a *close to* a_0). *Also,* $R = \frac{i}{2}\left\langle Q(a)^{-1}\frac{\partial}{\partial y}, \frac{\partial}{\partial y}\right\rangle$ *and is a linear second-order differential operator with respect to* y.

The meaning of the asymptotic series (7.4) lies in the property that by cutting off the expansion (7.4) after taking a sufficient number of terms we obtain a remainder of the form $O(\lambda^{-N})$ (uniformly in those a which are close to a_0), where N is an arbitrary number. We can obtain (7.4) by expanding, for example, the amplitude $g_1(y, a, \lambda)$ by Taylor's formula in y at the point $y = 0$ and then computing the model integrals that appear in each term and, finally, estimating the remainder.

Let us indicate an alternative way of obtaining the asymptotics of integrals of the form (7.1) that does not require a knowledge of the Morse coordinates y. We assume, as before, that there is a unique critical point $x(a)$ which is nondegenerate. We consider the quadratic part of the function $f(x, a)$ at the point $x(a)$:

$$f(x, a) = f(x(a), a) + \tfrac{1}{2}\langle Q(a)(x - x(a)), x - x(a)\rangle + r(x, a),$$

where $Q(a) = f''_{xx}(x(a), a)$ and the remainder $r(x, a)$ is determined by this formula and $r(x, a) = O(|x - x(a)|^3)$ as $x \to x(a)$.

Theorem 7.1′. *The asymptotic expansion*

$$I(a, \lambda) \sim (2\pi)^{n/2} e^{i\lambda f(x(a),a)} \lambda^{-n/2} |\det Q(a)|^{-1/2} \exp\left[\frac{\pi i}{4} \operatorname{sgn} Q(a)\right]$$

$$\times \sum_{k=0}^{\infty} \frac{\lambda^{-k}}{k!} [R^k(g(x, a, \lambda) e^{i\lambda r(x,a)})]|_{x=x(a)} \tag{7.5}$$

holds.

Here R is the same operator as in (7.4) and the expansion has analogous meaning. The expansion can be established by reducing it to (7.4) if $ge^{i\lambda r}$ is taken as the new amplitude and if it is verified beforehand that the discussion can be confined to a small neighbourhood of the point $x(a)$ by means of a cut-off function of the form $\varphi(\lambda^{\varkappa}(x - x(a)))$, where $0 < \varkappa < 1/2$, $\varphi \in C_0^\infty(\mathbb{R}^n)$ and $\varphi = 1$ in a neighbourhood of 0.

Let us cite an example to show how integrals of the form (7.1) arise. In $\mathbb{R}^3$, let there be a 2-dimensional surface S all the points of which are sources of radiation of a fixed frequency k with density $\alpha(y)\, dy$, where dy denotes the elementary surface area on S. If the propagation speed of the wave is taken to be unity, then the simplest sinusoidal outgoing wave is of the form $r^{-1}e^{ik(r-t)}$, where r denotes the distance from the source. Thus, under pointwise source at each point of the space, sinusoidal vibration with complex amplitude $r^{-1}e^{ikr}$ takes place, and all the sources situated on the surface give vibrations with complex amplitude

$$I(x, k) = \int_S \frac{e^{ik|x-y|}}{|x - y|} \alpha(y)\, dy \tag{7.6}$$

at the point $x \in \mathbb{R}^3$. To the passage to limit $k \to +\infty$, there corresponds transition from the wave optics to the geometric optics of the rays, and thus it is natural to try to obtain asymptotics of (7.6) as $k \to +\infty$. But this integral has the form (7.1) to within notation (here k is the parameter of the asymptotics which was earlier denoted by λ, and the point x in (7.6) plays the role of the parameter which was earlier played by a). The phase function $\varphi(y, x) = |y - x|$ has critical points precisely at those points y where the vector $x - y$ is proportional to the unit vector n_y which is normal to S at the point y. What is more, to a given x there may correspond several points y (or even infinitely many such points that may constitute a curve, for example). It is easy, however, to mention necessary and sufficient conditions on x under which all the corresponding critical points y are non-degenerate. Namely, x must not be a *focal point* of S. This means that x must not be a critical value of the map $S \times \mathbb{R}^1 \to \mathbb{R}^3$ which maps the point (y, s) into the point $y + sn_y$. In this case, the asymptotics of the integral (7.6) can be found according to the above scheme.

The problem of finding the asymptotics of integrals of the form (7.1) with degenerate critical points is a complicated one, and up to the present time it has not been investigated completely (see, for example, the book by Arnol'd, Varchenko, and Gusejn-Zade [1984] and the literature cited there).

7.2. Local Asymptotic Solutions of Hyperbolic Equations. Let us try to understand how wave optics goes over to geometric optics if the waves are very short, or equivalently the frequencies are very high. We first consider a plane wave with frequency ω and wave vector k:

$$u(t, x) = e^{i(\omega t - kx)} = e^{i\omega(t - x\cdot k/\omega)}.$$

We substitute for u into the wave equation

$$\Box u \equiv u_{tt} - a^2 \Delta u = 0, \tag{7.7}$$

and see that the equation is satisfied if and only if the vector k/ω has length $1/a$, where a is the speed of the wave propagation. In the case of the usual wave equation the quantity a is a constant, independent of ω. Thus for any vector k_0 of length $1/a$ and for any ω there is a plane wave of the form

$$u(t, x) = e^{i\omega(t - k_0\cdot x)}. \tag{7.8}$$

We now regard k_0 as fixed and let the frequency ω tend to $+\infty$. This is what is meant in the present case by the transition to geometric optics for the plane wave (7.8). Let us consider the phase function $\varphi = \omega(t - k_0\cdot x)$ of this wave and let us follow the surfaces $\varphi = \text{const}$ of the constant phase which are referred to as the *wave fronts*. We note that this concept of the wave front differs from that of wave front sets introduced in § 3, although there is a connection between them that will be mentioned below. For each t the wave front is the plane

$$\{x\colon k_0\cdot x = t + c\} \subset \mathbb{R}^n$$

that moves with changing t, with speed a in the direction of the vector k_0. Hence it is natural to refer to the straight lines going in the direction of k_0 as *rays*.

We now examine the general differential operator

$$A = a(x, D) = \sum_{|\alpha| \leq m} a_\alpha(x) D^\alpha, \quad x \in \Omega \subset \mathbb{R}^n,$$

with smooth coefficients and try to obtain, in analogy with the plane wave (7.8), the solution of the equation $Au = 0$ in the form

$$u(x) = u(x, \lambda) = e^{i\lambda S(x)}, \tag{7.9}$$

where λ is a large parameter (which plays the role of frequency), $\lambda \to +\infty$, and S is a real-valued smooth function. Substitution into the equation gives

$$Au(x) = e^{i\lambda S(x)}[\lambda^m a_m(x, S_x(x)) + O(\lambda^{m-1})],$$

where $a_m = a_m(x, \xi)$ is the principal symbol of A and $S_x = \partial S/\partial x = \operatorname{grad} S$. For the time being we are not concerned with the terms denoted by $O(\lambda^{m-1})$, which have the form of a polynomial in λ of degree not exceeding $m - 1$, with coefficients belonging to $C^\infty(\Omega)$. We assume here, and below, that the principal symbol a_m is real valued. Then it is clear that if the equation $Au = 0$ is satisfied up to $O(\lambda^{m-1})$ for large λ, the phase function $S = S(x)$ must satisfy the Hamilton-Jacobi equation

$$a_m(x, S_x(x)) \equiv 0, \tag{7.10}$$

which is also referred to as the *eikonal equation*. (The terminology *eikonal* has been taken from geometric optics where this term refers to the optical length of the path of the ray of light between two arbitrary points one of which lies in the space of objects and the other in the space of images.) This equation guarantees that the original equation $Au = 0$ is satisfied "to the first approximation".

To find a more precise solution of the equation, we seek the solution in the form of the formal series

$$u(x, \lambda) = e^{i\lambda S(x)} \sum_{j=0}^{\infty} b_j(x) \lambda^{-j}. \tag{7.11}$$

Applying the operator A to each term of this series, that is, to each function of the form $e^{i\lambda S(x)} b_j(x) \lambda^{-j}$, we obtain a sum of terms of the same form, but possibly with positive powers of λ. Hence by rearranging the terms, we see that the series Au can be written in the form

$$Au = e^{i\lambda S(x)} \sum_{j=0}^{\infty} v_j(x) \lambda^{m-j}. \tag{7.12}$$

We call the series for u the *asymptotic solution* or *rapidly oscillating asymptotic solution* if $Au = 0$, that is, if all the terms in (7.12) vanish. If we consider a finite sum

$$u_N(x, \lambda) = e^{i\lambda S(x)} \sum_{j=0}^{N} b_j(x) \lambda^{-j}$$

of this series, we obtain a function on Ω which depends on λ as a parameter and is such that

$$Au_N = O(\lambda^{-N-1+m})$$

as $\lambda \to +\infty$. Thus, the truncations u_N of the series which is an asymptotic solution are "almost solutions", and the larger the integer N the more accurate they are.

Since the equation is linear, it is clear that the asymptotic solution $\tilde{u}$ can be multiplied by a numerical factor. For example, if we multiply the solution by λ^{-k}, we again obtain an asymptotic solution, and so the series may start from $j = -k$ instead of $j = 0$. Further, we can assume that $b_0 \not\equiv 0$ in any of the subdomains of Ω because otherwise in such a subdomain we can start the series at the first non-zero term by taking a suitable power of λ outside the summation sign. Since the first term of the series for Au has the form

$$\lambda^m a_m(x, S_x(x)) b_0(x) e^{i\lambda S(x)},$$

we see that the phase function S must satisfy the eikonal equation. The computation of the next terms of Au gives

$$v_j = \sum_{l+k=j} (A_l b_k), \quad j = 0, 1, 2, \ldots, \quad 0 \leqslant l \leqslant m, \quad 0 \leqslant k \leqslant j,$$

where the A_l are linear differential operators of order l which depend on S and are defined by the formula

$$e^{-i\lambda S(x)} A(e^{i\lambda S(x)} f(x)) = \sum_{l=0}^{m} \lambda^{m-l} A_l f.$$

Furthermore, $A_0 = 0$ if S satisfies the eikonal equation and A_1 is given by the formula

$$\begin{aligned} A_1 &= \left.\frac{\partial a_m}{\partial \xi}\right|_{\xi=S_x(x)} \cdot D + \frac{1}{2i} \operatorname{Tr}\left(\frac{\partial^2 a_m}{\partial \xi^2} \cdot \frac{\partial^2 S}{\partial x^2}\right)\Bigg|_{\xi=S_x(x)} \\ &= \sum_{j=1}^{n} \left.\frac{\partial a_m}{\partial \xi_j}\right|_{\xi=S_x(x)} \cdot D_j + \frac{1}{2i} \sum_{j,k=1}^{n} \left.\frac{\partial^2 a_m}{\partial \xi_j \partial \xi_k} \frac{\partial^2 S}{\partial x_j \partial x_k}\right|_{\xi=S_x(x)} \\ &= i^{-1} L_1 + c(x), \end{aligned}$$

where L_1 denotes differentiation along the vector field

$$V(x) = \left.\frac{\partial a_m}{\partial \xi}\right|_{\xi=S_x(x)}.$$

The vector field $V(x)$ has a transparent geometric interpretation. The operator L_1 represents differentiation along rays which are projections onto the x-space of those bicharacteristics of the function a_m (that is, of solutions of the Hamiltonian system with the Hamiltonian a_m) that lie on the graph of the gradient of S. This implies that the trajectories of V are the vector functions $x(t)$ such that $(x(t), \xi(t))$, where $\xi(t) = S_x(x(t))$, is a solution of the Hamiltonian system

$$\begin{cases} \dot{x} = \dfrac{\partial a_m}{\partial \xi}, \\ \dot{\xi} = -\dfrac{\partial a_m}{\partial x}. \end{cases} \tag{7.13}$$

The equation $Au = 0$ implies that $v_j = 0$ ($j = 0, 1, 2, \ldots$). Note that $v_0 \equiv 0$ in view of the eikonal equation, and that each of the equations $v_{j+1} = 0$ ($j = 0, 1, 2, \ldots$) implies that

$$A_1 b_j = -\sum_{k=0}^{j-1} A_{j+1-k} b_k = f_j, \quad j = 0, 1, 2, \ldots, \tag{7.14}$$

where f_j depends only on $b_0, b_1, \ldots, b_{j-1}$.

The equations (7.14) are known as the *transport equations*. Taking into account the meaning of V, we can write each of these equations in the form

$$i^{-1} \frac{d}{dt} b_j(x(t)) + c(x(t)) b_j(x(t)) = f_j(x(t)). \tag{7.15}$$

These equations are linear first-order ordinary differential equations along rays for the functions b_j if $b_0, b_1, \ldots, b_{j-1}$ are already known. In particular, the function $b_j(x(t))$ is determined completely along the entire ray if $b_j(x(0))$ is known.

We see that an asymptotic solution of the equation $Au = 0$ can be constructed for a given phase function $S(x)$, which is a solution of the eikonal equation, if a smooth hypersurface Γ is given which is transversal to all the rays passing through its points. Moreover, the solution is constructed for any given initial data

$$b_j|_\Gamma = b_j^0, \quad j = 0, 1, 2, \ldots, \tag{7.16}$$

in a small neighbourhood of Γ. Furthermore, the solution exists and is unique in any neighbourhood U where the phase function S is defined. Also the rays constitute a good family in the sense that through every point $x \in U$ there passes exactly one ray $x(t)$, this ray intersecting Γ in exactly one point $x_0 = x(0)$, and this point x_0 as well as the value of t defined by $x(t) = x$ depend smoothly on x. Thus the description of the asymptotic solutions leads, in essence, to the solution of the Hamilton-Jacobi equation (7.10), which can also be solved by means of bicharacteristics and rays in accordance with the procedure laid out, for example, in Arnol'd [1979], Egorov and Shubin [1988], Hörmander [1983, 1985], Maslov and Fedoryuk [1976], Shubin [1978].

The eikonal equation (7.10) has a sufficient number of solutions when, for example, A is a strictly hyperbolic operator with respect to a certain direction which can be chosen, without loss of generality, to be the x_n-axis. Namely, let $\lambda_j = \lambda_j(x, \xi')$ ($j = 1, 2, \ldots, m$) be all the roots of $a_m(x, \xi', \lambda) = 0$, regarded as an equation in λ (here $\xi' \in \mathbb{R}^{n-1}$), which are chosen to be continuous in (x, ξ'). These roots are real and distinct for all $\xi' \neq 0$ in view of the definition of hyperbolicity. Then the phase function S, a solution of (7.10), actually satisfies one of the equations

$$\frac{\partial S(x)}{\partial x_n} - \lambda_j(x, S_{x'}(x)) = 0, \tag{7.17}$$

where $j = 1, \ldots, m$. Moreover, in a sufficiently small neighbourhood of the intersection of the hyperplane $x_n = 0$ with the domain Ω in question, the equation (7.17) has a unique solution S_j, such that

$$S_j|_{x_n=0} = S_j^0, \tag{7.18}$$

where $S_j^0 = S_j^0(x')$ is any sufficiently smooth function.

7.3. Cauchy Problem with Rapidly Oscillating Initial Data (Lax [1957], Maslov [1965], Maslov and Fedoryuk [1967]). Let us pose the following Cauchy problem for the hyperbolic equation of the form discussed above:

$$\begin{cases} Au = 0, \\ u|_{x_n=0} = e^{i\lambda S^0(x')} \sum_{j=0}^{\infty} c_j^{(0)}(x')\lambda^{-j}, \\ \left.\dfrac{\partial u}{\partial x_n}\right|_{x_n=0} = \lambda e^{i\lambda S^0(x')} \sum_{j=0}^{\infty} c_j^{(1)}(x')\lambda^{-j}, \\ \cdots\cdots\cdots\cdots\cdots\cdots \\ \left.\dfrac{\partial^{m-1} u}{\partial x_n^{m-1}}\right|_{x_n=0} = \lambda^{m-1} e^{i\lambda S^0(x')} \sum_{j=0}^{\infty} c_j^{(m-1)}(x')\lambda^{-j}, \end{cases} \tag{7.19}$$

where S^0 is a given real-valued smooth function such that $S^0_{x'}(x') \neq 0$ for all x' belonging to the domain of definition Ω' of S^0, and $\Omega' = \Omega \cap \{x: x_n = 0\}$. Also, $c_j^{(k)} \in C^\infty(\Omega')$. We remark that the presence of the factors $\lambda^0, \lambda^1, \ldots, \lambda^{m-1}$ in the initial data does not render the problem less general because we can always multiply u by λ^k since the equation is linear and, moreover, some terms at the start of the expansion of the initial data can be taken to be zero.

The meaning of the problem (7.19) is that we require solutions of the equation which are as accurate as possible (for large λ), with rapidly oscillating initial data that are truncations of the series occurring on the right-hand sides of (7.19).

The solution of this problem cannot, in general, be found in the form of the series (7.11), and we have to take a finite sum of such series with different phase functions S. Thus let us examine the sum

$$u(x, \lambda) = \sum_{r=1}^{m} u^r(x, \lambda), \tag{7.20}$$

where

$$u^r(x, \lambda) = e^{i\lambda S_r(x)} \sum_{j=0}^{\infty} b_j^{(r)}(x)\lambda^{-j}, \tag{7.21}$$

and the series for u^r is an asymptotic solution of the equation $Au = 0$. All the phase functions S_r are assumed to satisfy the eikonal equation (7.10) with the same initial condition

$$S_r|_{x_n=0} = S^0(x'). \tag{7.22}$$

Then all the initial data for u^r have the form

$$\left.\frac{\partial^k u^r}{\partial x_n^k}\right|_{x_n=0} = \lambda^k e^{i\lambda S^0(x')} \sum_{k=0}^{\infty} d_j^{(k)}(x')\lambda^{-j},$$

so that the problem (7.19) does make sense, and is to be understood in the sense of equality of coefficients for the various powers of λ. For the S_r we shall take m distinct phase functions which satisfy distinct equations (7.17), where S_r corresponds to the root λ_r of the characteristic equation.

Theorem 7.2 (Lax [1957]). *Under the above hypotheses, the Cauchy problem* (7.19) *is uniquely solvable in a small neighbourhood of the initial hyperplane Ω'.*

To prove this result, we note that each of the series u^r, being a solution of $Au = 0$, is uniquely determined by the initial data $u^r|_{x_n=0}$, which can be arbitrary. Therefore we need only choose these initial data so that the initial data of (7.19) are satisfied. This leads to the equations

$$\sum_{r=1}^{m} \left[\left(i\frac{\partial S_r}{\partial x_n}\right)^k b_j^{(r)}(x)\right]\Bigg|_{x_n=0} = f_j^{(k)}(x'); \quad k = 0, 1, \ldots, m-1; \quad j = 0, 1, \ldots, \tag{7.23}$$

where $f_j^{(k)}$ depends only on $b_0^{(r)}, \ldots, b_{j-1}^{(r)}$ $(r = 1, \ldots, m)$. But (7.23) is a linear system of equations in $b_j^{(r)}(x', 0)$ $(r = 1, \ldots, m)$ whose determinant is the non-zero Vandermonde determinant because all the quantities

$$\left.\frac{\partial S_r(x', x_n)}{\partial x_n}\right|_{x_n=0} = \lambda_r\left(x', 0, \frac{\partial S^0(x')}{\partial x'}\right), \quad r = 1, \ldots, m,$$

are distinct, in view of hyperbolicity.

A theorem analogous to Theorem 7.2 holds for hyperbolic systems too.

7.4. Local Parametrix of the Cauchy Problem and Propagation of Singularities of Solutions (Maslov and Fedoryuk [1976]). Let us describe the connection between Theorem 7.2 and the behaviour of singularities of the solutions to the Cauchy problem. The singularities can be conveniently described by means of the Fourier transformation. Let us therefore consider a solution $u = u(x)$ of the Cauchy problem

$$\begin{cases} Au = 0, \\ u|_{x_n=0} = 0, \\ \left.\dfrac{\partial u}{\partial x_n}\right|_{x_n=0} = 0, \\ \ldots\ldots\ldots\ldots\ldots \\ \left.\dfrac{\partial^{m-1} u}{\partial x_n^{m-1}}\right|_{x_n=0} = f(x'). \end{cases} \tag{7.24}$$

We note that, in the general case when none of the initial data is zero, the solution u is the sum of m solutions corresponding to those cases where only one of the initial data is not zero; each of these cases can be considered in the same way as the problem (7.24). Let $f \in \mathscr{E}'(\Omega')$. Since the propagation speed of perturbations is finite, we can assume that the problem (7.24) is considered throughout $\mathbb{R}^n$. We represent f by means of the Fourier transformation (in the generalized sense):

$$f(x') = (2\pi)^{-n+1} \int \tilde{f}(\xi') e^{ix' \cdot \xi'} \, d\xi'. \tag{7.25}$$

If $\tilde{f}(\xi') = O(|\xi'|^{-N})$ as $|\xi'| \to +\infty$, then $f \in C^{N-n-1}(\mathbb{R}^{n-1})$. Likewise, if we have to find the singularities of the solution u, it is sufficient to solve (7.24) up to functions whose Fourier transform with respect to x' decays rapidly enough as $|\xi'| \to +\infty$.

The formula (7.25) can be considered as the representation of f as a linear combination of the exponentials $x' \mapsto e^{ix' \cdot \xi'}$. Since the original problem is linear, it is sufficient to consider the solution of (7.24) in which $f(x')$ is replaced by $e^{ix' \cdot \xi'}$. To be more precise, let $\varphi \in C_0^\infty(\Omega')$, with $\varphi = 1$ in a neighbourhood of $\operatorname{supp} f$. Then (7.25) can be written as

$$f(x') = (2\pi)^{-n+1} \int \tilde{f}(\xi') \varphi(x') e^{ix' \cdot \xi'} \, d\xi',$$

so that it is sufficient to solve (7.24) with $f(x')$ replaced by $\varphi(x') e^{ix' \cdot \xi'}$. We now set $\lambda = |\xi'|$, $\eta = \xi'/|\xi'|$ to obtain

$$\varphi(x') e^{ix' \cdot \xi'} = e^{i\lambda S(x', \eta)} \varphi(x'),$$

where $S(x', \eta) = \eta \cdot x'$. Let $v(x, \eta, \lambda)$ denote a sufficiently long truncation of the asymptotic series which gives an asymptotic solution of the Cauchy problem (7.24) with $f(x')$ replaced by $e^{i\lambda S(x', \eta)} \varphi(x')$. Here η is regarded as a parameter and λ as a large parameter having the same meaning as above. We introduce a cut-off function $\psi = \psi(\xi')$ such that $\psi \in C_0^\infty(\mathbb{R}^{n-1})$ and $\psi = 1$ in a neighbourhood of zero. Then the function

$$u_1(x) = (2\pi)^{-n+1} \int (1 - \psi(\xi')) \tilde{f}(\xi') v(x, \xi'/|\xi'|, |\xi'|) \, d\xi' \tag{7.26}$$

is a solution of a problem that differs from (7.24) by the addition of sufficiently smooth functions on the right-hand sides. But then, in view of the well-known existence and uniqueness theorems of the Cauchy problem for hyperbolic equations, $u - u_1$ is also a sufficiently smooth function. This implies that (7.26) defines a parametrix for the Cauchy problem (7.24). Noting the form of the asymptotic solutions described above, we find that (7.26) can be written in the form

$$u_1(x) = \sum_{j=1}^{m} (\Phi_j f)(x), \tag{7.27}$$

where Φ_j is a Fourier integral operator of the form

$$\Phi_j f(x) = (2\pi)^{-n+1} \int e^{iS_j(x,\xi')} a(x, \xi') \tilde{f}(\xi')\, d\xi'$$

$$= (2\pi)^{-n+1} \int e^{i[S_j(x,\xi') - y'\cdot\xi']} a(x, \xi') f(y')\, dy'\, d\xi'. \tag{7.28}$$

Here the phase function S_j satisfies the eikonal equation (7.17) with initial condition $S_j|_{x_n=0} = x'\cdot\xi'$, and $a = a(x, \xi')$ is a classical symbol of order $-m+1$, being the sum of several smooth functions which are positive and homogeneous for $|\xi'| \geqslant 1$.

Further, the procedure for constructing the asymptotic solutions enables us to describe the singularities of the solution u_1. For simplicity, let $f(x') = \delta(x')$ so that $\tilde{f}(\xi') = 1$. Then the corresponding solution $u(x) = \mathscr{E}(x)$ is called the fundamental solution of the Cauchy problem for the equation $Au = 0$. We consider the corresponding solution $u_1 = \mathscr{E}_1$ of the form (7.26) which differs from $\mathscr{E}$ by a sufficiently smooth function. The solution $v(x, \eta, \lambda)$, which occurs in this formula, can be determined as a solution with initial data $f(x') = e^{i\lambda x'\cdot\eta}\varphi(x')$, where $\varphi \in C_0^\infty(\mathbb{R}^{n-1})$ and supp φ is contained in a sufficiently small ball with centre 0. But then v is given by a finite sum of the terms of the form $e^{i\lambda S_j(x,\eta)}\psi(x)$, where S_j is the phase function described above and ψ is obtained by solving the transport equations along the rays corresponding to the phase function S_j. Thus the support of v lies in the union of m cones of rays (with curvilinear generators) emanating from the origin. It can be seen easily that all these rays can be described as projections onto the x-space of all possible non-zero bicharacteristics of the principal symbol a_m which start above the point $x = 0$, that is, at the points of the form $(0, \xi', \lambda_j(0, \xi'))$.

A more detailed description of the singularities can be given easily in the language of wave fronts by analysing the integral (7.28) by the method of stationary phase with $\lambda = |\xi'|$. Thus, on multiplying $\Phi_j f(x)$ by a cut-off function $\varphi = \varphi(x)$ with a sufficiently small support, we obtain a distribution whose Fourier transform is given by the integral

$$\widetilde{\varphi\Phi_j f}(\eta) = (2\pi)^{-n+1} \int e^{i[S_j(x,\xi') - y'\cdot\xi' - x\cdot\eta]} a(x, \xi')\varphi(x) f(y')\, dy'\, d\xi'\, dx.$$

For $f(y') = \delta(y')$ the integral assumes the form

$$(2\pi)^{-n+1} \int e^{i[S_j(x,\xi') - x\cdot\eta]} a(x, \xi')\varphi(x)\, d\xi'\, dx$$

$$= (2\pi)^{-n+1}\lambda^{n-1} \int e^{i\lambda[S_j(x,\xi') - x\cdot\zeta]} a(x, \lambda\xi')\varphi(x)\, d\xi'\, dx,$$

where $\lambda = |\eta|$ and $\zeta = \eta/|\eta|$. The last integral decays rapidly with respect to λ outside the stationary phase points, that is, at those points where

$$\left|\frac{\partial S_j(x, \xi')}{\partial \xi'}\right| + \left|\frac{\partial S_j(x, \xi')}{\partial x} - \zeta\right| \neq 0.$$

We see that $WF(\Phi_j f)$ must be situated at those points (x, η) such that $\eta = \frac{\partial S_j(x, \xi')}{\partial x}$, where ξ' satisfies the condition $\frac{\partial S_j(x, \xi')}{\partial \xi'} = 0$. But these are precisely the points (x, η) that lie on the bicharacteristics starting from the points of the form $(0, \eta_0)$ (that is, above the point $x = 0$), because $\partial S_j/\partial \xi' = 0$ remains unchanged under motion along the bicharacteristic (ξ' is a parameter!) and $S_j|_{x_n=0} = x' \cdot \xi'$. In exactly the same way, the relation $\eta = \frac{\partial S_j(x, \xi')}{\partial x}$ also remains unchanged along the bicharacteristic, since the graph of the gradient of the function S_j, which is a solution of the Hamilton-Jacobi equation, is invariant with respect to the Hamiltonian flow; see the procedure for solving the Hamilton-Jacobi equation described in Egorov and Shubin [1988, § 3, Chap. 1]. Also, we can confine our attention only to null bicharacteristics because of the general result

$$WF(u) \subset \{(x, \xi): a_m(x, \xi) = 0\}$$

which is true for any solution of the equation $Au = 0$. This enables us to obtain a global assertion stating that the singularities of any solution are propagated along bicharacteristics (it is enough to establish this result for arbitrarily small segments of the bicharacteristics!) because any forward movement with respect to t can be regarded as a solution of the Cauchy problem with suitable initial data.

7.5. The Maslov Canonical Operator and Global Asymptotic Solutions of the Cauchy Problem (Maslov [1965, 1973], Maslov and Fedoryuk [1976], Vajnberg [1982]). In a majority of cases the asymptotic solution of the Cauchy problem (7.19) can be found only in a small neighbourhood of the initial surface. This is due to the fact that the Cauchy problem (7.17)–(7.18) for the eikonal equation can be solved only in a small neighbourhood of the initial surface even though the equation and the initial data are defined globally, an example being the case where x' varies on a compact manifold and $x_n \in \mathbb{R}$. The reason for this situation is the presence of *caustics*, that is, envelopes of families of rays which arise in the general situation from the solution of equations determining the rays. In order to overcome this difficulty, we can employ *the method of the Maslov canonical operator*, which we now describe briefly. From the analytical point of view, the essence of this method is that near the caustic we use, instead of asymptotic solutions of the form (7.11), the integrals of these solutions. More precisely, we use Fourier transforms of the asymptotic solution of the form (7.11) with respect to some or all of the variables, but given either in terms of variables ξ which are dual to x or in terms of variables $(x^{(1)}, \xi^{(2)})$, where $x = (x^{(1)}, x^{(2)})$ is a decomposition of the variables x into two groups and the variables $\xi^{(2)}$ are dual to $x^{(2)}$. It turns out that if the bicharacteristics of the equation in question are defined globally, then under certain topological conditions the asymptotic solution having the above-mentioned structure can be constructed and then these solutions can be glued to give the global asymptotic solution. In particular, in this manner we can obtain a global asymptotic solution of the Cauchy problem (7.19) for a strictly hyperbolic equation with rapidly oscillating initial data. This leads to the construction of the global parametrix of the Cauchy problem for such an equation, in accordance with the scheme outlined in § 7.4.

We first introduce a geometrical object that replaces the solution of the Hamilton-Jacobi equation

$$H\left(x, \frac{\partial S(x)}{\partial x}\right) = 0, \tag{7.29}$$

where H is a real-valued smooth function in $\mathbb{R}^{2n}$. To do this, we observe that the graph of the gradient of S, that is, the set of points $(x, S_x(x))$, is a *Lagrangian manifold* in $\mathbb{R}^{2n}$. In other words, this graph is an n-dimensional submanifold such that the restriction of the symplectic 2-form

$$\omega = \sum_{j=1}^{n} dx_j \wedge d\xi_j$$

to this submanifold is equal to 0, where $(x_1, \ldots, x_n, \xi_1, \ldots, \xi_n)$ denotes the coordinates in $\mathbb{R}^{2n} = \mathbb{R}^n_x \times \mathbb{R}^n_\xi$. Conversely, let $\Lambda \subset \mathbb{R}^{2n}$ be a Lagrangian manifold and let $p \in \Lambda$ be a point such that Λ is diffeomorphically projected onto $\mathbb{R}^n_x$ in a neighbourhood of this point. For this, it is necessary and sufficient that the differential of the projection $\pi\colon \Lambda \to \mathbb{R}^n_x$ is injective at p, by the Implicit Function Theorem. Then in a neighbourhood U of this point, Λ can be represented as the graph of the gradient of a function S which is defined uniquely up to an additive constant. Thus, S is determined by the formula

$$S(x) = S(x^0) + \int_{x^0}^{x} \xi(x)\, dx, \tag{7.30}$$

where $(x, \xi(x))$ is a point of the neighbourhood $U \subset \Lambda$ in question which lies above the point x, and the integral of the 1-form $\xi\, dx = \xi_1\, dx_1 + \cdots + \xi_n\, dx_n$ is taken along any path in $\pi(U)$ that joins x^0 and x. We note that the integral is independent of the path because the form $\xi\, dx$ is closed on Λ since Λ is a Lagrangian manifold. Thus every Lagrangian manifold Λ contained in the hypersurface $\Gamma = \{(x, \xi)\colon H(x, \xi) = 0\}$ is in essence a many-valued solution of the Hamilton-Jacobi equation (7.29). Let there be prescribed an initial function $S^0 = S^0(x')$ considered as a function on the hyperplane $x_n = 0$, and assume that the following condition holds.

Condition A. The equation $H\left(x', 0, \dfrac{\partial S^0(x')}{\partial x'}, \xi_n\right) = 0$ has a smooth solution $\xi_n = \xi_n(x')$ such that

$$H'_{\xi_n}\left(x', 0, \frac{\partial S^0(x')}{\partial x'}, \xi_n(x')\right) \neq 0.$$

Note that this condition is fulfilled automatically for equations of the form (7.17), and in this case the solution $\xi_n(x')$ is unique.

If Condition A holds, we can take all the bicharacteristics of H passing through the points $\left(x', 0, \frac{\partial S^0(x')}{\partial x'}, \xi_n(x')\right)$, and the points of these bicharacteristics will constitute the Lagrangian manifold Λ. In a number of important cases this manifold lies above the whole space $\mathbb{R}^n_x$, that is, the projection π: $\Lambda \to \mathbb{R}^n_x$ is surjective. This is the case, for example, with the Hamilton-Jacobi equations which are solved with respect to $\partial S/\partial x_n$ under certain conditions imposed on the growth of the Hamiltonian and, in particular, for equations of the form (7.17) which arise from uniformly strictly hyperbolic equations with coefficients having bounded derivatives.

We can generalise formula (7.30) by introducing the function

$$S(r) = S(r^0) + \int_{r^0}^{r} \xi\, dx \tag{7.31}$$

in a small neighbourhood U of any point r^0 on the Lagrangian manifold Λ. Here the integral is taken over any path that joins r^0 and r and lies in U. Again the integral is independent of the path because Λ is a Lagrangian manifold. The function $S = S(x)$, constructed above, was the image of the function (7.31) under the projection $\pi: \Lambda \to \mathbb{R}^n_x$. By a well-known lemma of Arnol'd [1979], in a neighbourhood of each of its points p^0 the Lagrangian manifold is diffeomorphically projected onto one of the Lagrangian coordinate planes in $\mathbb{R}^{2n}$, that is, onto one of the planes $(x^{(1)}, \xi^{(2)})$, where $(x^{(1)}, \xi^{(2)})$ are as described above. The image of S in (7.31) under this projection is a function $S(x^{(1)}, \xi^{(2)})$, which we use to construct a new function

$$S_U(x^{(1)}, \xi^{(2)}) = S(x^{(1)}, \xi^{(2)}) - \xi^{(2)} \cdot x^{(2)}(x^{(1)}, \xi^{(2)}),$$

where the function $x^{(2)}(x^{(1)}, \xi^{(2)})$ has been chosen so that there exists a point $(x^{(1)}, x^{(2)}(x^{(1)}, \xi^{(2)}), \xi^{(1)}(x^{(1)}, \xi^{(2)}), \xi^{(2)}) \in U \subset \Lambda$. The function S_U is called the *generating function* of the Lagrangian manifold Λ in a neighbourhood of the point p^0. Also, Λ can be locally recovered from the generating function $S_U(x^{(1)}, \xi^{(2)})$ as a set of points in $\mathbb{R}^{2n}$ for which

$$x^{(2)} = -\frac{\partial S_U(x^{(1)}, \xi^{(2)})}{\partial \xi^2}, \quad \xi^{(1)} = \frac{\partial S_U(x^{(1)}, \xi^{(2)})}{\partial x^{(1)}}.$$

We now assume for simplicity that the function $H = H(x, \xi)$ is homogeneous in ξ of degree m. We seek an asymptotic solution of the problem with rapidly oscillating initial data

$$\begin{cases} H(x, D_x)u = 0, \\ u|_{x_n=0} = e^{i\lambda S^0(x')}c(x'), \end{cases} \tag{7.32}$$

where $c \in C_0^\infty(\mathbb{R}^{n-1})$. As above, we assume that condition A holds. This enables us to construct a Lagrangian manifold Λ from the bicharacteristics such that the part lying above the hyperplane $x_n = 0$ has the form $\left(x', 0, \frac{\partial S^0(x')}{\partial x'}, \xi_n(x')\right)$, where the function $\xi_n(x')$ is given by condition A. For the coordinates of any point r on Λ we can choose the variables (t, y), where t is a parameter along the bicharacteristic which passes through r at time t and emanates (when $t = 0$) above the point $(0, y)$, where $y \in \mathbb{R}^{n-1}$. The usual coordinates x of the projection of the point r onto $\mathbb{R}^n_x$ will be functions of (t, y), that is, $x = x(t, y)$, and we can consider the Jacobian $J(t, y) = \det \frac{\partial x(t, y)}{\partial(t, y)}$. It does not vanish when $t = 0$ but vanishes at exactly those points where the differential of the canonical projection $\pi: \Lambda \to \mathbb{R}^n_x$ degenerates, that is, above the caustics.

Let the manifold Λ now be covered by the coordinate neighbourhoods U_j such that a set $(x^{(1)}, \xi^{(2)})$ can be chosen in each U_j as the local coordinates. For a function $\varphi \in C_0^\infty(U_j)$, we set

$$\begin{aligned} K(U_j)\varphi &= u(x, \lambda) \\ &= \left(-\frac{\lambda}{2\pi i}\right)^k \int \left|\frac{\partial(x^{(1)}, \xi^{(2)})}{\partial(t, y)}\right|^{-1/2} \exp\{i\lambda[S_{U_j}(x^{(1)}, \xi^{(2)}) \\ &\quad + x^{(2)} \cdot \xi^{(2)}]\} \varphi(r(x^{(1)}, \xi^{(2)}))\, d\xi^{(2)} \\ &= \left(-\frac{\lambda}{2\pi i}\right)^k \int \left|\frac{\partial(x^{(1)}, \xi^{(2)})}{\partial(t, y)}\right|^{-1/2} e^{i\lambda S(r(x^{(1)}, \xi^{(2)}))} \\ &\quad \times \varphi(r(x^{(1)}, \xi^{(2)}))\, d\xi^{(2)}, \end{aligned} \tag{7.33}$$

where S_{U_j} is the generating function of the Lagrangian manifold Λ in $(x^{(1)}, \xi^{(2)})$ coordinates, S denotes the function (7.31) on Λ and k denotes the number of variables $\xi^{(2)}$. The operator $K(U_j): C_0^\infty(U_j) \to C^\infty(\mathbb{R}^n_x \times \mathbb{R}_+)$ is known as a *precanonical operator* on U_j associated with the coordinates $(x^{(1)}, \xi^{(2)})$. If $k = 0$, that is, if $\xi^{(2)}$ are absent and U_j ia diffeomorphically projected into $\mathbb{R}^n_x$, we find that $K(U_j)\varphi$ coincides with the function

$$e^{i\lambda S(r(x))} J(x)^{-1/2} \varphi(r(x)) \tag{7.34}$$

up to an insignificant numerical factor. The Jacobian appears here because the first transport equation (namely, the equation (7.14) with $j = 0$ and $a_m = H$) can be written in the form

$$J^{-1/2} \frac{d}{dt}(J^{1/2} b_0) - \frac{1}{2} \sum_{j=1}^n \frac{\partial^2 H}{\partial x_j \partial \xi_j} b_0 = 0 \tag{7.35}$$

by the well-known Liouville formula. Here the values of all the functions are taken along the given bicharacteristic. This gives the leading term u_0 of the asymptotic solution (7.11) in the form

$$u_0(x, \lambda) = b_0^0(y) \sqrt{\frac{J(0, y)}{J(t, y)}} \exp\left[i\lambda S(r(x)) - \frac{1}{2}\int_0^t \sum_{j=1}^n \frac{\partial^2 H}{\partial x_j \partial \xi_j} dt\right], \tag{7.36}$$

where $t = t(x)$, $y = y(x)$, $\xi = \xi(x)$ and $r(x)$ are the coordinates on the Lagrangian manifold Λ in U_j, when expressed in terms of x.

If we now assume that in a neighbourhood U_j the set $(x^{(1)}, \xi^{(2)})$ giving local coordinates on Λ can be chosen in various ways, that is, if $(\tilde{x}^{(1)}, \tilde{\xi}^{(2)})$ is another such set and $\tilde{K}(U_j)$ is the corresponding precanonical operator, then the method of stationary phase yields the following connection between $K(U_j)$ and $\tilde{K}(U_j)$:

$$\tilde{K}(U_j)\varphi = e^{\gamma i\pi/2} K(U_j)\varphi + O(\lambda^{-1}), \tag{7.37}$$

where

$$\gamma = \left[\operatorname{inerdex} \frac{\partial x^{(2)}(x^{(1)}, \xi^{(2)})}{\partial \xi^{(2)}} - \operatorname{inerdex} \frac{\partial \tilde{x}^{(2)}(\tilde{x}^{(1)}, \tilde{\xi}^{(2)})}{\partial \tilde{\xi}^{2}}\right] \bmod 4. \tag{7.38}$$

Here we use the notation that, for a symmetric matrix A, inerdex A denotes the negative inertia index, which is the number of negative eigenvalues of the matrix of the corresponding quadratic form; in the present case mod γ 4 does not depend on the choice of the point in question.

Let us now try to glue the precanonical operators $K(U_j)$ to obtain the *canonical operator*

$$K_\Lambda\colon C_0^\infty(\Lambda) \to C^\infty(\mathbb{R}^n \times \mathbb{R}_+) \bmod O(\lambda^{-1}). \tag{7.39}$$

To do this, we set

$$(K_\Lambda \varphi)(x, \lambda) = \sum_j c_j K(U_j)(e_j(r)\varphi(r))(x), \tag{7.40}$$

where $\{e_j\}_{j=1}^\infty$ is a partition of unity on Λ which is subordinated to the covering U_j, and the c_j are constants chosen so that this definition depends on neither the choice of the covering U_j nor the choice of the coordinates in each neighbourhood U_j. It turns out that this can always be achieved in the case of the so-called *quantized Lagrangian manifolds*, that is, for those manifolds Λ for which, first, $\int_l \xi\, dx = 0$ for any closed path l on Λ and, secondly, ind $l = 0$, where ind l denotes the *Maslov index* of this path. The Maslov index of any path (not necessarily closed) on Λ can be defined, for example, by covering this path by a set $U_{i_0}, U_{i_1}, \ldots, U_{i_s}$ of coordinate neighbourhoods described above, such that $U_{i_k} \cap U_{i_{k+1}} \neq \varnothing$ $(k = 0, 1, \ldots, s-1)$. Then for each pair $U_{i_k}, U_{i_{k+1}}$ we introduce on $U_{i_k} \cap U_{i_{k+1}}$ the number $\gamma(U_{i_k}, U_{i_{k+1}})$ given by the formula (7.38), in which $(x^{(1)}, \xi^{(2)})$ denotes the coordinates on U_{i_k} and $(\tilde{x}^{(1)}, \tilde{\xi}^{(2)})$ denotes the coordinates on $U_{i_{k+1}}$. Finally we set

$$\operatorname{ind} l = \sum_{k=0}^{s-1} \gamma(U_{i_k}, U_{i_{k+1}}) \bmod 4, \tag{7.41}$$

this definition being independent of the choice of the arbitrary elements.

We remark that, in particular, every simply connected Lagrangian manifold is quantized.

In the case of a quantized manifold Λ, the quantities c_j in the definition (7.40) of K_Λ can be taken as $c_j = \exp(-i\pi\gamma_j/2)$, for example, where γ_j is the Maslov index of the path joining a fixed point $r_0 \in \Lambda$ to any point $r \in U_j$. We then find that

$$c_i K(U_i)\varphi = c_j K(U_j)\varphi + O(\lambda^{-1}),$$

if $\varphi \in C_0^\infty(U_i \cap U_j)$, and this ensures that our canonical operator K_Λ is well defined.

Let us return to the Cauchy problem (7.32). It turns out that this problem has an asymptotic solution of the form

$$u(x, \lambda) = K_\Lambda\left(\sum_{j=0}^{\infty} \lambda^{-j} b_j(r)\right)(x, \lambda), \tag{7.42}$$

with $b_j \in C^\infty(\Lambda)$, wherever the Lagrangian manifold Λ, constructed just now, is defined (under the assumption that it is quantized). The functions b_j here are found by solving the transport equations on Λ. These equations, which are obtained by substituting the series for $u(x, \lambda)$ into the equation $H(x, D_x)u = 0$ and then equating to zero the coefficients of various powers of λ, have the form

$$R_1 b_j = f_j. \tag{7.43}$$

Here

$$R_1 = \frac{d}{dt} - \frac{1}{2}\sum_{j=1}^{n} \frac{\partial^2 H(x, \xi)}{\partial x_j \partial \xi_j} \tag{7.44}$$

is a first-order differential operator on Λ in which d/dt denotes differentiation along bicharacteristics and the point (x, ξ) is taken on Λ, and the functions f_j are expressed in terms of $b_0, \ldots, b_{j-1}$. By assigning suitable initial conditions for b_j on the hyperplane $x_n = 0$ which guarantee that the initial condition of the problem (7.32) is fulfilled, we can successively solve the transport equations (7.43) (explicitly, once we know Λ!) and find the desired asymptotic solution.

We now examine the Cauchy problem (7.19) with a uniformly strictly hyperbolic equation of order m in $\mathbb{R}^n$ whose coefficients possess derivatives of all orders. Then we have m Hamilton-Jacobi equations (7.17) that define m Lagrangian manifolds $\Lambda_1, \ldots, \Lambda_m$. The presence of a distinguished variable x_n and the simple structure of the initial data ensure that the manifolds are simply connected and, hence, that they are quantized. Thus m Maslov canonical operators $K_{\Lambda_1}, \ldots, K_{\Lambda_m}$ are defined.

Theorem 7.3 (Maslov; see Maslov and Fedoryuk [1976], Vajnberg [1982]). *Under the above hypotheses, the Cauchy problem* (7.19) *has a global asymptotic solution of the form*

$$u = \sum_{j=1}^{m} K_{\Lambda_j}\left(\sum_{k=0}^{\infty} \lambda^{-k} b_{k,j}\right), \tag{7.45}$$

where the functions $b_{k,j} \in C^\infty(\Lambda_j)$ are found by solving the transport equations on Λ_j with suitable initial conditions.

In view of the procedure described in § 7.4, this theorem gives a way of constructing the global parametrix of the Cauchy problem for the equation in question.

There is a direct connection between the Maslov canonical operator and the Fourier integral operators, which appeared later (see a description of this connection in Mishchenko, Sternin and Shatalov [1978], and Nazajkinskij, Oshmyan, Sternin and Shatalov [1981]).

The method described above and its improved variants give us a way of constructing asymptotic solutions of a wide variety of problems of mathematical physics and, in particular, semiclassical approximations in a number of important quantum-mechanical problems concerning asymptotic problems in spectral theory. A detailed account of these questions can be found in the monographs by Babich and Buldyrev [1972], Fedoryuk [1977], Guillemin and Sternberg [1977], Hörmander [1983, 1985], Ivrii [1984], Leray [1978], Maslov [1965, 1973, 1977, 1987], Maslov and Fedoryuk [1976], Mishchenko, Sternin and Shatalov [1978], Shubin [1978], Taylor [1981], Trèves [1980]. These monographs and the literature cited there also contain the details of the constructions described here together with the necessary proofs.

§ 8. Asymptotics of Eigenvalues of Self-adjoint Differential and Pseudodifferential Operators

In the present section we discuss simple problems concerning the asymptotics of eigenvalues of self-adjoint operators which arise in analysis, geometry and mathematical physics.

8.1. Variational Principles and Estimates for Eigenvalues (Berezin and Shubin [1983], Courant and Hilbert [1937], Glazman [1963], Shubin [1978]). We have already dealt with eigenvalue problems associated with elliptic boundary-value problems (see Egorov and Shubin [1988, § 3, Chap. 2]). The simplest of these is the problem of finding eigenvalues and eigenfunctions for the Laplace operator in a bounded domain $\Omega \subset \mathbb{R}^n$ with the Dirichlet boundary condition:

$$\begin{cases} -\Delta\psi = \lambda\psi & \text{in } \Omega, \\ \psi|_{\partial\Omega} = 0. \end{cases} \tag{8.1}$$

In its generalized formulation, this problem consists in investigating the eigenvalues of the operator generated by the quadratic form

$$[u, u] = \int_\Omega |\nabla u|^2 \, dx \tag{8.2}$$

over the functions $u \in \mathring{H}^1(\Omega)$. This means that a function ψ is said to be an eigenfunction of the problem (8.1), corresponding to the eigenvalue λ, if $\psi \in \mathring{H}^1(\Omega)$, $\psi \neq 0$ and

$$[\psi, v] = \lambda(\psi, v) \tag{8.3}$$

for any $v \in \mathring{H}^1(\Omega)$, where the sesqui-linear forms $[\cdot, \cdot]$ and $(\cdot, \cdot)$ define Hilbert scalar products in $\mathring{H}^1(\Omega)$ and $L_2(\Omega)$ respectively:

$$[u, v] = \int_\Omega \nabla u \cdot \overline{\nabla v}\, dx, \quad (u, v) = \int_\Omega u\bar{v}\, dx.$$

Incidentally, it is enough that the identity (8.3) holds for functions $v \in C_0^\infty(\Omega)$. When the domain Ω has smooth boundary (of class C^∞), the generalized formulation is equivalent to the classical formulation (8.1). It also follows automatically that $\psi \in C^\infty(\overline{\Omega})$.

We assume that the eigenvalues of (8.1) are arranged in increasing order:

$$\lambda_1 \leqslant \lambda_2 \leqslant \lambda_3 \leqslant \cdots,$$

where each eigenvalue is repeated according to its multiplicity. It can be shown that λ_1 is non-repeated so that actually $\lambda_1 < \lambda_2$, but this observation will not play any role in our discussion below.

There is a variational principle, known as *Courant's Minimax Principle*, which expresses the eigenvalues λ_j in terms of the quadratic form (8.2):

$$\lambda_j = \max_{\substack{L \subset H_1 \\ \dim L = j-1}} \min_{u \in L^\perp \setminus \{0\}} \frac{[u, u]}{(u, u)}, \quad j = 1, 2, \ldots, \tag{8.4}$$

where $H_1 = \mathring{H}^1(\Omega)$, L is a linear subspace of H_1, and the maximum is taken over all such subspaces of dimension $j - 1$. Also, $L^\perp$ is the orthogonal complement of L in $L_2(\Omega)$, and it is understood that the minimum is only taken over those u that lie in $L^\perp \cap H_1$. When $j = 1$, this formula has to be understood without the first maximum, that is,

$$\lambda_1 = \min_{u \in H_1 \setminus \{0\}} \frac{[u, u]}{(u, u)}. \tag{8.4$_1$}$$

Noting that $C_0^\infty(\Omega)$ is dense in $\mathring{H}^1(\Omega)$, we can easily transform (8.4) to the form

$$\lambda_j = \sup_{\substack{L \subset D_0 \\ \dim L = j-1}} \inf_{u \in L^\perp \setminus \{0\}} \frac{(Au, u)}{(u, u)}, \quad j = 1, 2, \ldots, \tag{8.5}$$

where $D_0 = C_0^\infty(\Omega)$, $A = -\Delta$ and the minimum is taken over all $u \in L^\perp \cap D_0$. As before, the orthogonal complement is taken in $L_2(\Omega)$.

We now introduce the *distribution function of eigenvalues*:

$$N(\lambda) = \#\{j\colon \lambda_j \leqslant \lambda\} = \sum_{\lambda_j \leqslant \lambda} 1 = \max\{j\colon \lambda_j \leqslant \lambda\}.$$

(In this first of two alternative ways of expressing $N(\lambda)$ it is immaterial whether the eigenvalues have been arranged in increasing order or not.) This function is right-continuous, that is, $N(\lambda + 0) = N(\lambda)$. Then we also have a variational principle for $N(\lambda)$, known as *Glazman's Lemma*:

$$N(\lambda) = \max_{\substack{L \subset H_1 \\ [u,u] \leqslant \lambda(u,u),\, u \in L}} \dim L. \tag{8.6}$$

An analogue of (8.5) has the form

$$N(\lambda - 0) = \max_{\substack{L \subset D_0 \\ (Au,u) < \lambda(u,u),\, u \in L\setminus\{0\}}} \dim L. \tag{8.7}$$

The variational principles (8.4)–(8.7) lie at the foundation of the standard numerical methods for finding the eigenvalues λ_j. They can also be used effectively for investigating the qualitative behaviour of the eigenvalues and, in particular, for finding their asymptotics.

Suppose that, for example, there are two domains Ω and Ω' in $\mathbb{R}^n$ such that $\Omega \subset \Omega'$. We denote by $\lambda_j(\Omega)$ and $\lambda_j(\Omega')$ respectively the eigenvalues of the Dirichlet problems in Ω and Ω', and by $N_\Omega(\lambda)$ and $N_{\Omega'}(\lambda)$ the corresponding distribution functions. Since $C_0^\infty(\Omega) \subset C_0^\infty(\Omega')$, it evidently follows from (8.7) that

$$N_\Omega(\lambda) \leqslant N_{\Omega'}(\lambda), \quad \lambda \in \mathbb{R}, \tag{8.8}$$

or equivalently

$$\lambda_j(\Omega) \geqslant \lambda_j(\Omega'), \quad j = 1, 2, \ldots . \tag{8.9}$$

Thus the eigenvalues of the Dirichlet problem decrease when the domain is enlarged.

As an example, we use this fact to compare the eigenvalues for Ω with the eigenvalues for a cube

$$K_l = \{x : 0 \leqslant x_j \leqslant l, j = 1, \ldots, n\}.$$

For K_l, the eigenvalues

$$\lambda_{k_1 \ldots k_n} = \left(\frac{\pi}{l}\right)^2 (k_1^2 + \cdots + k_n^2), \quad k_j \in N,$$

and the corresponding eigenfunctions

$$\psi_{k_1 \ldots k_n} = \sin \frac{\pi k_1 x_1}{l} \ldots \sin \frac{\pi k_n x_n}{l}$$

are found by the method of separation of variables. It can be seen easily that

$$N_{K_l}(\lambda) = (2\pi)^{-n} l^n \omega_n \lambda^{n/2} + O(\lambda^{(n-1)/2}), \tag{8.10}$$

where ω_n is the volume of the unit sphere in $\mathbb{R}^n$. By including a given bounded domain Ω within a sufficiently large cube, and again by taking a sufficiently small cube lying in Ω, we find, in view of (8.8), that there is a constant $C > 0$ such that

$$C^{-1} \lambda^{n/2} \leqslant N_\Omega(\lambda) \leqslant C \lambda^{n/2}. \tag{8.11}$$

We now examine the eigenvalue problem with the Neumann boundary condition:

$$\begin{cases} -\Delta\psi = \lambda\psi & \text{in } \Omega, \\ \left.\dfrac{\partial\psi}{\partial n}\right|_{\partial\Omega} = 0. \end{cases} \tag{8.12}$$

The generalized formulation is obtained by arguments similar to those described in Egorov and Shubin [1988, § 6, Chap. 2]. Thus ψ is said to be an eigenfunction corresponding to the eigenvalue λ if $\psi \in H^1(\Omega)$, $\psi \neq 0$ and the identity (8.3) holds for any function $v \in H^1(\Omega)$. The variational principles (8.4) and (8.6) remain valid but now we take $H_1 = H^1(\Omega)$ (and not $\mathring{H}^1(\Omega)$ as we did in the case of the Dirichlet boundary conditions). By defining $D_0 = \left\{u\colon u \in C^\infty(\Omega), \left.\dfrac{\partial u}{\partial n}\right|_{\partial\Omega} = 0\right\}$ we can again use the formulations (8.5) and (8.7).

We denote the eigenvalues of (8.12) by $'\lambda_j = {'\lambda_j}(\Omega)$ $(j = 1, 2, \ldots)$. Just as in the case of the Dirichlet problem, it is assumed that

$$'\lambda_1 \leqslant {'\lambda_2} \leqslant {'\lambda_3} \leqslant \cdots.$$

We note that $'\lambda_1 = 0$ and this eigenvalue is simple. Hence $'\lambda_2 > 0$. Finally, let $'N(\lambda) = {'N_\Omega}(\lambda)$ be the distribution function of the eigenvalues $'\lambda_j$. In the case of the cube K_l, the eigenvalues of (8.12) are

$$\lambda_{k_1 \ldots k_n} = \left(\frac{\pi}{l}\right)^2 (k_1^2 + \cdots + k_n^2), \quad k_j \in \mathbb{Z}_+,$$

and the corresponding eigenfunctions are $\psi_{k_1 \ldots k_n}(x) = \cos\dfrac{\pi k_1 x_1}{l} \cdots \cos\dfrac{\pi k_n x_n}{l}$. Thus the formula (8.10) remains valid for the problem (8.12).[1] Furthermore, since $\mathring{H}^1(\Omega) \subset H^1(\Omega)$, a comparison of (8.5) for the Dirichlet and Neumann problems yields the inequality

$$N_\Omega(\lambda) \leqslant {'N_\Omega}(\lambda), \tag{8.13}$$

or

$$\lambda_j(\Omega) \geqslant {'\lambda_j}(\Omega), \quad j = 1, 2, \ldots. \tag{8.14}$$

8.2. Asymptotics of the Eigenvalues of the Laplace Operator in a Euclidean Domain. Under minimal additional conditions on the domain Ω, we can obtain the asymptotic formula

$$N_\Omega(\lambda) = (2\pi)^{-n} \operatorname{mes} \Omega \cdot \omega_n \cdot \lambda^{n/2}(1 + o(1)) \tag{8.15}$$

instead of the estimate (8.11), where mes Ω denotes the n-dimensional Lebesgue measure of Ω. H. Weyl, in 1912, was the first to establish such asymptotics. One of the proofs of this result consists of approximating Ω by a union of cubes whose vertices are the points of a lattice of integers that has been suitably contracted.

[1] The monotonicity properties (8.8), (8.9) are not true for the Neumann conditions, and (8.11) is true only if the boundary is sufficiently regular.

A lower estimate is obtained if these cubes are open, disjoint and lie inside Ω. To obtain an upper estimate, we take a similar union of cubes which contains Ω, but now we consider the Neumann condition on the boundary of each cube. Thus (8.15) is at least obtained for those bounded domains Ω with $\operatorname{mes}(\partial\Omega) = 0$.

By carrying out the above variational arguments more precisely, Courant (see Courant and Hilbert [1931, 1937]) obtained the formula

$$N_\Omega(\lambda) = (2\pi)^{-n} \operatorname{mes} \Omega \cdot \omega_n \lambda^{n/2}(1 + O(\lambda^{-1/2} \ln \lambda)), \tag{8.16}$$

with a more precise estimate for the remainder term. This formula is true if, for example, the boundary $\partial\Omega$ has the property that $\operatorname{mes}(\partial\Omega)_\varepsilon = O(\varepsilon)$ as $\varepsilon \to +0$, where $(\partial\Omega)_\varepsilon$ is an ε-neighbourhood of $\partial\Omega$ in $\mathbb{R}^n$. Finally, the following theorem gives an estimate for the remainder that cannot be improved.

Theorem 8.1 (Seeley [1978]). *Let Ω be a bounded domain with smooth boundary in $\mathbb{R}^n$. Then, as $\lambda \to \infty$,*

$$N_\Omega(\lambda) = (2\pi)^{-n} \operatorname{mes} \Omega \cdot \omega_n \lambda^{n/2}(1 + O(\lambda^{-1/2})). \tag{8.17}$$

This theorem cannot be obtained from the variational methods, and to prove this result we require the method of hyperbolic equations which is the most precise of the known Tauberian methods; this method will be dealt with later.

Theorem 8.1 remains valid without change for the eigenvalues of the Neumann problem.

As far back as 1912, H. Weyl conjectured that the second term of the asymptotic formula for $N(\lambda)$ contains an $(n-1)$-dimensional measure ("area") of the boundary $\partial\Omega$. An answer to the problem of the second term of the asymptotic formula was first obtained by Ivrii in 1980 for domains having smooth boundary under the following condition regarding the billiard trajectories of Ω, where the billiard trajectory in Ω is taken with the usual reflections at the boundary.[2]

Condition (A). The set of periodic points of the billiard in Ω has measure zero.

This condition implies that the set of vectors tangent to Ω that are initial conditions for periodic billiard trajectories has measure zero in $\Omega \times \mathbb{R}^n$ (that is, the $2n$-dimensional Lebesgue measure is zero). Alternatively, we can state that the set of periodic points of the geodesic flow on Ω has measure zero, the geodesic flow being prescribed on unit vectors tangent to Ω.

Theorem 8.2 (Ivrii [1980, 1984]). *Suppose that the domain Ω, with smooth boundary, satisfies condition* (A). *Then for the eigenvalues of the Dirichlet problem,*

$$N(\lambda) = (2\pi)^{-n} \operatorname{mes} \Omega \cdot \omega_n \cdot \lambda^{n/2} - \frac{1}{4}(2\pi)^{-n+1} \operatorname{mes}_{n-1}(\partial\Omega)\lambda^{(n-1)/2} + o(\lambda^{(n-1)/2}), \tag{8.18}$$

[2] Melrose [1980] independently obtained the second term for manifolds with concave boundary.

as $\lambda \to \infty$, *where* $\mathrm{mes}_{n-1}(\partial\Omega)$ *denotes the volume of the boundary with respect to the Riemannian metric that is induced by the standard Euclidean metric in* $\mathbb{R}^n$. *Under the same assumptions,*

$$'N(\lambda) = (2\pi)^{-n} \,\mathrm{mes}\, \Omega \cdot \omega_n \cdot \lambda^{n/2} + \frac{1}{4}(2\pi)^{-n+1} \,\mathrm{mes}_{n-1}(\partial\Omega)\lambda^{(n-1)/2} + o(\lambda^{(n-1)/2}) \tag{8.19}$$

for the eigenvalues of the Neumann problem.

Vasil'ev [1986] showed that condition (A) is satisfied for all strictly convex domains with analytic boundary, while Petkov and Stojanov [1985–1986] showed that this condition is satisfied for generic domains. Thus Theorem 8.2 is valid in both these cases for Ω.

A survey of several variants and generalizations of Theorems 8.1 and 8.2 can be found in Ivrii [1984].[3]

We note that Vasil'ev [1986] established analogues of Theorems 8.1 and 8.2 in the case of self-adjoint elliptic boundary-value problems for operators of arbitrary order m (see Egorov and Shubin [1988, §3, Chap. 2]). The corresponding asymptotic formulae are of the form

$$N(\lambda) = c_1 \lambda^{n/m} + O(\lambda^{(n-1)/m}), \tag{8.20}$$

$$N(\lambda) = c_1 \lambda^{n/m} + c_2 \lambda^{(n-1)/m} + o(\lambda^{(n-1)/m}), \tag{8.21}$$

where the constants c_1 and c_2 are expressed in terms of the coefficients of the operator and the boundary conditions. The second of the two formulae is valid when an analogue of condition (A) holds for the Hamiltonian flow which is defined by the principal symbol of the operator in question with suitable reflection on the boundary; in particular, it is true when a convexity condition, which Vasil'ev calls the Hamiltonian convexity, is satisfied.

We also note the results of Ivrii and Fedorova [1986]. By combining the method of hyperbolic equations with a perturbation technique, they extended Theorems 8.1 and 8.2 to a number of irregular cases where the irregularity is due either to the domain being unbounded, or to non-smoothness of the boundary, or to some degeneracy. Finally, in some cases Gureev and Safarov [1988] dispensed with condition (A) and its analogues by modifying the second term of (8.21) and obtaining a term of the form $Q(\lambda)\lambda^{(n-1)/m}$, where Q is an almost-periodic function.[4]

We also note that the asymptotic formulae for $N(\lambda)$, as $\lambda \to \infty$, can always be written as asymptotic formulae for λ_j as $j \to \infty$. Thus, the formula

$$N(\lambda) \sim c\lambda^{\alpha}, \quad \lambda \to \infty,$$

where $c > 0$ and $\alpha > 0$, is equivalent to

$$\lambda_j \sim c^{-1/\alpha} j^{1/\alpha}, \quad j \to \infty.$$

---

[3] See also forthcoming books by V. Ivrii [in preparation] and by Yu.G. Safarov and D.G. Vassiliev [in preparation]

[4] See also Safarov [1988a, b], Safarov and Vasil'ev [1992]

In particular, from (8.15) we obtain the asymptotic formula

$$\lambda_j \sim (2\pi)^2(\text{mes}\,\Omega)^{-2/n}\omega_n^{-2/n}j^{2/n}, \quad j \to \infty \tag{8.22}$$

for the eigenvalues of the Dirichlet problem in a bounded domain Ω with $\text{mes}(\partial\Omega) = 0$. The formula

$$N(\lambda) = c\lambda^{\alpha}(1 + O(\lambda^{-\delta})) \quad \text{as } \lambda \to \infty,$$

where $c > 0$, $\alpha > 0$, and $\delta > 0$, is equivalent to

$$\lambda_j = c^{-1/\alpha}j^{1/\alpha}(1 + O(j^{-\delta/\alpha})).$$

This enables us to formulate Theorem 8.1 in the form

$$\lambda_j = (2\pi)^2(\text{mes}\,\Omega)^{-2/n}\omega_n^{-2/n}j^{2/n}(1 + O(j^{-1/n})), \quad j \to \infty. \tag{8.23}$$

Similarly, under the conditions imposed on the domain, Theorem 8.2 gives two terms of the asymptotics for λ_j as $j \to \infty$.

8.3. General Formula of Weyl Asymptotics and the Method of Approximate Spectral Projection. Let us describe informally the idea which is essentially borrowed from quantum mechanics and which in many cases enables us to write correctly the principal term of the asymptotics for eigenvalues.

Let a correspondence be prescribed between the operators A from a class $\mathfrak{A}$ and the functions $a = a(v)$ on a fixed space M with measure dv. These functions will be referred to as the *symbols*. The symbol of A will be denoted by $\sigma(A)$. The correspondence $A \to \sigma(A)$ must be linear and such that $\sigma(A_1 A_2)$ coincides with $\sigma(A_1)\sigma(A_2)$ up to some lower terms. We express this by writing

$$\sigma(A_1 A_2) \sim \sigma(A_1)\sigma(A_2). \tag{8.24}$$

Heuristically, we expect that if A is a self-adjoint operator, then for a sufficiently well-behaved function $f\colon \mathbb{R} \to \mathbb{R}$

$$\sigma(f(A)) \sim f(\sigma(A)). \tag{8.25}$$

We note that, if f is a polynomial, then this result is indicated by (8.24). In particular, let us take for f the function χ_λ defined by

$$\chi_\lambda(x) = \begin{cases} 1 & \text{for } x \leqslant \lambda, \\ 0 & \text{for } x > \lambda. \end{cases}$$

Then $\chi_\lambda(A) = E_\lambda$ is the spectral projection of A (if A is semibounded from below with discrete spectrum, E_λ is the projection operator onto the subspace spanned by the eigenfunctions corresponding to the eigenvalues $\lambda_j \leqslant \lambda$). Clearly,

$$N(\lambda) = \text{Tr}\, E_\lambda. \tag{8.26}$$

From what has been said above, we may expect that

$$\sigma(E_\lambda) \sim \chi_\lambda(\sigma(A)). \tag{8.27}$$

We note that for any function $a\colon M \to \mathbb{R}$

$$\chi_\lambda(a)(v) = \begin{cases} 1 & \text{for } a(v) \leqslant \lambda \\ 0 & \text{for } a(v) > \lambda. \end{cases} \tag{8.28}$$

We assume, finally, that if the operator P has a trace, then

$$\operatorname{Tr} P \sim \int_M \sigma(P)(v)\, dv, \tag{8.29}$$

where the $\sim$ sign must be understood as an asymptotic equivalence when a large or a small parameter is present.

For a self-adjoint operator A with symbol $\sigma(A) = a$, the formulae (8.26)–(8.29) lead to the relation

$$N(\lambda) \sim \operatorname{mes}\{v\colon a(v) \leqslant \lambda\} = \int_M \chi_\lambda(a(v))\, dv. \tag{8.30}$$

This formula is the pre-image of all the asymptotic formulae for $N(\lambda)$ (with respect to λ as $\lambda \to \infty$ or some other parameter such as a small parameter h in the semi-classical problems). We now cite examples illustrating different situations.

Example 8.1. Let $X = \mathbb{R}^n$. In $\mathbb{R}^n$, we consider pseudodifferential operators $A = a^w(x, D_x)$ defined by the Weyl symbols (see § 1). The precise meaning of the condition (8.24) is contained in the composition theorem. The trace of the operator P with the usual symbol, or with the Weyl symbol, $p(x, \xi) = p(y)$ is given by the formula

$$\operatorname{Tr} P = (2\pi)^{-n} \int p(y)\, dy = (2\pi)^{-n} \int p(x, \xi)\, dx\, d\xi \tag{8.31}$$

in the case of a sufficiently smooth and sufficiently rapidly decaying symbol p. Therefore if we prescribe the measure $dv = dv(y)$ by the formula $dv(y) = (2\pi)^{-n}\, dy$, we obtain the necessary correspondence in accordance with the above scheme, and the expected asymptotic formula for the function $N(\lambda)$ in the case of a self-adjoint operator A with Weyl symbol $a(x, \xi)$ has the form

$$N(\lambda) \sim (2\pi)^{-n} \operatorname{mes}\{(x, \xi)\colon a(x, \xi) \leqslant \lambda\} = (2\pi)^{-n} \int_{a(x,\xi) \leqslant \lambda} dx\, d\xi, \tag{8.32}$$

where mes denotes the Lebesgue measure on $\mathbb{R}^{2n}$. We can also consider the usual symbol in place of the Weyl symbol, but then we replace $a(x, \xi)$ by $\operatorname{Re} a(x, \xi)$ in (8.32). This case can also be included within the framework of the above scheme if $\sigma(A)$ is regarded as the principal part of the symbol $a(x, \xi)$ in some natural sense (for example, as the homogeneous leading part); $\sigma(A)$ will automatically be real valued. In particular, for the Schrödinger operator $A = -\Delta + q(x)$, where $a(x, \xi) = \xi^2 + q(x)$, the formula (8.32), on integration with respect to ξ, yields

$$N(\lambda) \sim (2\pi)^{-n} \omega_n \int_{\mathbb{R}^n} (\lambda - q(x))_+^{n/2}\, dx, \tag{8.33}$$

where $(\lambda - q(x))_+ = \max(0, \lambda - q(x))$.

Example 8.2. Let X be an n-dimensional compact manifold without boundary. We consider the algebra of all classical pseudodifferential operators on X, and by $\sigma(A)$ we understand the principal symbol of A. Then we must take $M = T^*X$ and $dv = (2\pi)^{-n}\, dx\, d\xi$, where $dx\, d\xi$ is the canonical measure on T^*X defined in each coordinate neighbourhood by the local coordinates x and the corresponding dual coordinates ξ. The relation (8.24) becomes an equality, and the formula (8.31) shows that something akin to (8.29) is valid, because we can decompose the pseudodifferential operator on X into a sum of operators whose kernels are concentrated in the direct product of a pair of coordinate neighbourhoods. This leads to the expected result (8.30) in which a is the principal symbol of A.

Example 8.3. If X is a manifold with boundary (for example, a domain in $\mathbb{R}^n$), we may expect that the boundary does not affect the principal term of the asymptotics for $N(\lambda)$. In this context the significance of formula (8.15) becomes clear, because the principal symbol of the Laplace operator is $-|\xi|^2$ and

$$(2\pi)^{-n} \operatorname{mes}\{(x, \xi)\colon x \in \Omega, |\xi|^2 < \lambda\} = (2\pi)^{-n} \operatorname{mes} \Omega \cdot \omega_n \cdot \lambda^{n/2}.$$

Let us cite a few more simple but important generalizations of the type (8.30) and the idea put forward above. Suppose that the values of the symbol $a(v)$ are not numbers but are themselves operator functions (for example, matrix functions) with values in a space of operators in some linear space, possibly depending on v, which is simpler than the original space. We denote the trace of $\sigma(P)(v)$ by $\operatorname{tr} \sigma(P)(v)$ in order to distinguish it from the trace Tr of the operators in question. Instead of (8.29), we now assume that the relation

$$\operatorname{Tr} P \sim \int_M \operatorname{tr} \sigma(P)(v)\, dv \tag{8.29'}$$

holds. Then we may expect that the asymptotic formula

$$N(\lambda) \sim \int_M \operatorname{tr} E_\lambda(a(v))\, dv = \sum_j \operatorname{mes}\{v\colon \lambda_j(a(v)) \leqslant \lambda\} \tag{8.30'}$$

holds in place of (8.30). Here $E_\lambda(a(v))$ is the spectral projection of the operator $a(v)$, where $a(v)$ is assumed to be self-adjoint, and $\lambda_j(a(v))$ are the eigenvalues of $a(v)$. When all the spaces where the operators $a(v)$ act are one-dimensional, the formula (8.30′) becomes (8.30). The simplest examples where (8.30′) is applicable are provided by the matrix operators in Examples 8.1 and 8.3 or by the operators in sections of vector bundles in Example 8.2. We now present a more complicated case. In considering operators on a manifold X which degenerate on the boundary Y, it proves useful to take a different operator symbol for $a(v)$ instead of the usual symbol. This operator symbol is defined for $v \in T^*Y\backslash 0$ (the cotangent bundle space of Y without the zero section) acting in the space $L_2(\mathbb{R}^+)$ as a pseudodifferential operator whose symbol is obtained from the symbol of the original operator by fixing the point $x' \in Y$ and the dual variable $\xi' = v$.

Although the above *a priori* discussions leading to the formulae (8.30) and (8.30′) are only heuristic, they can be caried out rigourously in many cases. This was first done by Tulovskij and Shubin (Shubin [1978, Chap. 4]) in the context of Example 8.1 for operators in $\mathbb{R}^n$ with Weyl symbols belonging to classes G_ρ^m (see § 1). Then main idea is as follows. If we take a suitably smoothed function $\tilde{\chi}_\lambda$, instead of the characteristic function χ_λ, then the operator $\mathscr{E}_\lambda$ with symbol $\tilde{\chi}_\lambda(a(x, \xi))$ possesses properties that are close to those of the spectral projection E_λ. For example, we may assume that $\mathscr{E}_\lambda^* = \mathscr{E}_\lambda$ and, as $\lambda \to \infty$, the operator $\mathscr{E}_\lambda^2 - \mathscr{E}_\lambda$ is small in the trace-class norm in comparison to the function

$$V(\lambda) = (2\pi)^{-n} \operatorname{mes}\{(x, \xi): a(x, \xi) \leqslant \lambda\}$$

which gives the asymptotics assumed. This implies that, as $\lambda \to \infty$, the eigenvalues of the operator $\mathscr{E}_\lambda$ are concentrated asymptotically near the points 0 and 1. The subspace spanned by the eigenvectors of $\mathscr{E}_\lambda$ corresponding to the eigenvalues close to 1 should be used as a test subspace L in the variational principle (8.6) or (8.7) by first establishing an estimate of the form

$$\mathscr{E}_\lambda(A - \lambda I)\mathscr{E}_\lambda \leqslant C\lambda^{1-\varepsilon} \tag{8.34}$$

with some $\varepsilon > 0$ and $C > 0$ that are independent of λ. This gives a lower estimate for $N(\lambda)$ in terms of a quantity approximately equal to $\operatorname{Tr} \mathscr{E}_\lambda \sim V(\lambda)$. Similarly, by considering the orthogonal complement of the same L and using an estimate of the form

$$(I - \mathscr{E}_\lambda)(A - \lambda I)(I - \mathscr{E}_\lambda) \geqslant -C\lambda^{1-\varepsilon}, \tag{8.35}$$

we can obtain an upper estimate for $N(\lambda)$. These two estimates together yield the asymptotics for $N(\lambda)$. The operator $\mathscr{E}_\lambda$ that imitates the properties of the spectral projection E_λ can naturally be called the *approximate spectral projection*. The method of finding asymptotics by means of the approximate spectral projection has been developed by several authors, but the most significant contribution has been made by Levendorskij (see Levendorskij [1985, 1990] and the works cited there). He applied this method to a large class of problems and, in particular, to non-elliptic and degenerate operators and systems and to various other problems with a small parameter, such as semi-classical problems and problems in the theory of shells. This method also enables us to obtain an estimate for the remainder term in the asymptotics although the result is not as precise as the one obtained by the method of hyperbolic equations.

We also note that in the asymptotic problem for the eigenvalues of the operator with Weyl h-symbol $a(x, \xi)$ (that is, for the operator with Weyl symbol $a(x, h\xi)$) the asymptotics for the distribution function $N_h(\lambda)$, as $h \to +0$ with λ fixed, must be of the form

$$N_h(\lambda) \sim (2\pi h)^{-n} \operatorname{mes}\{(x, \xi): a(x, \xi) < \lambda\}, \tag{8.36}$$

in accordance with the formulae of § 1 and the idea developed above. This formula can also be justified under suitable assumptions by the method of the approximate spectral projection which was first done by V. Roitburd (see Shubin [1978, Supplement 2]), and by other methods.

8.4. Tauberian Methods. Tauberian methods are based on the study of certain functions of eigenvalues from which the application of so-called Tauberian theorems enable us to draw conclusions regarding the asymptotic behaviour of the eigenvalues themselves. Let us cite one example explaining the situation.

We consider the *ζ-function* $\zeta_A(z)$ of a positive self-adjoint operator A with eigenvalues $\lambda_1 \leqslant \lambda_2 \leqslant \lambda_3 \leqslant \cdots$ which satisfy the condition that $\lambda_j \geqslant cj^\varepsilon$ with some $c > 0$ and $\varepsilon > 0$ or equivalently, $N(\lambda) \leqslant c_1 \lambda^\alpha$ for some $c_1 > 0$ and $\alpha > 0$. The ζ-function is defined by

$$\zeta_A(z) = \sum_{j=1}^{\infty} \lambda_j^z = \int_0^\infty \lambda^z \, dN(\lambda). \tag{8.37}$$

This integral converges for z in a half-plane $\operatorname{Re} z < -\mu_0$, and it defines a holomorphic function of z in this half-plane.

Let us assume for a moment that $N(\lambda)$ has the asymptotic expansion

$$N(\lambda) = \sum_{k=0}^{r-1} c_k \lambda^{\alpha_k} + O(\lambda^{\alpha_r}), \tag{8.38}$$

as $\lambda \to \infty$, where $\alpha_0 > \alpha_1 > \cdots > \alpha_{r-1} > \alpha_r$. Substituting this expression into (8.37) and replacing the integral by a similar integral with limits from 1 to ∞ we find that

$$\zeta_A(z) = \sum_{k=0}^{r-1} \frac{c_k \alpha_k}{z - \alpha_k} + \rho_r(z), \tag{8.39}$$

where $\rho_r(z)$ is a holomorphic function for $\operatorname{Re} z > \alpha_r$. Thus we see that if $N(\lambda)$ has the expansion (8.38), then the function ζ_A has a meromorphic extension to the half-plane $\{z\colon \operatorname{Re} z > \alpha_r\}$ with simple poles at the points α_k and the residues at these poles are $c_k \alpha_k$. In this way, all the terms of the expansion (8.38), corresponding to the exponents $\alpha_k > 0$, can be recovered from the poles and residues of ζ_A.

The hypothetical expansion (8.38) is almost never valid in a real analytic situation. For an elliptic operator on a closed manifold X with dimension $n \geqslant 2$ we have, in general, only one term of such an expansion. For example, for the Laplace-Beltrami operator Δ on the unit sphere $S^n \subset \mathbb{R}^{n+1}$, the eigenvalues $\mu_k = -k(k+n-1)$ $(k = 0, 1, 2, \ldots)$ have such a high multiplicity $N_k = \binom{n+k}{n} - \binom{n+k-2}{n}$ that the second term of the asymptotics (8.38) for the distribution function $N(\lambda)$ corresponding to the operator $A = -\Delta$ cannot exist. What is striking is that the ζ-function is still meromorphic in the whole complex plane for the most general case:

Theorem 8.3 (Seeley [1967]). *Let A be a self-adjoint classical pseudodifferential operator of order $m > 0$ on an n-dimensional closed manifold X. Then the ζ-function $\zeta_A = \zeta_A(z)$ has a meromorphic extension to the whole complex z-plane with possible simple poles at the points $z_j = (j - n)/m$ $(j = 0, 1, 2, \ldots)$ which form an arithmetical progression. In fact, the point $z = 0$ is not a pole, and if A is a differential operator, then the points $z = 0, 1, 2, \ldots$ are also not poles.*

The residues at these poles and the value at the point zero (or the values at the points 0, 1, 2, ..., if A is a differential operator) can be expressed in the form of certain integrals over S^*X (over the fibres of the cotangent spheres) of expressions that are obtained from recurrence formulae in terms of the homogeneous components of the symbol. For example, the residue r_0 at the first pole $z_0 = -n/m$ is expressed in terms of the principal symbol a_m by a formula that can be written in any local coordinates in the form

$$r_0 = -(2\pi)^{-n} m^{-1} \int_X dx \int_{|\xi|=1} a_m^{-n/m}(x, \xi)\, dS_\xi, \tag{8.40}$$

after using a partition of unity, where dS_ξ is the standard measure on the unit sphere $\{\xi: |\xi| = 1\}$.

Returning to the connection between (8.38) and (8.39), we find that the coefficient c_0 in the principal term of the asymptotics (8.38) has the form $c_0 = r_0/\alpha_0 = r_0/z_0 = -r_0 m/n$. It can be verified easily that this result agrees with the Weyl formula of Example 8.3.

The proof of Theorem 8.3 is based on the description of the structure of the complex powers of A given in the language of the pseudodifferential operators in § 1. If we compute $\zeta_A(z) = \operatorname{Tr} A^z$ using a partition of unity and local coordinates on X, then passing to homogeneous components of the approxmation of the symbol of A^z and finally a polar coordinates on the set $\{(x, \xi): |\xi| \geqslant 1\}$, we find that everything reduces to a meromorphic continuation of the integrals

$$\int_1^{+\infty} r^{mz-j+n-1}\, dr = -\frac{m^{-1}}{z - z_j}.$$

Theorem 8.3 is also valid for a non-self-adjoint operator A if there is a ray in $\mathbb{C}$ that passes through 0 and does not contain values of the principal symbol a_m. In this case, we take $\zeta_A(z) = \operatorname{Tr} A^z$, or define $\zeta_A(z)$ by the first equation in (8.37). Results of the type in Theorem 8.3 are also valid for an operator on a compact manifold with boundary if by A we mean a self-adjoint operator defined by an elliptic formally self-adjoint differential operator L with homogeneous elliptic boundary conditions, provided that $L - \lambda I$ with these conditions satisfies the ellipticity condition with the parameter λ in $(-\infty, 0]$ (Seeley [1969a, b]). The case of pseudodifferential boundary-value problems is much more complicated (see Grubb [1986], Rempel and Schulze [1982]).

In place of the ζ-function, we can consider the kernel $K_z = K_z(x, y)$ of the complex power A^z as a function of z for fixed $x, y \in X$. If $\operatorname{Re} z < -n/m$, then such a kernel is a continuous function of x and y on $X \times X$ and a holomorphic function of z. Moreover, a statement similar to Theorem 8.3 holds for the value $K_z(x, x)$ on the diagonal, that is, the function $z \mapsto K_z(x, x)$ can be extended meromorphically to the whole complex plane $\mathbb{C}$ with possible poles at the same points $z_j = (j - n)/m$ $(j = 0, 1, 2, \ldots)$ which constitute an arithmetical progression, excluding the point $z = 0$ which is not a pole (and also excluding integers $z \geqslant 0$ if A is a differential operator). The residues at these poles and the value at

$z = 0$ are given by formulae similar to those mentioned earlier. For example, the residue at the first pole $z_0 = -n/m$ is obtained by dropping the integration with respect to x in (8.40). When $x \neq y$, the function $z \mapsto K_z(x, y)$ can be extended to an entire function of z which vanishes for $z = 0$ (and for all integers $z \geqslant 0$ if A is a differential operator).

We now introduce the *spectral function of the operator* A, this being the kernel $e(\lambda, x, y)$ of the spectral projection E_λ. This function is defined for every self-adjoint operator A in $L_2(X)$, but if A is semibounded below and has discrete spectrum, then the function can be expressed in terms of the eigenvalues λ_j and normalized eigenfunctions by the formula

$$e(\lambda, x, y) = \sum_{\lambda_j \leqslant \lambda} \psi_j(x)\overline{\psi_j(y)}.$$

In particular, this implies that, under the hypotheses of Theorem 8.3, $e(\lambda, \cdot, \cdot) \in C^\infty(X \times X)$ for each fixed λ. The function $\lambda \mapsto e(\lambda, x, x)$ is increasing in λ for each fixed x. Moreover, the kernel $e(\lambda', \lambda'', x, y) = e(\lambda', x, y) - e(\lambda'', x, y)$ is hermitian and positive for $\lambda' > \lambda''$, that is, the matrix $\|e(\lambda', \lambda'', x_i, x_j)\|_{i,j=1}^N$ is hermitian and positive for every set of points $x_1, \ldots, x_N \in X$. Using this fact for $N = 2$, we can show easily that the function $\lambda \mapsto e(\lambda, x, y)$ is of bounded variation of each bounded interval of the λ-axis.

We now note that on account of the formula

$$A^z = \int \lambda^z \, dE_\lambda,$$

the kernels $K_z(x, y)$ and $e(\lambda, x, y)$ are connected by the formula

$$K_z(x, y) = \int \lambda^z \, de(\lambda, x, y), \tag{8.41}$$

where the differential d is taken with respect to λ. Here the integral can be understood in a weak sense, but for $\operatorname{Re} z < -n/m$ the integral converges as a Steiltjes integral for every fixed x and y. The formula (8.41) can be used for obtaining information about the spectral function on the basis of properties of the kernel K_z in accordance with the above scheme but subject to the comments given below.

Theorem 8.3 can also be used to study and compute the index of A due to the *Atiyah-Bott formula*

$$\operatorname{ind} A = \zeta_{I+A^*A}(z) - \zeta_{I+AA^*}(z)$$

which follows from the fact that all the non-zero eigenvalues of the operators AA^* and A^*A coincide, and have the same multiplicity. The most convenient approach is to take $z = 0$ on the right-hand side because the value of this side at $z = 0$ is expressed in terms of the symbol of the operator A. In this way we obtain an expression for the index in terms of integrals of expressions which contain the symbol and its derivatives. One of the proofs of the celebrated Atiyah-Singer index theorem is based on the simplification of the above complicated formula (which, in particular, contains lower terms of the symbol) with the help of the theory of invariants (see Atiyah, Bott and Patodi [1973], Berline, Getzler, Vergne [1992]).

We turn finally to the question of what information about the asymptotics of $N(\lambda)$ can be derived from Theorem 8.3. Here the following theorem plays a fundamental role (see, for example, Shubin [1978]).

Theorem 8.4 (Ikehara's Tauberian theorem). *Let the function* $\zeta_A(z)$ *be defined by* (8.37) *in terms of the non-decreasing function* $N(\lambda)$ *which is zero if* $\lambda < 0$. *Suppose that the integral in* (8.37) *converges for* $\operatorname{Re} z < -k_0$, *where* $k_0 > 0$, *and that the function* $\zeta_A(z) + \dfrac{r_0}{z + k_0}$ *is continuous in the closed half-plane* $\operatorname{Re} z \leqslant -k_0$. *Then, as* $\lambda \to \infty$,

$$N(\lambda) = r_0 k_0^{-1} \lambda^{k_0}(1 + o(1)).$$

Thus Theorem 8.3 yields the Weyl asymptotics

$$N(\lambda) = c_0 \lambda^{n/m}(1 + o(1)) \tag{8.42}$$

in the case of an elliptic operator A of order m which is self-adjoint and semi-bounded below on a closed manifold X. We note that the condition $A > 0$ can be made to hold by replacing A by $A + cI$, where $c > 0$ is sufficiently large. However, in the transition from $\zeta_A(z)$ to $N(\lambda)$ in Theorem 8.4, all the information regarding the ζ-function is lost, except the information concerning the first pole, and we fail to utilise it for improving the asymptotics (8.42). Likewise, in the non-self-adjoint case, if the principal symbol is complex-valued, we fail to utilise the information regarding the ζ-function (and other known methods) for obtaining even the principal term of the asymptotics of the eigenvalues.

Using the meromorphic extension of the function $z \to K_z(x, x)$, we can easily derive the asymptotics of the spectral function $e(\lambda, x, x)$ from Theorem 8.4. Thus, under the hypotheses of Theorem 8.3, we have

$$e(\lambda, x, x) = c_0(x)\lambda^{n/m}(1 + o(1)), \tag{8.43}$$

as $\lambda \to \infty$, where $c_0(x)$ is expressed in local coordinates in the form

$$c_0(x) = (2\pi)^{-n} \operatorname{mes}\{\xi\colon a_m(x, \xi) < 1\}. \tag{8.44}$$

The asymptotics (8.43) is uniform in x and, by integrating (8.43), we obtain the asymptotics (8.42) for $N(\lambda)$. The holomorphicity of the function $z \mapsto K_z(x, y)$ for $x \neq y$ does not give any asymptotics nor even an estimate for $e(\lambda, x, y)$, because the function $\lambda \mapsto e(\lambda, x, y)$ is not monotonic in λ.

Instead of using the ζ-function $\zeta_A(z)$, we can straightaway use the trace $\operatorname{Tr}(A - zI)^{-1}$ of the resolvent, this trace being defined if $m > n$; otherwise A must be replaced by A^k with a sufficiently large k. This gives information regarding the behaviour of $N(\lambda)$ from the behaviour of its *Steiltjes transform*

$$\operatorname{Tr}(A - zI)^{-1} = \int (\lambda - z)^{-1}\, dN(\lambda).$$

Another variant of the use of the Tauberian technique is to consider the function

$$\theta_A(t) = \sum_{j=1}^{\infty} e^{-t\lambda_j} = \operatorname{Tr} e^{-tA} = \int e^{-\lambda t}\, dN(\lambda).$$

In using this function, we usually study the operator e^{-tA} as an operator solution of the Cauchy problem for the parabolic equation

$$\frac{\partial u}{\partial t} = -Au, \quad u|_{t=0} = u_0.$$

The kernel $K = K(t, x, y)$ of the operator e^{-tA}, which is a fundamental solution of the Cauchy problem, is usually constructed by the Levi method (see Egorov and Shubin [1988, § 5, Chap. 2]). Under the hypotheses of Theorem 8.3, the function $\theta_A(t)$ has an asymptotic expansion as $t \to +0$, from which we can again obtain the asymptotics (8.42). Historically, this method was the first Tauberian method by means of which Carleman [1935] first established the asymptotic formula (8.43) for the spectral function of the Laplace operator in a domain. A detailed account of the parabolic equation method and the resolvent method can be found in the paper by Rozenblyum, Solomyak and Shubin on the spectral theory of differential operators to be published in one of the volumes of the present series.

8.5. The Hyperbolic Equation Method. The hyperbolic equation method, developed by Levitan [1952], is a Tauberian method which uses the Fourier transformation and its analogues. This method enables us to obtain the most precise information on the asymptotic distribution of the spectrum, though for a narrower class of problems. Let us describe the method in its simplest form.

Let A be a classical self-adjoint elliptic pseudodifferential operator of order $m > 0$ on an n-dimensional closed manifold X with principal symbol $a_m = a_m(x, \xi) \geqslant 0$. By replacing A by $A + CI$, if necessary, where the number $C > 0$ is sufficiently large, we can assume that $A > 0$. Instead of A itself, we use the operator $B = A^{1/m}$ which is also a classical pseudodifferential operator of order 1 with principal symbol $b_1 = a_m^{1/m}$. The spectral distribution functions $N_A(\lambda)$ and $N_B(\lambda)$ of A and B are connected by the obvious relation $N_A(\lambda^m) = N_B(\lambda)$, and so it is sufficient to study the asymptotics of $N_B(\lambda)$. We consider the Fourier transform of the corresponding measure

$$\sigma(t) = \int e^{-i\lambda t}\, dN_B(\lambda) = \sum_j e^{-i\mu_j t}, \tag{8.45}$$

where μ_j are the eigenvalues of B, and the sum and integral are understood in the sense of distributions. We now introduce the operator $U(t) = e^{-iBt}$ which is an operator solution of the Cauchy problem

$$\frac{\partial u}{\partial t} = -iBu, \quad u|_{t=0} = u_0, \tag{8.46}$$

where $u = u(t, x)$, $t \in \mathbb{R}$, $x \in X$, and $u_0 = u_0(x)$. This problem is the simplest hyperbolic pseudodifferential Cauchy problem and its operator solution $U(t)$ is a Fourier integral operator which depends smoothly on t (see § 4). The structure of this operator enables us to describe the singularities of its kernel $U(t, x, y)$. In particular, it can be shown easily that if

$$\mathrm{WF}'(U) = \{((x, \xi), y, \eta): (x, y, \xi, -\eta) \in \mathrm{WF}(U)\},$$

then

$$\mathrm{WF}'(U(t, \cdot, \cdot)) \subset \{((x, \xi), \Phi^t(x, \xi)): (x, \xi) \in T^*X \backslash 0\}, \tag{8.47}$$

where Φ^t is the Hamiltonian flow defined on T^*X by the Hamiltonian $b_1 = a_m^{1/m}$. This implies, in particular, that if $u = u(t, x)$ is a solution of the Cauchy problem (8.46), then

$$\mathrm{WF}(u(t, \cdot)) = \Phi^t(\mathrm{WF}(u_0)), \tag{8.48}$$

and this gives important information concerning the propagation of singularities. We note that formally

$$\sigma(t) = \operatorname{Tr} U(t) = \int_X U(t, x, x)\, dx. \tag{8.49}$$

In fact, this formula can be given a precise meaning after multiplication by an arbitrary function $\varphi = \varphi(t) \in C_0^\infty(\mathbb{R})$ and integration with respect to t, and in this regularised form it is easily established. Since $U(t, x, y)$ is a smooth function of t with values in $\mathcal{D}'(X \times X)$, the function σ is smooth for those t_0 such that, for nearby values of t, the restriction $U(t, x, x)$ to the diagonal $\Delta = \{(x, x)\} \subset X \times X$ depends continuously on the distribution $U(t, \cdot, \cdot)$ (and this restriction $U(t, x, x)$ must be understood as an element of $\mathcal{D}'(X)$ that depends on t). But this evidently holds if the microlocal condition for the existence of restriction to the diagonal is satisfied (see § 3):

$$\mathrm{WF}(U(t_0, \cdot, \cdot)) \cap N^*\Delta = \varnothing, \tag{8.50}$$

where $N^*\Delta$ denotes the conormal bundle of the diagonal, that is,

$$N^*\Delta = \{(x, x, \xi, -\xi): x \in X\} \subset T^*(X \times X).$$

It follows from (8.47) that, if (8.50) is not satisfied, then there is a point $(x, \xi) \in T^*X \backslash 0$ such that $\Phi^{t_0}(x, \xi) = (x, \xi)$, that is, t_0 is the period of the Hamiltonian flow Φ^t (in this case, (x, ξ) is known as the *periodic point of the flow* Φ^t). In this way, we find that

$$\operatorname{sing\,supp} \sigma \subset \{t: \exists (x, \xi) \in T^*X \backslash 0,\ \Phi^t(x, \xi) = (x, \xi)\}. \tag{8.51}$$

This relation was first obtained independently by Chazarain and Colin de Verdiere (see Trèves [1980]). It is referred to as the *Poisson relation*, because in the simplest case of the operator $A = -d^2/dx^2$ on the circle $S^1 = \mathbb{R}/T\mathbb{Z}$, in which $T > 0$, this relation follows from the Poisson summation formula

$$\sum_{n\in Z} e^{-2\pi i\xi T/n} = T\sum_{n\in Z}\delta(\xi - Tn). \tag{8.52}$$

It is clear from the Poisson relation (8.51) that to study the asymptotics of $N(\lambda)$ as $\lambda\to\infty$, it is most important to know the behaviour of $\sigma(t)$ near the periods of the Hamiltonian flow. The most important point here is the point $t=0$. The singularity of σ, as $t\to 0$, is referred to as the *big singularity*. From its study we can obtain important information concerning the asymptotics of the function $N_B(\lambda)$. We note that this singularity is isolated, and this fact is an important feature of the hyperbolic equation method. We cloose a function $\rho\in S(\mathbb{R})$ such that $\tilde{\rho}\in C_0^\infty(\mathbb{R})$, $\tilde{\rho}(0)=1$, $\rho(\lambda)>0$ and $\rho(0)>0$. Then $\tilde{\rho}\sigma$ is the Fourier transform of the function

$$(\rho * dN_B)(\lambda) = \int \rho(\lambda-\mu)\, dN_B(\mu).$$

If supp $\tilde{\rho}$ is sufficiently small, we may assume that $\tilde{\rho}\sigma$ has a singularity only at 0 and, by analysing this singularity of $\tilde{\rho}\sigma$, we can easily find the complete asymptotic expansion

$$(\rho * dN_B)(\lambda) = c_0\lambda^{n-1} + c_1\lambda^{n-2} + c_2\lambda^{n-3} + \cdots. \tag{8.53}$$

Levitan's Tauberian theorem enables us to derive from (8.53) the asymptotic formula

$$N_B(\lambda) = c_0 n^{-1}\lambda^n + O(\lambda^{n-1}), \tag{8.54}$$

which is equivalent to the asymptotic formula

$$N(\lambda) = c_0 n^{-1}\lambda^{n/m} + O(\lambda^{(n-1)/m}). \tag{8.55}$$

This formula in the above context was obtained by Hörmander [1968], who was the first to apply Fourier integral operators to spectral theory. The estimate for the remainder term in (8.55) cannot be improved in the general case, as is shown by examining the Laplace-Beltrami operator on the standard unit sphere. Under some additional conditions of a geometrical nature, we can also obtain the second term of the asymptotics. This was first done by Duistermaat and Guillemin [1975]. The first step of their method consists of replacing $\tilde{\rho}$ by $\tilde{\rho}_T$, where $\tilde{\rho}_T(t)=\tilde{\rho}(t/T)$ and $T\to\infty$. Next, to obtain the asymptotics of $\rho_T * dN_B(\lambda)$ as $\lambda\to\infty$, we need to know the singularities of the function $\tilde{\rho}_T\sigma$, and for this we need information concerning the singularities of σ on a segment with length T. The necessary information is obtained if the set of periodic points of the flow Φ^t has measure zero. Further, it is sufficient to consider, in place of periodic points, *absolutely periodic points*, that is, the points $(x,\xi)\in T^*X\backslash 0$ such that the graph of the map Φ^t at the point $((x,\xi),\Phi^t(x,\xi))$ has contact of infinite order with the diagonal. Under this assumption, the formula

$$\rho_T * dN_B(\lambda) = c_0\lambda^{n-1} + c_1\lambda^{n-2} + o(\lambda^{n-2}) \tag{8.56}$$

holds for any $T>0$. From (8.56) and with the help of a certain refinement of Levitan's Tauberian theorem, we can obtain the formula

$$N_B(\lambda) = c_0 n^{-1} \lambda^n + c_1 (n-1)^{-1} \lambda^{n-1} + o(\lambda^{n-1}) \tag{8.57}$$

or

$$N(\lambda) = c_0 n^{-1} \lambda^{n/m} + c_1 (n-1)^{-1} \lambda^{(n-1)/m} + o(\lambda^{(n-1)/m}). \tag{8.58}$$

Here the coefficient c_1 is expressed in terms of the subprincipal symbol of the operator A and is equal to zero if A is a differential operator.

The asymptotic formulae (8.55) and (8.58) were extended to the case of manifolds X with boundary by Ivrii and Vasil'ev, and their results were stated above in § 8.2. They also use some analogues of the hyperbolic equation method just described but with essential modifications.

The estimate $o(\lambda^{(n-1)/m})$ in (8.58) for the remainder term can be improved in certain cases. Namely, for the Laplace-Beltrami operator on closed manifolds with negative curvature, Berard [1977] found that $o(\lambda^{(n-1)/2})$ can be replaced by $O(\lambda^{(n-1)/2}/\log \lambda)$. Volovoj [1986, 1990a, b] extended this result and proved that $o(\lambda^{(n-1)/m})$ can be replaced by $O(\lambda^{(n-1)/m}/\log \lambda)$ for a class of operators of an arbitrary order $m > 0$ (on a closed manifold), and under certain assumptions Volovoj also established an asymptotic formula with the estimate $O(\lambda^{(n-1)/m-\varepsilon})$, where $\varepsilon > 0$, for the remainder. These results take into account a more precise dependence on T in asymptotics of the type (8.56); this is done by studying the behaviour of $U(t)$ as $|t| \to \infty$. The assumptions include estimates on the measure of the set of the trajectories of the Hamiltonian flow that are nearly closed.

For the spectral function $e(\lambda, x, y)$ of second-order operators in $\mathbb{R}^n$ with coefficients which are constant in a neighbourhood of infinity, as well as for the *Hill operator* (that is, the Schrödinger operator with periodic potential on $\mathbb{R}^1$), we can even obtain a complete asymptotic expansion in powers of λ (see Popov and Shubin [1983], Schenk and Shubin [1985], and Vajnberg [1983]).

Bibliographical Comments

The theory of pseudodifferential operators appeared in its present form in the mid-1960s. Its appearance was preceded by a long development and perfection of the theory of singular integral and integro-differential operators (see Agranovich [1965], Calderon [1958], Calderon and Zygmund [1952], Mikhlin [1962], and Stein [1970]). But only the formulation of the basic elements in the language of Fourier transformations in the work of Kohn and Nirenberg [1965] enabled the theory to reach its final form, and subsequently to display breadth and power as it penetrated practically all branches of the theory of linear partial differential equations. Later in the course of the natural development of the theory of pseudodifferential operators, the ideas were combined with concepts from asymptotic theory (Maslov [1965]) to produce the theory of Fourier integral operators, a direct ancestor of which is the theory of the Maslov canonical operator (see Leray [1978], Maslov [1965], Maslov and Fedoryuk [1976], Mishchenko, Sternin and Shatalov [1978], Nazajkinskij, Oshmyan, Sternin and Shatalov [1981], Vajnberg [1982]). Together with the concept of the wave front, introduced by Sato [1970] and Hörmander [1971] and very widely used, these ideas came to form one of the fundamental directions in the modern development of the theory of partial differential equations and constitute what is referred to as *microlocal analysis*.

The various aspects of microlocal analysis and its applications have been discussed in Agranovich [1969, 1971], Atiyah, Bott and Patodi [1973], Beals and Fefferman [1973], Duistermaat and Hörmander [1972], Egorov [1975, 1984, 1985], Eskin [1973], Fedoryuk [1971], Friedrichs [1968], Grubb [1986], Guillemin and Sternberg [1977], Hörmander [1966, 1983, 1985], Kumano-go [1982], Melrose [1978], Mishchenko, Sternin and Shatalov [1978], Nazajkinskij, Oshmyan, Sternin and Shatalov [1981], Nirenberg [1970, 1973], Olejnik and Radkevich [1969], Plamenevskij [1986], Reed and Simon [1972–1978], Seeley [1967], Shubin [1978], Taylor [1981], and Trèves [1980]. An account of symplectic geometry can be found in Arnol'd [1979] and in the survey paper of Arnol'd and Givental published in vol. 4 of the present series. The method of stationary phase with degenerate critical points has been discussed by Arnol'd, Varchenko and Gusejn-Zade [1984] and by Duistermaat [1974].

The appearance of the theory of pseudodifferential operators was stimulated by the requirements of index theory (see Fedosov [1974a], Palais [1965], and Rempel and Schulze [1982]).

The asymptotic methods and their various applications have been discussed in Babich and Buldyrev [1972], Fedoryuk [1977, 1986], Ivrii [1984], Leray [1978], Maslov [1965, 1973, 1977, 1987], Maslov and Fedoryuk [1976], Mishchenko, Sternin and Shatalov [1978], Nazajkinskij, Oshmyan, Sternin and Shatalov [1981], and Vajnberg [1975, 1982].

The papers by Birman and Solomyak [1974, 1977] contain a survey of the classical questions connected with the asymptotics of the spectrum and, for further development in this direction, the reader is referred to Ivrii [1984] and to the survey paper by Rozenblyum, Solomyak and Shubin to be published in the present series. An account of the geometry of the spectrum can be found in Berard's book [1986].

We also note that the commentary which accompanies the works of Petrovskij [1986] contains a survey of the most diverse questions in the modern theory of partial differential equations, a theory greatly indebted to Petrovskij himself.

References*

Agranovich, M.S.

[1965] Elliptic singular integro-differential operators. Usp. Mat. Nauk *20*, No. 5, 3–120. English transl.: Russ. Math. Surv. *20*, No. 5, 1–121, Zbl.149,361

[1969] Boundary-value problems for systems of first-order pseudodifferential operators. Usp. Mat. Nauk *24*, No. 1, 61–125. English transl.: Russ. Math. Surv. *24*, No. 1, 59–126 (1969), Zbl.175,108

[1971] Boundary-value problems for systems with a parameter. Mat. Sb. *84*, No. 1, 27–65. English transl.: Math. USSR–Sb. *13*, 25–64, Zbl.207,108

Agranovich, M.S., Vishik, M.I.

[1964] Elliptic problems with a parameter and parabolic problems of general type. Usp. Mat. Nauk *19*, No. 3, 53–161. English transl.: Russ. Math. Surv. *19*, No. 3, 53–157, Zbl.137,296

Arnol'd, V.I.

[1979] Mathematical Methods in Classical Mechanics. Second edition, Moscow: Nauka. English transl.: Graduate Texts in Math. *60*, New York-Berlin-Heidelberg: Springer-Verlag 1978, Zbl.386.70001

*For the convenience of the reader, references to reviews in Zentralblatt für Mathematik (Zbl.), compiled using the MATH database, have, as far as possible, been included in this bibliography.

Arnol'd, V.I., Varchenko, A.N., Gusejn-Zade, S.M.
[1984] Singularities of Differentiable Maps. Monodromy and Asymptotics of Integrals. Moscow: Nauka. English transl.: Monographs in Mathematics *83*, Boston, MA: Bikhäuser 1988, Zbl.545.58001

Atiyah, M., Bott, R., Patodi, V.K.
[1973] On the heat equation and the index theorem. Invent. Math. *19*, 279–330. Errata ibid *28* (1975), 277–280, Zbl.257.58008; Zbl.301.58018

Babich, V.M., Buldyrev, V.S.
[1972] Asymptotic Methods in Short Wave Diffraction Problems. Moscow: Nauka. English transl.: Springer Ser. Wave Phenom. *4* 1991, Berlin-Heidelberg-New York: Springer-Verlag, Zbl.255.35002

Balaban, T.
[1969] On the mixed problem for a hyperbolic equation. Bull. Acad. Polon. Sci. Math. Astron. Phys. *17*, No. 4, 231–235, Zbl.176,404

Beals, R., Fefferman, C.
[1973] On local solvability of linear partial differential equations. Ann. Math., II. Ser. *97*, No. 3, 482–498, Zbl.256.35002

Bérard, P.H.
[1977] On the wave equation on a compact Riemannian manifold with conjugate points. Math. Z. *155*, 249–276, Zbl.341.35052
[1986] Spectral Geometry: Direct and Inverse Problems. Lect. Notes Math. *1207*, Berlin-Heidelberg-New York: Springer-Verlag, Zbl.608.58001

Berezin, F.A., Shubin, M.A.
[1983] The Schrödinger Equation. Moscow: Izd. Moskov. Gos. Univ. English transl.: Dordercht: Kluwer Academic Publishers 1991, Zbl.546.35002

Berline, N., Getzler, E., Vergne, M.
[1992] Heat Kernels and Dirac Operators.
Berlin-Heidelberg-New York: Springer-Verlag. Zbl.744.58001

Bers, L., John, F., Schechter, M.
[1964] Partial Differential Equations. Proc. Summer Semin. Boulder, New York: Interscience. Zbl.126,2

Birman, M.Sh., Solomyak, M.Z.
[1974] Quantitative analysis in Sobolev's embedding theorems and its application to the spectral theory. In the book: Desyataya Mat. Shkola, Kiev, 5–189. English transl.: Am. Math. Soc., Transl., II. Ser. *114* (1980), Zbl.426.46019
[1977] Asymptotics of the spectrum of differential equations. Itogi Nauki Tekh., Ser. Mat. Anal. *14*, 5–58. English transl.: J. Sov. Math. *12*, 247–283 (1979), Zbl.417.35061

Calderon, A.P.
[1958] Uniqueness in the Cauchy problem for partial differential equations. Amer. J. Math. *80*, 16–36, Zbl.80,303
[1963] Boundary-value problems for elliptic equations. Outlines of the joint Soviet American symposium on partial differential equations, Novosibirsk, 303–304.

Calderon, A.P., Zygmund, A.
[1952] On the existence of certain singular integrals. Acta Math. *88*, No. 1–2, 85–139, Zbl.47,102

Carleman, T.
[1935] Propriétés asymptotiques des fonctions fondamentales des membranes vibrantes. C.R. 8-ème Congr. des Math. Scand., Stockholm, 1934. Lund, 34–44, Zbl.12,70

Cordes, H.O.
[1979] Elliptic Pseudodifferential Operators. An abstract theory. Lect. Notes Math. *756*, Berlin-Heidelberg-New York: Springer-Verlag, Zbl.417.35004

Courant, R.
[1962] Methods of Mathematical Physics, Vol. II: Partial Differential Equations. New-York–London, Zbl.99,295

Courant, R., Hilbert, D.

[1931, 1937] Methoden der Mathematischen Physik, vols 1 and 2. Berlin: Springer. Zbl.1,5; Zbl.17,397. English transl.: 1953/1962; Reprint 1989, New York: Wiley

Duistermaat, J.J.

[1974] Oscillatory integrals, Lagrange immersions and unfolding of singularities. Commun. Pure Appl. Math. *27*, No. 2, 207–281, Zbl.276.35010

Duistermaat, J.J., Guillemin, V.

[1975] The spectrum of positive elliptic operators and periodic bicharacteristics. Invent. Math. *29*, No. 1, 39–79, Zbl.307.35071

Duistermaat, J.J., Hörmander, L.

[1972] Fourier integral operators. II. Acta Math. *128*, No. 3–4, 183–269, Zbl.232.47055

Egorov, Yu.V.

[1975] Subelliptic operators. Usp. Mat. Nauk *30*, No. 3, 57–104. English transl.: Russ. Math. Surv. *30*, No. 3, 59–118 (1975), Zbl.309.35068

[1984] Linear Differential Equations of Principal Type. Moscow: Nauka. English transl.: New York: Contemp. Sov. Math. (1986), Zbl.574.35001

[1985] Lectures on Partial Differential Equations. Supplementary Chapters. Moscow: Izdat. Mosk. Univ., Zbl.615.35001

Egorov, Yu.V., Shubin, M.A.

[1988] Linear Partial Differential Equations. Foundations of Classical Theory. Itogi Nauki Tekh., Ser. Sovrem. Probl. Mat., Fundam. Napravleniya *30*. English transl.: Encycl. Math. Sci. *30*, Berlin-Heidelberg-New York: Springer-Verlag 1992, Zbl.738.35001

Èskin, G.I. (= Ehskin, G.I.)

[1973] Boundary-Value Problems for Elliptic Pseudodifferential Equations. Moscow: Nauka, Zbl.262.35001

[1984] Initial-boundary value problem for hyperbolic equations. Proc. Int. Congr. Math., Warsaw 1983, 1165–1173, Zbl.571.35066

Fedoryuk, M.V.

[1971] The method of stationary phase and pseudodifferential operators. Usp. Mat. Nauk *26*, No. 1, 67–112. English transl.: Russ. Math. Surv. *26*, No. 1, 65–115, Zbl.221.47036

[1977] The Saddle-Point Method. Moscow: Nauka, Zbl.463.41020

[1986] Asymptotic Methods in Analysis. Itogi Nauki Tekh., Ser. Sovrem. Probl. Mat., Fundam. Napravleniya *13*, 93–210. English transl.: Encycl. Math. Sci. *13*, Berlin-Heidelberg-New York: Springer-Verlag 1989, Zbl.655.41034

Fedosov, B.V.

[1974a] Analytic formulae for the index of elliptic operators. Tr. Mosk. Mat. O-va. *30*, 159–241. English transl.: Trans. Mosc. Math. Soc. *30*, 159–240 (1976), Zbl.349.58006

[1974b] Analytic formula for the index of elliptic boundary-value problem. I, II, Mat. Sb. *93*, No. 1, 62–89; *95*, No. 4, 525–550. English transl.: Math. USSR, Sb. *22*, 61–90 (1975); *24*, 511–535 (1976), Zbl.306.58016; Zbl.312.58010

[1976] Analytic formulae for the index of elliptic boundary-value problems. III. Mat. Sb. Nov. Ser. *101*, No. 3, 380–401. English transl.: Math. USSR, Sb. *30*, 341–359 (1978), Zbl.349.58007

Friedlander, F.G.

[1976] The wave front set of the solution of a simple initial-boundary value problem with glancing rays. Math. Proc. Camb. Philos. Soc. *79*, 145–159, Zbl.319.35053

Friedrichs, K.O.

[1968] Pseudodifferential Operators. An Introduction. New York: New York Univ. Press. (1970) 2nd rev. ed, New York: Courant Inst. Math. Sci., Zbl.226.47028

Friedrichs, K.O., Lax, P.D.

[1965] Boundary-value problems for first order operators. Commun. Pure Appl. Math. *18*, 365–388, Zbl.178,114

Glazman, I.M.
[1963] Direct Methods for Qualitative Spectral Analysis of Singular Differential Operators. Moscow: Fizmatgiz. English transl.: London Oldbourne Press 1965, Zbl.143,365
Gokhberg, I.Ts., Krejn, M.G.
[1965] Introduction to Theory of Non-self-adjoint Operators. Moscow: Nauka. English transl.: Trans. Math. Monographs *18* (1969), AMS Providence, Zbl.138,78
Grubb, G.
[1986] Functional Calculus of Pseudodifferential Boundary Problems. Prog. Math. *65*, Boston: Birkhäuser, Zbl.622.35001
Guillemin, V., Sternberg, S.
[1977] Geometric Asymptotics. Providence, R.I.: Amer. Math. Soc., Zbl.364.5301
Gureev, T.E., Safarov, Yu.G.
[1988] Precise asymptotics of the spectrum for the Laplace operator on manifolds with periodic geodesics. (Preprint LOMI, E-1-86) Tr. Mat. Inst. Steklova *179*, 36–53. English transl.: Proc. Steklov. Inst. Math. *179*, 35–53, Zbl.701.58058
Hersh, R.
[1963] Mixed problems in several variables. J. Math. Mech. *12*, 317–334, Zbl.149,66
Hörmander, L.
[1963] Linear Partial Differential Operators. Berlin-Heidelberg-New York: Springer-Verlag, Zbl.108,93
[1966] Pseudodifferential operators and non-elliptic boundary problems. Ann. Math. II. Ser. *83*, 129–209, Zbl.132,74
[1968] The spectral function of an elliptic operator. Acta Math. *121*, 193–218, Zbl.164,132
[1971] Fourier integral operators. I. Acta Math. *127*, No. 1–2, 79–123, Zbl.212,466
[1983] The Analysis of Linear Partial Differential Operators. Vols 1 and 2. Berlin-Heidelberg-New York: Springer-Verlag, Zbl.521.35001, Zbl.521.35002
[1985] The Analysis of Linear Partial Differential Operators. Vols 3 and 4. Berlin-Heidelberg-New York: Springer-Verlag, Zbl.601.35001, Zbl.612.35001
Ikawa, M.
[1969] A mixed problem for hyperbolic equations of second order with a first order derivative boundary condition. Publ. Res. Inst. Math. Soc. *5*, No. 2, 119–147, Zbl.207,101
Ivrij, V.Ya (= Ivrii, V. Ya)
[1980] On the second term of the spectral asymptotic expansion of the Laplace-Beltrami operator on manifolds with boundary. Funkts. Anal. Prilozh, *14*, No. 2, 25–34. English transl.: Funct. Anal. Appl. *14*, 98–106 (1980), Zbl.435.35059
[1984] Precise Spectral Asymptotics for Elliptic Operators Acting in Fiberings over Manifolds with Boundary. Lect. Notes Mathematics *1100*, Berlin-Heidelberg-New York: Springer-Verlag, Zbl.565.35002
Ivrij, V.Ya., Fedorova, S.I.
[1986] Dilations and asymptotics of eigenvalues of spectral problems with singularities. Funkts. Anal. Prilozh. *20*, No. 4, 29–34. English transl.: Funct. Anal. Appl. *20*, 277–281, Zbl.628.35077
Ivrii, V.
[in preparation] Semiclassical Microlocal Analysis and Precise Spectral Asymptotics. Berlin-Heidelberg-New York: Springer-Verlag.
Kajitani, K.
[1974] A necessary condition for the well-posedness of hyperbolic mixed problem with variable coefficients. J. Math. Kyoto Univ. *14*, 231–242, Zbl.291.35058
Kohn, J.J., Nirenberg, L.
[1965] An algebra of pseudodifferential operators. Commun. Pure Appl. Math. *18*, No. 1–2, 269–305, Zbl.171,351
Kreiss, H.O.
[1958] Über sachgemäße Cauchyprobleme für Systeme von linearen partiellen Differen-

tialgleichungen. Kungliga Tekniska Högskolans Handlingar, Stockholm, No. 127, 1–31, Zbl.84,298
[1970] Initial-boundary value problem for hyperbolic systems. Commun. Pure Appl. Math. *23*, 277–298, Zbl.193,69

Kumano-go, H.
[1982] Pseudodifferential Operators. Cambridge, Mass.: MIT Press, Zbl.489.35003

Lax, P.D.
[1957] Asymptotic solutions of oscillatory initial-value problems. Duke Math. J. *24*, No. 4, 627–646, Zbl.83,318

Leray, J.
[1978] Analyse Lagrangienne et mecanique quantique; une structure mathematique apparentee aux developpements asymptotiques et a l'indice de Maslov. Strasbourg, I.R.M.A. English transl.: Cambridge, Mass.: MIT Press (1981), Zbl.483.35002

Levendorskij, S.Z.
[1985] The method of approximate spectral projector. Izv. Akad. Nauk SSSR, Ser. Mat. *49*, No. 6, 1177–1228. English transl.: Math. USSR, Izv, *27*, 451–502 (1986), Zbl.614.35021

Levendorskij, S.Z.
[1990] Asymptotic Distribution of Eigenvalues of Differential Operators. Transl. from the Russian. Mathematics and its Applications (Soviet Series). English transl.: Dordrecht: Kluwer Academic Publishers. Zbl.721.35049

Levitan, B.M.
[1952] On the asymptotic behaviour of the spectral function of a self-adjoint differential equation of second order. Izv. Akad. Nauk SSSR, Ser. Mat. *16*, No. 4, 325–352, Zbl.48,324

Maslov, V.P.
[1965] Theory of Perturbations and Asymptotic Methods. Moscow: Izd. Moskv. Gos. Univ. French translation: Paris: Dunod 1972, Zbl.247.47010
[1973] The Operator Methods. Moscow: Nauka. English transl.: Moscow: Mir Publ. 1976, Zbl.288.47042
[1977] The Complex WKB Method in Nonlinear Equations. Moscow: Nauka, Zbl.449.58001
[1987] The Asymptotic Methods for Solving Pseudodifferential Equations. Moscow: Nauka, Zbl.625.35001

Maslov, V.P., Fedoryuk, M.V.
[1976] Quasi-classical Approximation for Equations of Quantum Mechanics. Moscow: Nauka. English transl.: Dordrecht-Boston-London: D. Reidel 1981, Zbl.449.58002

Melrose, R.B.
[1976] Equivalence of glancing hypersurfaces. Invent. Math. *37*, 165–191, Zbl.354.53033
[1978] Airy operators. Commun. Partial Differ. Equations *3*, No. 1, 1–76, Zbl.384.35052

Melrose, R.
[1980] Weyl's conjecture for manifolds with concave boundary. Geometry of the Laplace Operator. Proc. Symp. Pure Math. *36* (1980) 257–274. Zbl.436.58024

Mikhlin, S.G.
[1962] Multidimensional Singular Integrals and Integral Equations. Moscow: Fizmatgiz. English transl.: Pergamon Press 1965, Zbl.105,303

Mishchenko, A.S., Sternin, B.Yu., Shatalov, V.E.
[1978] Lagrangian Manifolds and Method of Canonical Operator. Moscow: Nauka, Zbl.727.58001. English transl.: Berlin-Heidelberg-New York: Springer-Verlag 1990

Mizohata, S.
[1965] The Theory of Partial Differential Equations (in Japanese) Tokyo: Iwanami Shoten. English transl.: London: Cambridge Univ. Press 1973, Zbl.263.35001

Nazajkinskij, V.E., Oshmyan, V.G., Sternin, B.Yu., Shatalov, V.E.
[1981] Fourier integral operators and canonical operator. Usp. Mat. Nauk *36*, No. 2, 81–140. English transl.: Russ. Math. Surv. *36*, No. 2, 93–161 (1981), Zbl. 507.58043

Nirenberg, L.
[1970] Pseudodifferential operators. Global Analysis, Proc. Symp. Pure Math., Providence, R.I.: Amer. Math. Soc. *16*, 149–167, Zbl.218.35075
[1973] Lectures on linear partial differential equations. Amer. Math. Soc. Regional Conf. Ser. Math., No. 17, 1–58, Zbl.267.35001

Olejnik, O.A., Radkevich, E.V.
[1969] Equations of the second order with a non-negative characteristic form. Itogi Nauki, Ser. Mat., Mat. Anal. 1969, Zbl.217,415

Oshima, T.
[1978] On analytic equivalence of glancing hypersurfaces. Sci. Papers College Gen. Ed. Univ. Tokyo *28*, 51–57, Zbl.382.53026

Palais, R.S.
[1965] Seminar on the Atiyah-Singer Index Theorem. Princeton: Princeton Univ. Press, Zbl.137,170

Petkov, V.M., Stojanov, L.N.
[1985/86] Propriétés génériques de l'application de Poincaré et des géodésiques périodiques généralisées. Sémin., Equations Dériv. Partielles, 1985/86, Exp. No. 11, Zbl.611.58048

Petrovskij, I.G.
[1986] Selected Works. Systems of Partial Differential Equations. Algebraic Geometry. Moscow: Nauka, Zbl.603.01018

Plamenevskij, B.A.
[1986] Algebras of Pseudodifferential Operators. Moscow: Nauka, Zbl.615.47038. English transl.: Dordrecht: Kluwer 1989

Popov G.S., Shubin, M.A.
[1983] Asymptotic expansion of spectral function for second-order elliptic operators in $\mathbb{R}^n$. Funkts. Anal. Prilozh. *17*, No. 3, 37–45. English transl.: Funct. Anal. Appl. *17*, 193–200, Zbl.533.35072

Povzner, A.Ya., Sukharevskij, I.V.
[1960] Discontinuities of the Green's function of a mixed problem for the wave equation. Mat. Sb. *51*, No. 1, 3–26, Zbl.156,114. English transl.: Am. Math. Soc., Transl., II. Ser. *47* (1965), 131–156

Rauch, J.
[1972] L_2 is a continuable initial conditions for Kreiss' mixed problems. Commun. Pure Appl. Math. *25*, No. 3, 265–285, Zbl.226.35056

Reed, M., Simon, B.
[1972–'78] Methods of Modern Mathematical Physics, Vols. 1, 2, 3, 4. New York: Acad. Press. I: Zbl.242.46001, I: Zbl.459.46001 rev. ed. (1980), II. Zbl.308.47002, III: Zbl.405.47007, IV: Zbl.401.47001

Rempel, S., Schulze, B.-W.
[1982] Index theory of elliptic boundary problems. Berlin: Akademie-Verlag, Zbl.504.35002

Safarov, Yu.G.
[1988a] Exact asymptotics of the spectrum of a boundary value problem and periodic billiards. Izv. Akad. Nauk *52*, No. 6, 1230–1251. English transl.: Math. USSR Izvestiya *33*, No. 3 (1989). Zbl.682.35082
[1988b] Asymptotics of the spectral function of a positive elliptic operator without the non-trap condition. Funkts. Anal. Prilozh. *22*, No. 3, 53–65. English transl.: Funct. Anal. Appl. *22*, No. 3, 213–223 (1988). Zbl.679.35074

Safarov, Yu.G., Vasil'ev, D.G.
[1992] The asymptotic distribution of eigenvalues of differential operators. In: "Spectral Theory of Operators" (ed. S. Gindikin), Transl., Ser. 2, Amer. Math. Soc. *150*, 55–111. Zbl.751.58040
[in preparation] The asymptotic distribution of eigenvalues of partial differential operators. Providence: Amer. Math. Soc.

Sakamoto, R.
[1970a] Mixed problems for hyperbolic equations. I. J. Math. Kyoto Univ., *10*, No. 2, 349–373, Zbl.203,100
[1970b] Mixed problems for hyperbolic equations. II. ibid, No. 3, 403–417, Zbl.206,401
Sato, M.
[1970] Regularity of hyperfunction solutions of partial differential equations. Actes Congr. Inte. Math. Nice 1970, *2*, 785–794, Zbl.232.35004
Schenk, D., Shubin, M.A.
[1985] Asymptotic expansion of density of states and of spectral function of the Hill operator. Mat. Sb., Nov. Ser. *128*, No. 4, 474–491, Zbl.604.34015. English transl.: Math. USSR, Sb. *56*, 473–490 (1987)
Seeley, R.T.
[1967] Complex powers of an elliptic operator. Proc. Symp. Pure Math., Am. Math. Soc. *10*, 288–307, Zbl.159,155
[1969a] The resolvent of an elliptic boundary problem. Am. J. Math. *91*, No. 4, 889–920, Zbl.191,118
[1969b] Analytic extension of the trace associated with elliptic boundary problems. Am. J. Math. *91*, No. 4, 963–983, Zbl.191,119
[1978] A sharp asymptotic remainder estimate for the eigenvalues of the Laplacian in a domain of $\mathbb{R}^3$. Adv. Math. *29*, No. 2, 244–269, Zbl.382.35043
Shubin, M.A.
[1978] Pseudodifferential Operators and Spectral Theory. Moscow: Nauka. English transl.: Berlin-Heidelberg-New York: Springer-Verlag 1987, Zbl.451.47064
Stein, E.M.
[1970] Singular Integrals and Differentiability Properties of Functions. Princeton: Princeton Univ. Press, Zbl.207,135
Taylor, M.E.
[1981] Pseudodifferential Operators. Princeton: Princeton Univ. Press, Zbl.453.47026
Trèves, F.
[1980] Introduction to pseudodifferential and Fourier integral operators. Vols. 1, 2. New York: Plenum Press, Zbl.453.47027
Vajnberg, B.R.
[1975] On the short wave asymptotic behaviour of solutions of stationary problems and the asymptotic behaviour as $t \to \infty$ of solutions of non-stationary problems. Usp. Mat. Nauk *30*, No. 2, 3–55. English transl.: Russ. Math. Surv. *30*, No. 2, 1–58, Zbl.308.35011
[1982] Asymptotic Methods in Equations of Mathematical Physics. Moscow: Izd. Moskov. Gos. Univ., Zbl.518.35002. English transl.: New York: Gordon & Breach 1989
[1983] A complete asymptotic expansion of spectral function of elliptic operators in $\mathbb{R}^n$. Vestn. Mosk. Univ., Ser. I Mat. Mekh. 1983, No. 4, 29–36. English transl.: Mosc. Univ. Math. Bull. *38*, No. 4, 32–39, Zbl.547.35040
Vasil'ev, D.G.
[1986] The asymptotics of the spectrum of a boundary-value problem. Tr. Mosk. Mat. O-va., *49*, 167–237. English transl.: Trans. Mosc. Math. Soc., 173–245 (1987), Zbl.623.58024
Volovoj, A.V.
[1986] Sharper estimates for the remainder term in the two-member asymptotics for distribution function of eigenvalues of the elliptic operator on a compact manifold. Usp. Mat. Nauk *41*, No. 4, 185
Volovoy, A.V.
[1990a] Improved two-term asymptotics for the eigenvalue distribution function of an elliptic operator on a compact manifold. Commun. Partial Differ. Equations *15*, No. 11, 1509–1563. Zbl.724.35081
[1990b] Verification of the Hamiltonian flow conditions associated with Weyl's conjecture. Ann. Glob. Anal. Geom. *8*, No. 2, 127–136. Zbl.721.58048

II. Linear Partial Differential Equations with Constant Coefficients

A.I. Komech

Translated from the Russian
by P.C. Sinha

Contents

Preface

The present survey paper is devoted to the methods of solution and investigation of equations of the form

$$Pu(x) := P(\partial_x)u(x) := \sum_{|\alpha| \leqslant m} p_\alpha \partial_x^\alpha u(x) = f(x), \quad x \in \mathbb{R}^n. \tag{0.1}$$

Here $\alpha = (\alpha_1, \ldots, \alpha_n)$ are multi-indices, $\alpha_k = 0, 1, 2, \ldots$ for $k = 1, \ldots, n$; $p_\alpha \in \mathbb{C}$ and

$$\partial_x^\alpha u(x) = \frac{\partial^{|\alpha|} u(x)}{\partial x_1^{\alpha_1} \ldots \partial x_n^{\alpha_n}}, \quad |\alpha| = \alpha_1 + \cdots + \alpha_n.$$

The basic tools for investigating these equations are the Fourier transformation and the theory of generalized functions.

Equations of the form (0.1) are used in mathematical physics to describe various phenomena in the theories of elasticity, acoustics, electrodynamics and quantum mechanics. The investigation of the properties of solutions of these equations, as well as finding formulae for the solutions themselves, attracted the attention of Euler, d'Alembert, Laplace, Fourier, Bessel, Poisson, Kirchhoff, Heaviside, and Green back in the 18th–19th centuries. From the beginning of the 20th century and right up to the forties, significant results in this direction were obtained by Hadamard, Riesz, Herglotz, Petrovskij, Gårding, and Leray. During the following decades, a sufficiently general theory of such equations was developed in the works of Schwartz, Gel'fand, Shilov, Malgrange, Ehrenpreis, Hörmander, and others. This was achieved by means of the theory of generalized functions, the appearance and development of which was closely associated with the investigation of the equations of mathematical physics.

The basic method of investigating the equations (0.1) is to represent $u(x)$ and $f(x)$ in terms of Fourier integrals, that is, representations in terms of harmonics of the form

$$u(x) = \int e^{-ix\xi} \tilde{u}(\xi)\, d\xi, \quad f(x) = \int e^{-ix\xi} \tilde{f}(\xi)\, d\xi;$$
$$x\xi = x_1\xi_1 + \cdots + x_n\xi_n \quad \text{for } x, \xi \in \mathbb{R}^n. \tag{0.2}$$

In fact, the exponentials $e^{-ix\xi}$ are eigenfunctions of the operator P of the form (0.1). Thus

$$Pe^{-ix\xi} = \tilde{P}(\xi) e^{-ix\xi},$$

where

$$\tilde{P}(\xi) = \sum_{|\alpha| \leqslant m} p_\alpha (-i\xi)^\alpha = P(-i\xi), \tag{0.3}$$

and therefore $\tilde{P}(\xi)$, the symbol of P, is an eigenvalue corresponding to the eigenfunction $e^{-ix\xi}$. Consequently, the Fourier representation (0.2) is the decomposition in terms of the eigenfunctions of P, and so reduces P to diagonal form, after which (0.1) can be solved easily. For example, if $f(x) = \tilde{f}(\xi)e^{-ix\xi}$, that is, if $f(x)$ contains only one harmonic, then $u(x)$ can also be sought in the form $\tilde{u}(\xi)e^{-ix\xi}$. Substituting $u(x)$ into (0.1) and taking into account (0.3), we obtain

$$\tilde{P}(\xi)\tilde{u}(\xi) = \tilde{f}(\xi). \tag{0.4}$$

Hence $\tilde{u}(\xi) = \tilde{f}(\xi)/\tilde{P}(\xi)$ if $\tilde{P}(\xi) \neq 0$. If, however, $\tilde{P}(\xi) = 0$, then $\tilde{u}(\xi)$ is an arbitrary quantity when $\tilde{f}(\xi) = 0$, while $\tilde{u}(\xi)$ does not exist if $\tilde{f}(\xi) \neq 0$. A similar analysis is possible with the help of (0.2) in the general case of an arbitrary function $f(x)$. This again leads to (0.4), which holds for all ξ and, consequently,

$$\tilde{u}(\xi) = \frac{\tilde{f}(\xi)}{\tilde{P}(\xi)}, \tag{0.5}$$

formally for all ξ. Substituting into (0.2), we find that, formally,

$$u(x) = \int e^{-ix\xi} \frac{\tilde{f}(\xi)}{\tilde{P}(\xi)}\, d\xi. \tag{0.6}$$

The main difficulty in carrying out this programme lies in giving a precise meaning to the formal computations (0.2) and (0.4)–(0.6) and in justifying the division in (0.5). These problems arise due to the fact that the operator P, which is defined on functions in the whole space $\mathbb{R}^n$, has a continuous spectrum and its eigenfunctions $e^{-ix\xi}$ do not lie in $L_2(\mathbb{R}^n)$, that is, the eigenfunctions are "generalized" eigenfunctions in the sense of Gel'fand-Kostyuchenko-Levitan (see Gel'fand and Shilov [1958c]). Consequently, the "coefficients" $\tilde{u}(\xi)$ and $\tilde{f}(\xi)$ in (0.2) are also generalized functions. In view of these remarks, the theory of generalized functions turns out to be the fundamental tool for investigating equations of the form (0.1) and, accordingly, we start our survey with a brief account of this theory in Chap. 1.

In Chap. 1 we recall the basic rules of operations with generalized functions. We construct the solutions of (0.1) with an arbitrary right-hand side $f(x)$ in terms of the convolution of f and a fundamental solution of (0.1).

In Chap. 2 we introduce the Fourier transformation of generalized functions by means of the closure property with respect to continuity. The Payley-Wiener theory of the Fourier transformation in the complex domain (see Paley and Wiener [1934]) and a connection between the convolution and the Fourier transformation are discussed. This theory has important applications, for example, in justifying the dispersion relations in quantum field theory (Bogolyubov and Shirkov [1973]), and in constructing the Heaviside operational calculus (see § 3, Chap. 3, and Doetsch [1967]).

Chap. 3 deals with the existence and smoothness of the solutions of (0.1). A plan for solving the division problem (0.5) is put forward, and the theory of solvability for equations (0.1) in a cone and the operational calculus are discussed. The fundamental concepts of the theory of partial differential equations, namely, the concepts of characteristics, bicharacteristics, the wave front, etc., are introduced and are discussed in detail. We also give an account of the results on the propagation of singularities for real equations of principal type as well as an account of the classical "Weyl lemma" on the smoothness of generalized solutions of elliptic equations.

By a unified method, which generalizes the Riesz method [1949] of complex powers, in Chap. 4 we compute in an explicit form the fundamental solutions of all real second-order equations with a non-degenerate quadratic form. For homogeneous operators of the form $P = \sum_{i=1}^{n} \pm \frac{\partial^2}{\partial x_i^2}$, with $n \neq 2$, the fundamental solutions are obtained in the form of an analytic continuation of the function $C\rho^\lambda \left(\text{where } \rho = \left(\sum_i \pm x_i^2 \right)^{1/2} \right)$ with respect to λ to the point $\lambda = -(n-2)$. For non-homogeneous operators of the form $P = \sum_{i=1}^{n} \pm \frac{\partial^2}{\partial x_i^2} + q$, with $q \neq 0$, (and, in particular, for the Klein-Gordon operator) the fundamental solutions are obtained in the form of an analytic continuation of the Bessel function $C\rho^\nu J_\nu(\rho)$ to the point $\nu = -\frac{n-2}{2}$ (for $n \neq 2$). These fundamental solutions in terms of improper integrals ("the Hadamard gallows") were constructed back in 1932 by Hadamard [1932]. However, in the modern language of distribution theory these fundamental solutions were first constructed for $q = 0$ by Leray [1953] and Riesz [1949] (see Schwartz [1966]), and for $q \neq 0$ they have been constructed in the present paper. These fundamental solutions play an important role in the problems of modern mathematical physics (see Bogolyubov and Shirkov [1973]). On this account we devote much attention to the methods by which they are computed and the ways in which they are used.

Chap. 5 discusses the basic theory of boundary-value problems in a half-space due to Petrovskij [1937a, b; 1938], Gel'fand and Shilov [1953, 1958c], Krejn [1967]. We present sufficient conditions, which are close to being necessary, to be imposed on the boundary-value problem to ensure a unique solution for any boundary data. We also present necessary conditions to be imposed on the differential operator under which such boundary-value problems for the operator do exist. In the treatment of these problems there appears the natural classification of the equations into hyperbolic, parabolic and elliptic types which plays a fundamental role in the modern theory of partial differential equations, although this classification is not exhaustive. We mention the "proper" formulation of boundary-value problems for equations of hyperbolic, parabolic and elliptic types.

Chap. 6, written by Vasil'ev, is devoted to a survey of the theory of lacunae for hyperbolic equations as developed by Petrovskij, and Atiyah, Bott and Gårding.

We denote by $x_k \xrightarrow{X} x$ a sequence of elements of a linear topological space X that converges to $x \in X$. By continuous maps of these spaces we mean, for simplicity, sequentially continuous maps, although we do not always mention it explicitly.

Chapter 1
Generalized Functions and Fundamental Solutions of Differential Equations

§ 1. Generalized Functions and Operations on them

1.1. Differentiation of Generalized Functions. We denote by $\mathscr{D} = \mathscr{D}(\mathbb{R}^n) = C_0^\infty(\mathbb{R}^n)$ the *space of test functions* or the *space of basic complex-valued functions* $\varphi(x) \in C^\infty(\mathbb{R}^n)$ which are equal to zero for $|x| > R$, where R may depend on φ. Convergence in $\mathscr{D}$ is defined in the following manner. By definition, a sequence $\varphi_k \to \varphi$ in $\mathscr{D}$ if

1) the derivatives $\varphi_k^{(\alpha)}$ of any order $\alpha = (\alpha_1, \ldots, \alpha_n)$ converge uniformly on $\mathbb{R}^n$ to $\varphi^{(\alpha)}(x)$; thus, for all α

$$\varphi_k^{(\alpha)}(x) \rightrightarrows \varphi^{(\alpha)}(x), \quad x \in \mathbb{R}^n,$$

2) for some R, independent of k,

$$\varphi_k(x) = 0 \quad \text{for } |x| > R.$$

We denote by $\mathscr{D}' \equiv \mathscr{D}'(\mathbb{R}^n)$ the set of all linear functions on $\mathscr{D}$ that are continuous with respect to the convergence introduced on $\mathscr{D}$. We can also introduce in $\mathscr{D}'$ the concept of *weak convergence*. Namely, by definition, a sequence $u_k(x) \in \mathscr{D}'$ converges weakly to $u(x) \in \mathscr{D}'$ as $k \to \infty$ if

$$\langle u_k, \varphi \rangle \underset{k\to\infty}{\rightarrow} \langle u, \varphi \rangle \quad \text{for all } \varphi \in \mathscr{D}. \tag{1.1}$$

We then write $u_k \overset{\mathscr{D}'}{\rightarrow} u$.

Example 1.1. If $u(x) \in L_1^{\text{loc}}(\mathbb{R}^n)$, then the functional of $\varphi \in \mathscr{D}$ of the form

$$u\colon \varphi \mapsto u(\varphi) = \langle u(x), \varphi(x) \rangle := \int_{\mathbb{R}^n} u(x)\varphi(x)\, dx \tag{1.2}$$

is an element of $\mathscr{D}'(\mathbb{R}^n)$.

Definition 1.1. The functionals $u \in \mathscr{D}'$ are called *generalized functions* or *distributions.*

In analogy with (1.2), we define a "scalar product" for $u \in \mathscr{D}'$ by

$$\langle u(x), \varphi(x) \rangle := u(\varphi), \quad \varphi \in \mathscr{D}. \tag{1.3}$$

We remark that here $u(x)$ is merely a notation and does not denote the value at the points x of a function u. Roughly speaking, for a generalized function u the "observable" quantities are the values $u(\varphi)$ for $\varphi \in \mathscr{D}$ and not $u(x)$ for $x \in \mathbb{R}^n$. Generalized functions will often be simply referred to as functions.

Example 1.2. The celebrated *Dirac δ-function* is defined by the formula

$$\langle \delta(x), \varphi(x) \rangle = \varphi(0), \quad \varphi \in \mathscr{D}(\mathbb{R}^n), \tag{1.4}$$

(see Dirac [1958]). Sometimes we also write $\delta_n(x)$ for the function $\delta(x)$.

Example 1.1 shows that the space $L_1^{\text{loc}}(\mathbb{R}^n)$ of "classical" functions is mapped into $\mathscr{D}'(\mathbb{R}^n)$ in a natural manner. It turns out that to distinct functions $u(x) \in L_1^{\text{loc}}(\mathbb{R}^n)$ there correspond (in accordance with formula (1.2)) distinct functionals $u \in \mathscr{D}'$. Therefore the map $L_1^{\text{loc}}(\mathbb{R}^n) \to \mathscr{D}'(\mathbb{R}^n)$ defined by (1.2) is an *embedding*, that is, $L_1^{\text{loc}}(\mathbb{R}^n)$ can be identified with a subset of $\mathscr{D}'(\mathbb{R}^n)$, and it can be assumed that

$$L_1^{\text{loc}}(\mathbb{R}^n) \subset \mathscr{D}'(\mathbb{R}^n).$$

On the other hand, it turns out (Gel'fand and Shilov [1958a], Shilov [1965]) that $L_1^{\text{loc}}(\mathbb{R}^n)$ does not coincide with $\mathscr{D}'(\mathbb{R}^n)$, because, for example, $\delta \in \mathscr{D}'$ but $\delta \notin L_1^{\text{loc}}(\mathbb{R}^n)$. Thus the space $\mathscr{D}'(\mathbb{R}^n)$ is larger than $L_1^{\text{loc}}(\mathbb{R}^n)$.

What is striking is that such an extension of the space $L_1^{\text{loc}}(\mathbb{R}^n)$ is closed with respect to the operation of differentiation.

Definition 1.2. For any generalized function $u(x)$ and any multi-index $\alpha = (\alpha_1, \ldots, \alpha_n)$ the *derivative* $\partial_x^\alpha u(x) = u^{(\alpha)}(x)$ is defined by the identity

$$\langle u^{(\alpha)}(x), \varphi(x) \rangle = (-1)^{|\alpha|} \langle u(x), \varphi^{(\alpha)}(x) \rangle, \quad \varphi \in \mathscr{D}. \tag{1.5}$$

It turns out (Gel'fand and Shilov [1958a], Shilov [1965]) that $u^{(\alpha)}(x) \in \mathscr{D}'$ for all $u \in \mathscr{D}'$ and all multi-indices $\alpha = (\alpha_1, \ldots, \alpha_n)$, and that the operator $\partial^\alpha \colon u \mapsto u^{(\alpha)}$ is (sequentially) continuous in $\mathscr{D}'$ with respect to the weak convergence (1.1). If $u \in C^\infty(\mathbb{R}^n)$, the derivatives $\partial_x^\alpha u(x)$ coincide with the usual derivatives in the sense of classical analysis. Since $C^\infty(\mathbb{R}^n)$ is dense in $\mathscr{D}'(\mathbb{R}^n)$, such a continuous extension of the operator ∂_x^α to $\mathscr{D}'(\mathbb{R}^n)$ is unique.

Example 1.3. Let $\theta(x)$ be the *Heaviside function* ("the step"):

$$\theta(x) = \begin{cases} 1, & x > 0, \\ 0, & x < 0. \end{cases} \tag{1.6}$$

Then $\langle \theta'(x), \varphi(x) \rangle = -\langle \theta, \varphi' \rangle = -\int_0^\infty \varphi'(x)\, dx = \varphi(0)$, that is,

$$\theta'(x) = \delta(x). \tag{1.7}$$

Example 1.4. The derivatives of $\delta(x)$. For $\varphi \in \mathscr{D}(\mathbb{R})$

$$\langle \delta'(x), \varphi(x) \rangle = -\varphi'(0); \quad \langle \delta''(x), \varphi(x) \rangle = \varphi''(0). \tag{1.8}$$

An extremely important role in applications is played by the following simple *differentiation formula for piecewise smooth functions.*

Let $u(x) \in C^1(\mathbb{R} \setminus \{a\})$, where $a \in \mathbb{R}$. We denote by

$$\{u'(x)\} = u'(x), \quad x \neq a, \tag{1.9}$$

a function which is equal to the usual derivative of $u(x)$ at those points $x \in \mathbb{R}$ where it exists. We assume that the limits $u(a \pm 0)$ exist, and we further assume (although it is not necessary to do so) that the limits $u'(a \pm 0)$ exist.

Lemma 1.1. *In the sense of the differentiation of generalized functions,*

$$u'(x) = \{u'(x)\} + h\delta(x - a), \tag{1.10}$$

where

$$h = u(a + 0) - u(a - 0)$$

denotes the jump of the function $u(x)$ at the point $x = a$.

Proof. For $\varphi(x) \in \mathcal{D}(\mathbb{R})$, we have

$$\langle u', \varphi \rangle = -\langle u, \varphi' \rangle = -\int_{-\infty}^{\infty} u\varphi' \, dx = -\int_{-\infty}^{a} u\varphi' \, dx - \int_{a}^{\infty} u\varphi' \, dx$$

$$= -u\varphi|_{-\infty}^{a} - u\varphi|_{a}^{\infty} + \int_{x \neq a} u'\varphi \, dx = h\varphi(a) + \langle \{u'\}, \varphi \rangle. \quad \blacksquare$$

Example 1.5. It can be derived easily from (1.10) that $\theta'(x) = \delta(x)$, which agrees with (1.7). Furthermore, $|x|'' = 2\delta(x)$, and

$$\left(\frac{d}{dx} + \lambda\right)(\theta(x)e^{-\lambda x}) = \delta(x); \quad \left(-\frac{d^2}{dx^2} + m_0^2\right)\frac{e^{-m_0|x|}}{2m_0} = \delta(x), \quad m_0 \neq 0;$$
$$\left(\frac{d^2}{dx^2} + \omega^2\right)\frac{\sin \omega|x|}{2\omega} = \delta(x), \quad \omega \neq 0. \tag{1.11}$$

1.2. Change of Variables in Generalized Functions. Let $m \geqslant n$. Let $h: \mathbb{R}^m \to \mathbb{R}^n$ be a smooth map of rank n at each point $y \in \mathbb{R}^m$. Then the pre-image of each point $x \in \mathbb{R}^n$ is a smooth submanifold $\Gamma_x = h^{-1}(x)$ in $\mathbb{R}^m$ of dimension $v = m - n$. On Γ_x a smooth measure $\omega_x(dy)$ is defined with the following property. For any function $f(y) \in C(\mathbb{R}^m)$ vanishing outside a bounded set (each function has its own set) in $\mathbb{R}^m$,

$$\int f(y) d^m y = \int \left(\int_{\Gamma_x} f(y)\omega_x(dy)\right) d^n x, \tag{1.12}$$

where $d^n x$ and $d^m y$ are forms of the volume element on $\mathbb{R}^n$ and $\mathbb{R}^m$ respectively. Such a measure can be constructed easily (Hörmander [1983, 1985]). It can be shown that

$$\omega_x(dy) = \frac{\sigma_v(dy)}{G(y)}, \tag{1.13}$$

where $\sigma_v(dy)$ denotes the v-dimensional Euclidean measure on Γ_x, and $G(y)$ denotes the n-dimensional volume of a parallelopiped that is spanned by the vectors grad $h_1(y), \ldots,$ grad $h_n(y)$ $(h(y) = (h_1(y), \ldots, h_n(y)))$. Thus,

$$G(y) = \left(\sum_{1 \leqslant k_1 < \cdots < k_n \leqslant m} \left| \frac{\partial(h_1, \ldots, h_n)}{\partial(y_{k_1}, \ldots, y_{k_n})} \right|^2 \right)^{1/2}. \tag{1.14}$$

Since h has rank n at each point $y \in \mathbb{R}^m$, it follows that $G(y) \neq 0$.

Example 1.6. If $m = n$, then h: $\mathbb{R}^n \to \mathbb{R}^n$ is a diffeomorphism, $\Gamma_x = h^{-1}(x)$ is a single point, $\sigma_0(\Gamma_x) = 1$ and

$$G(y) = \left| \frac{\partial(h_1, \ldots, h_n)}{\partial(y_1, \ldots, y_n)} \right| \tag{1.15}$$

is the Jacobian of h at the point y.

Example 1.7. If $n = 1$, we have $h = h_1(y)$ and grad $h(y) \neq 0$ for $y \in \mathbb{R}^m$. Then

$$G(y) = |\operatorname{grad} h(y)|. \tag{1.16}$$

Example 1.8. If $m = 3$ and $n = 2$, then $G(y) = |[\operatorname{grad} h_1(y), \operatorname{grad} h_2(y)]|$ is the modulus of the "vector" product.

Proposition 1.1. *Let* rank $\dfrac{\partial h(y)}{\partial y} = n$ *for all* $y \in \mathbb{R}^m$. *Then the map* $u(x) \mapsto u(h(y))$ *acts continuously from* $C(\mathbb{R}^n)$ *into* $C(\mathbb{R}^m)$ *and can be extended to an operator from* $\mathscr{D}'(\mathbb{R}^n)$ *into* $\mathscr{D}'(\mathbb{R}^m)$ *that is (sequentially) continuous in the convergence* (1.1).

Proof. We take a function $u(x) \in C(\mathbb{R}^n)$. With $\varphi(y) \in \mathscr{D}(\mathbb{R}^n)$, we find from (1.12) that

$$\begin{aligned} \langle u(h(y)), \varphi(y) \rangle &= \int u(x) \left(\int_{\Gamma_x} \varphi(y) \omega_x(dy) \right) d^n x \\ &= \langle u(x), \overline{\varphi}(x) \rangle. \end{aligned} \tag{1.17}$$

Here

$$\overline{\varphi}(x) = \int_{\Gamma_x} \varphi(y) \omega_x(dy). \tag{1.18}$$

It can be shown easily that

$$\overline{\varphi}(x) \in \mathscr{D}(\mathbb{R}^n). \tag{1.19}$$

For example, $\varphi(y)$ can be decomposed into a sum of functions having sufficiently small supports and in each support the coordinates $y_1, \ldots, y_m$ can be chosen so that $y_k(x) = h_k(x)$ for $k = 1, \ldots, m$. Therefore the map $u(x) \to u(h(y))$ is continuous on $C(\mathbb{R}^n)$ with respect to the convergence (1.1) and admits a closed extension with respect to (sequential) continuity on $\mathscr{D}'(\mathbb{R}^n)$. ∎

By taking the closure with respect to continuity, from (1.17) we have

Definition 1.3. For $u(x) \in \mathscr{D}'(\mathbb{R}^n)$,

$$\langle u(h(y)), \varphi(y) \rangle = \langle u(x), \overline{\varphi}(x) \rangle, \quad \varphi \in \mathscr{D}(\mathbb{R}^m). \tag{1.20}$$

Example 1.9. If, as in Example 1.6, $h: \mathbb{R}^n \to \mathbb{R}^n$ is a diffeomorphism, then $\overline{\varphi}(x) = \varphi(h^{-1}(x))/G(h^{-1}(x))$, and

$$\langle u(h(y)), \varphi(y)\rangle = \left\langle u(x), \frac{\varphi(h^{-1}(x))}{G(h^{-1}(x))}\right\rangle. \tag{1.21}$$

Remark 1.1. The assumption that $h(y)$ has rank n at each point $y \in \mathbb{R}^m$ is not necessary for the composition $u(h(y))$ to be meaningful. The condition $\overline{\varphi} \in \mathscr{D}(\mathbb{R}^n)$ of (1.19) may be satisfied for all $\varphi \in \mathscr{D}(\mathbb{R}^m)$ for some maps $h(y)$ that do not have rank n everywhere. Further, the condition that $\overline{\varphi} \in \mathscr{D}(\mathbb{R}^n)$ is also not necessary. Roughly speaking, the function $\overline{\varphi}(x)$ must be smooth only at those points x_0 where $u(x)$ has singularities, and the map $h(y)$ must have rank n only at the points $y_0 = h^{-1}(x_0)$.

Example 1.10. If $h(y) \in C^\infty(\mathbb{R}^m)$ and grad $h(y) \neq 0$ for $y \in \mathbb{R}^m$, then, by Definition 1.3 and (1.16),

$$\langle \delta_1(h(y)), \varphi(y)\rangle = \langle \delta_1(x), \overline{\varphi}(x)\rangle = \overline{\varphi}(0) = \int_{\Gamma_0} \frac{\varphi(y)\sigma_{m-1}(dy)}{|\text{grad } h(y)|} \tag{1.22}$$

for $\varphi \in \mathscr{D}(\mathbb{R}^m)$; here $\Gamma_0 = h^{-1}(0) = \{y \in \mathbb{R}^m: h(y) = 0\}$.

For $h(y) = |y|^2 - \omega^2$, with $\omega \in \mathbb{R}\backslash 0$, it is natural to define

$$\langle \delta_1(|y|^2 - \omega^2), \varphi(y)\rangle = \int_{|y|=|\omega|} \frac{\varphi(y)\sigma_{m-1}(dy)}{2|\omega|}, \tag{1.23}$$

even though grad $h(y) = 0$ at $y = 0$.

Note that the right-hand side of (1.22) is valid for any $\varphi \in \mathscr{D}(\mathbb{R}^m)$ provided that $h \in C^\infty(\mathbb{R}^m)$, grad $h(y) \neq 0$ for $y \neq 0$, $m \geqslant 2$, and

$$\int_{\Gamma_0^R\backslash 0} \frac{\sigma_{m-1}(dy)}{|\text{grad } h(y)|} < \infty \quad \forall R > 0; \quad \Gamma_0^R = \{y \in \Gamma_0: |y| \leqslant R\}. \tag{1.24}$$

Definition 1.4. With condition (1.24), we define $\delta_1(h(y))$ by the formula (1.22), assuming that $\sigma_{m-1}(\{0\}) = 0$ if $0 \in \Gamma_0$.

Example 1.11. If $h(y) = y_1^2 + \cdots + y_k^2 - y_{k+1}^2 - \cdots - y_m^2$ is a non-degenerate quadratic form, then grad $h(y) = 0$ if $y = 0 \in \Gamma_0$. However, (1.24) is satisfied for $m \geqslant 3$.

Note that (1.22) implies that

$$\delta(a(y)h(y)) = \frac{1}{a(y)}\delta(h(y)), \tag{1.25}$$

if $a(y) \in C^\infty(\mathbb{R}^m)$ and $a(y) \neq 0$ for $y \in \mathbb{R}^m$.

Remark 1.2. The condition (1.24) enables us to define $\delta_1(h(y))$ but this condition cannot be used for defining $u(h(y))$ if $u \in \mathscr{D}'(\mathbb{R})$ is an arbitrary function. For each concrete $u \in \mathscr{D}'$ one can find its own (1.24)-type condition on $h(y)$ under which $u(h(y))$ is defined and depends continuously in some sense on the map h from the class in question. For example, the generalized function $\delta_1(h(y))$, defined by (1.22), depends continuously on $h(\cdot) \in C^1(\mathbb{R}^m)$ in the class of functions $h(y)$ for which grad $h(y) \neq 0$ with $y \in \Gamma_0$.

Remark 1.2'. We note that the expression (1.22) coincides with the result obtained by the "usual" formal integration performed with the δ-function as an integration with respect to the measure $\delta(h)\, dh = \delta_{(0)}(dh)$ concentrated at the point $h = 0$. Namely, in a neighbourhood of the surface Γ_0 we choose the coordinates $(h(y), y')$, where y' are local coordinates on Γ_0. This is possible provided that grad $h(y) \neq 0$ for $y \in \Gamma_0$ and $h(y) \in C^1(\mathbb{R}^m)$. Then $d^m y = \dfrac{dh\, \sigma_{m-1}(dy')}{|\text{grad } h(y)|}$ and therefore, formally,

$$\int \delta(h(y))\varphi(y)\, dy = \int_{\Gamma_0} \left(\int_R \delta(h)\varphi(h, y')\, dh \right) \frac{\sigma_{m-1}(dy')}{|\text{grad } h(y)|}$$

$$= \int_{\Gamma_0} \frac{\varphi(0, y')\sigma_{m-1}(dy')}{|\text{grad } h(y)|}. \tag{1.26}$$

Example 1.12. Let the map $h\colon \mathbb{R}^2 \to \mathbb{R}$ be a projection. In particular, let $h(y_1, y_2) = y_1$. Then

$$\langle \delta_1(y_1), \varphi(y) \rangle = \langle \delta_1(x), \overline{\varphi}(x) \rangle = \overline{\varphi}(0) = \int_{-\infty}^{\infty} \varphi(0, y_2)\, dy_2 \tag{1.27}$$

for $\varphi \in \mathcal{D}(\mathbb{R}^2)$. Similarly,

$$\langle \delta_1(y_2), \varphi(y) \rangle = \int_{-\infty}^{\infty} \varphi(y_1, 0)\, dy_1. \tag{1.28}$$

Example 1.13. Let $m = n$. Let $x = h(y) = y - z$ be the shift by the vector $z \in \mathbb{R}^n$. Then, by (1.21),

$$\langle u(y - z), \varphi(y) \rangle = \langle u(x), \varphi(x + z) \rangle. \tag{1.29}$$

In particular,

$$\langle \delta(y - z), \varphi(y) \rangle = \varphi(z), \quad \varphi \in \mathcal{D}(\mathbb{R}^n). \tag{1.30}$$

Remark 1.3. It is clear from (1.30) that $\delta(y - z)$ is formally the integral kernel of the identity operator in $\mathcal{D}(\mathbb{R}^n)$. This is what prompted Dirac in 1927 to denote such a "function" by the letter δ, in analogy with δ_{ij}, the Kronecker δ-notation, which constitutes the matrix of the identity operator in $\mathbb{R}^N$: $\sum_j \delta_{ij}\varphi_j = \varphi_i$ for $\varphi = (\varphi_1, \varphi_2, \ldots, \varphi_N) \in \mathbb{R}^N$, in a similar way to (1.30) (see Dirac [1958]).

Example 1.14. If $x = h(y) = ay$, where $a \in \mathbb{R}\backslash 0$, then

$$\langle u(ay), \varphi(y) \rangle = \frac{1}{|a|^n} \left\langle u(x), \varphi\left(\frac{x}{a}\right) \right\rangle \tag{1.31}$$

for $\varphi \in \mathcal{D}(\mathbb{R}^N)$. In particular,

$$\langle \delta_n(ay), \varphi(y) \rangle = \frac{1}{|a|^n}\varphi(0) \Rightarrow \delta_n(ay) = \frac{1}{|a|^n}\delta_n(y). \tag{1.32}$$

Thus $\delta_n(y)$ is a homogeneous, positive generalized function of order $-n$, and $\delta_n(y)$ is an even function.

Example 1.15. If $x = h(y) = Cy$, where $C \in GL_{\mathbb{R}}(n)$ is a non-degenerate real matrix, we set $u_C(y) = u(Cy)$:

$$\langle u_C(y), \varphi(y) \rangle = \langle u(Cy), \varphi(y) \rangle = \frac{1}{|\det C|}\langle u(x), \varphi(C^{-1}x) \rangle. \tag{1.33}$$

In particular, if $|\det C| = 1$, then

$$\langle \delta_n(Cy), \varphi(y) \rangle = \varphi(0) \Rightarrow \delta_n(Cy) = \delta_n(y). \tag{1.34}$$

This implies, for example, that $\delta_n(y)$ is invariant under rotations $C \in O(n)$ and under the *Lorentz transformations* (which preserve the quadratic form $y_1^2 - y_2^2 - \cdots - y_n^2$).

1.3. Support of a Generalized Function. The *support* of a function $\varphi(x) \in \mathscr{D}(\mathbb{R}^n)$ is defined to be the closure of the set of points $x \in \mathbb{R}^n$ where $\varphi(x) \neq 0$:

$$\operatorname{supp} \varphi := \overline{\{x \in \mathbb{R}^n : \varphi(x) \neq 0\}}. \tag{1.35}$$

Clearly, supp φ is a compact set. For an open domain $\Omega \subset \mathbb{R}^n$, we set $\mathscr{D}(\Omega) = C_0^\infty(\Omega) = \{\varphi \in \mathscr{D}(\mathbb{R}^n): \operatorname{supp} \varphi \subset \Omega\}$. By definition, $\varphi_k \xrightarrow{\mathscr{D}(\Omega)} \varphi$ if $\varphi_k \xrightarrow{\mathscr{D}(\mathbb{R}^n)} \varphi$ and $\operatorname{supp} \varphi_k \subset K$, where K is a compact set in Ω which is independent of k. The embedding $\mathscr{D}(\Omega) \subset \mathscr{D}(\mathbb{R}^n)$ is continuous and the dual map $\mathscr{D}'(\mathbb{R}^n) \to \mathscr{D}'(\Omega)$ is defined to be the operation of restricting the generalized functions $u(x)$ to the open set Ω: $u \mapsto u|_\Omega$. To give the detailed definition,

$$\langle u|_\Omega, \varphi \rangle_\Omega = \langle u, \varphi \rangle, \quad \varphi \in \mathscr{D}(\Omega), \tag{1.36}$$

where $\langle \cdot, \cdot \rangle_\Omega$ denotes the duality between $\mathscr{D}'(\Omega)$ and $\mathscr{D}(\Omega)$.

Definition 1.5. For $u, v \in \mathscr{D}'(\mathbb{R}^n)$, we say that $u(x) = v(x)$ for $x \in \Omega$ if $u|_\Omega = v|_\Omega$, that is, if

$$\langle u, \varphi \rangle = \langle v, \varphi \rangle, \quad \varphi \in \mathscr{D}(\Omega). \tag{1.37}$$

Proposition 1.2 (Gel'fand and Shilov [1958a], Schwartz [1966], Shilov [1965]). *If $\{\Omega_i\}$ is a family of open sets $\Omega_i \subset \mathbb{R}^n$ and $u|_{\Omega_i} = 0$ for all i, then $u|_\Omega = 0$, where $\Omega = \bigcup_i \Omega_i$.*

Thus there exists a largest domain $\Omega(u) \subset \mathbb{R}^n$ where $u(x) = 0$, namely

$$\Omega(u) = \bigcup_{\Omega: u|_\Omega = 0} \Omega.$$

Definition 1.6. The *support* of a function $u \in \mathscr{D}'(\mathbb{R}^n)$ is the set

$$\operatorname{supp} u := \mathbb{R}^n \backslash \Omega(u). \tag{1.38}$$

If supp $u \subset K$, where K is a set in $\mathbb{R}^n$, then we say that u is *concentrated* on K. We write $\mathscr{D}'_K = \{u \in \mathscr{D}'(\mathbb{R}^n)\colon \operatorname{supp} u \subset K\}$. If K is closed in $\mathbb{R}^n$, then $\mathscr{D}'_K$ is a closed subspace of $\mathscr{D}'$ and, by definition, the convergence in $\mathscr{D}'_K$ coincides with that in $\mathscr{D}'$.

Example 1.16. We have supp $\theta(x) = \overline{\mathbb{R}}_+$, where $\mathbb{R}_+ = \{x \in \mathbb{R}\colon x > 0\}$, and

$$\operatorname{supp} \delta(x-a) = \operatorname{supp} \delta^{(\alpha)}(x-a) = \{a\}, \quad a \in \mathbb{R}^n. \tag{1.39}$$

Clearly, supp u is a closed set. If $u \in \mathscr{D}(\mathbb{R}^n)$, the supports (1.35) and (1.38) coincide.

Definition 1.7. We denote by $\mathscr{E}' = \mathscr{E}'(\mathbb{R}^n)$ the space of distributions $u(x) \in \mathscr{D}'(\mathbb{R}^n)$ which have compact support (see Gel'fand and Shilov [1958a], Shilov [1965]). By definition, $u_k \overset{\mathscr{E}'}{\to} u$ if $u_k \overset{\mathscr{D}'}{\to} u$ and supp $u_k \subset K$, where K is a compact set in $\mathbb{R}^n$ which is independent of k. A distribution $u(x) \in \mathscr{E}'(\mathbb{R}^n)$ is referred to as a *distribution with compact support.*

If $u(x) \in \mathscr{E}'$, the scalar product (1.3) can be defined for all $\varphi(x) \in C^\infty(\mathbb{R}^n)$, and even for those $\varphi(x) \in C^\infty(\Omega)$ which are smooth in a neighbourhood Ω of the set supp u. To do this, we take a "cut-off" function $\psi(x) \in \mathscr{D}(\Omega)$ which is unity in a neighbourhood of the set supp u and then write

$$\langle u(x), \varphi(x)\rangle = \langle u(x), \psi(x)\varphi(x)\rangle, \tag{1.40}$$

where $\psi(x)\varphi(x) \equiv 0$ for $x \in \mathbb{R}^n \backslash \Omega$. Here $\psi(x)\varphi(x) \in \mathscr{D}(\mathbb{R}^n)$ and the definition (1.40) does not depend on the choice of $\psi(x)$ with the required properties. It can be shown easily that $\mathscr{E}'$ is the dual space of $\mathscr{E} \equiv C^\infty(\mathbb{R}^n)$.

Lemma 1.2 (Gel'fand and Shilov [1958a], Yosida [1965]). *If $u \in \mathscr{E}'(\mathbb{R}^n)$ and* supp $u = K$, *then for each* $\varepsilon > 0$

$$|\langle u, \varphi\rangle| \leqslant C\|\varphi\|_{C^{(m)}(K_\varepsilon)}, \quad \varphi \in C^\infty(\mathbb{R}^n), \tag{1.41}$$

for some $C > 0$ and some integer m. Here K_ε denotes an ε-neighbourhood of K.

Then m is known as the *order* (*of singularity*) *of the generalized function u*, and the least such m is known as the exact order. For example, $\delta(x)$ has order $m = 0$ while $\delta^{(\alpha)}(x)$ has order $m = |\alpha|$.

Example 1.17. The series $u(x) = \sum_{j=0}^{\infty} \delta^{(j)}(x - j)$, $x \in \mathbb{R}$, converges in $\mathscr{D}'(\mathbb{R})$ but supp u is not a compact set and the order of u is ∞ (that is, it is not finite).

Lemma 1.2 yields immediately

Corollary 1.1 (Gel'fand and Shilov [1958a], Shilov [1965]). *If $u \in \mathscr{D}'(\mathbb{R}^n)$ and* supp $u = \{a\}$, *where $a \in \mathbb{R}^n$, then*

$$u(x) = \sum_{|\alpha| \leqslant m} C_\alpha \delta^{(\alpha)}(x-a)$$

for some $m < \infty$ and $C_\alpha \in \mathbb{C}$ (compare with (1.39)).

We now extend the definition (1.40).

Definition 1.8. If $u(x) \in \mathscr{D}'(\mathbb{R}^n)$ and $\varphi(x) \in \mathscr{E} = C^\infty(\mathbb{R}^n)$, where $K := \operatorname{supp} u \cap \operatorname{supp} \varphi$ is a bounded set of $\mathbb{R}^n$, then we write

$$\langle u(x), \varphi(x) \rangle = \langle u(x), \psi(x)\varphi(x) \rangle, \tag{1.42}$$

where $\psi(x) \in \mathscr{D}(\mathbb{R}^n)$ and $\psi(x) = 1$ in a neighbourhood of K.

Evidently, the formula (1.42) does not depend on the choice of $\psi(x)$ with the required properties.

1.4. Singular Support of Generalized Functions. As for supp u, we define the *singular support of a generalized function* $u(x) \in \mathscr{D}'(\mathbb{R}^n)$ by

$$\operatorname{sing\,supp} u = R^n \backslash \Omega^\infty(u), \tag{1.43}$$

where

$$\Omega^\infty(u) = \bigcup_{\Omega:\, u|_\Omega \in C^\infty(\Omega)} \Omega$$

is the largest domain in which $u \in C^\infty$.

For a closed set $K \subset \mathbb{R}^n$, we set $\mathscr{D}'S_K = \{u \in \mathscr{D}'(\mathbb{R}^n)\colon \operatorname{sing\,supp} u \subset K\}$ or, in other words, $\mathscr{D}'S_K = \{u \in \mathscr{D}'\colon u|_{\mathbb{R}^n \backslash K} \in C^\infty(\mathbb{R}^n \backslash K)\}$. By definition, $u_m \xrightarrow{\mathscr{D}'S_K} u$ if $u_m \xrightarrow{\mathscr{D}'} u$ and

$$u_m|_{\mathbb{R}^n \backslash K} \xrightarrow{C^\infty(\mathbb{R}^n \backslash K)} u|_{\mathbb{R}^n \backslash K}.$$

Remark 1.4. If $\operatorname{sing\,supp} u \subset K$, where $K \subset \mathbb{R}^n$, then the substitution $u(h(y))$ is defined under milder restrictions on h than those mentioned earlier. Thus, in order to define $u(h(y))$ it is enough to require that the map $h \in C(\mathbb{R}^m, \mathbb{R}^n) \cap C^\infty(h^{-1}(\Omega), \Omega)$ and $\operatorname{rank} \dfrac{\partial h}{\partial y} = n$ only at the points $y \in h^{-1}(\Omega)$, where Ω is a neighbourhood of K. The map $u(x) \mapsto u(h(y))$ is continuous as a map $\mathscr{D}'S_K \to \mathscr{D}'(\mathbb{R}^m)$. Therefore, if $u_\lambda(x)$ is a continuous function of the parameter $\lambda \in \mathbb{C}$ with values in $\mathscr{D}'S_K$, then $u_\lambda(h(y))$ is also a continuous function of $\lambda \in \mathbb{C}$ with values in $\mathscr{D}'(\mathbb{R}^m)$. The same is true for differentiable, holomorphic, and meromorphic functions of a parameter (see Gel'fand and Shilov [1958a]).

Example 1.18. Let $u(x) = \delta_1(x)$ and $h(y) = |y|^2 - R^2$ for $y \in \mathbb{R}^m$ and $R \in \mathbb{R}$ with $R \neq 0$. Then (1.20), (1.18), (1.16) and (1.13) imply that

$$\langle \delta_1(|y|^2 - R^2), \varphi(y) \rangle = \frac{1}{2R} \int_{|y|=R} \varphi(y)\, dy. \tag{1.44}$$

We note that $\operatorname{grad}(|y|^2 - R^2) = 0$ if $y = 0$. Therefore $u(|y|^2 - R^2)$ is not meaningful for all $u \in \mathscr{D}'(\mathbb{R})$.

1.5. The Convolution of Generalized Functions. The *convolution* of two functions u and v is defined to be the function

$$(u * v)(x) := \int_{R^n} u(x - y)v(y)\, dy = \int_{\mathbb{R}^n} u(z)v(x - z)\, dz, \quad x \in \mathbb{R}^n. \tag{1.45}$$

The properties of the convolution of test functions.

1) It follows from (1.45) that convolution is commutative, bilinear, and associative. Thus

$$u * v = v * u, \quad u * (\alpha v + \beta w) = \alpha u * v + \beta u * w, \quad (u * v) * w = u * (v * w) \quad (1.46)$$

for u, v and $w \in \mathscr{D}(\mathbb{R}^n)$ and $\alpha, \beta \in \mathbb{C}$.

2) The principal properties of the convolution are the following translation and differentiation formulae: for u and $v \in \mathscr{D}$

$$T_a(u * v) = (T_a u) * v = u * T_a v,$$

where $T_a u(x) := u(x + a)$, $a \in \mathbb{R}^n$, and

$$\partial_x^\alpha (u * v) = (\partial_x^\alpha u) * v = u * \partial_x^\alpha v, \quad \alpha = (\alpha_1, \ldots, \alpha_n). \quad (1.47)$$

3) The support of the convolution is contained in the algebraic sum of the supports, that is,

$$\operatorname{supp}(u * v) \subset \operatorname{supp} u + \operatorname{supp} v \equiv \{x = z + y \in \mathbb{R}^n : z \in \operatorname{supp} u, y \in \operatorname{supp} v\}. \quad (1.48)$$

4) The map $(u, v) \mapsto u * v$ is continuous as a map on $\mathscr{D} \times \mathscr{D} \to \mathscr{D}$.

The *convolution of generalized functions* is defined by the taking the closure with respect to continuity of the convolution of the test functions.

We recall (Schwartz [1966]) that the *direct product* of the generalized functions $u(z) \in \mathscr{D}'(\mathbb{R}^n)$ and $v(y) \in \mathscr{D}'(\mathbb{R}^m)$ is defined as the functional $u \times v$ which acts on test functions of the form $\varphi(z)\psi(y)$ according to the formula

$$\langle u \times v, \varphi(z)\psi(y) = \langle u, \varphi \rangle \langle v, \psi \rangle, \ \varphi \in \mathscr{D}(\mathbb{R}^n), \ \psi \in \mathscr{D}(\mathbb{R}^m). \quad (1.49)$$

Proposition 1.3 (Schwartz [1966]). *For all $u \in \mathscr{D}'(\mathbb{R}^n)$ and all $v \in \mathscr{D}'(\mathbb{R}^m)$ there exists a unique distribution $u \times v \in \mathscr{D}'(\mathbb{R}^n \times \mathbb{R}^m)$ satisfying the identity* (1.49). *The map $(u, v) \mapsto u \times v$ is continuous as a map on $\mathscr{D}'(\mathbb{R}^n) \times \mathscr{D}'(\mathbb{R}^m) \to \mathscr{D}'(\mathbb{R}^n \times \mathbb{R}^m)$. Furthermore,*

$$\operatorname{supp} u \times v = \operatorname{supp} u \times \operatorname{supp} v. \quad (1.50)$$

In order to decide whether the operation of forming the convolution of distributions is closed, we note that (1.45) yields the following identity: for $u, v \in \mathscr{D}(\mathbb{R}^n)$ and $\varphi \in \mathscr{D}(\mathbb{R}^n)$

$$\langle u * v, \varphi \rangle = \iint u(x - y) v(y) \varphi(x) \, dx \, dy$$

$$= \langle u(z) \times v(y), \varphi(y + z) \rangle. \quad (1.51)$$

This expression can be defined for $u, v \in \mathscr{D}'(\mathbb{R}^n)$, with $\varphi \in \mathscr{D}(\mathbb{R}^n)$, by means of Definition 1.8 provided that

for all $\varphi \in \mathscr{D}(\mathbb{R}^n)$ the set $B_\varphi := \operatorname{supp}(u \times v) \cap \operatorname{supp}(\varphi(y + z))$
is bounded in $\mathbb{R}^{2n}$. (1.52)

In view of (1.50), this condition is equivalent to the following condition concerning the sets $U = \operatorname{supp} u$ and $V = \operatorname{supp} v$:

$$\text{for all } R > 0 \text{ the set } B_R := (U \times V) \cap \{(z, y) \in \mathbb{R}^{2n}: |y + z| < R\rangle \text{ is bounded in } \mathbb{R}^{2n}. \tag{1.52'}$$

Definition 1.9 (Schwartz [1966]). If $u, v \in \mathscr{D}'(\mathbb{R}^n)$ and the sets $U = \operatorname{supp} u$ and $V = \operatorname{supp} v$ satisfy (1.52′), then the convolution $u * v$ is defined by the identity

$$\langle u * v, \varphi \rangle = \langle u(z) \times v(y), \varphi(z + y)\rangle, \quad \varphi \in \mathscr{D}(\mathbb{R}^n), \tag{1.53}$$

where the right-hand side is defined in accordance with (1.42).

Example 1.19. Since $\operatorname{supp} \delta(x) = \{0\}$, it follows that the convolution $\delta * u$ is defined for all $u \in \mathscr{D}'(\mathbb{R}^n)$ and

$$\langle \delta * u, \varphi \rangle = \langle \delta(z) \times u(y), \varphi(z + y)\rangle = \langle u(y), \varphi(y)\rangle \Rightarrow \delta * u = u. \tag{1.54}$$

The condition (1.52) is equivalent to the following condition which is geometrically more obvious:

$$\text{for all } R > 0, \quad (x - U) \cap V \subset D_R \quad \text{for } |x| < R, \tag{1.52''}$$

where D_R is a bounded set in $\mathbb{R}^n$.

The conditions (1.52″) and (1.52′) are satisfied, in particular, in the following important practical cases.

1) U or V is a bounded set.

2) $U \subset K$ and $V \subset K$, where K is a (convex) cone in $\mathbb{R}^n$ that does not contain straight lines. By a *cone* here (and in what follows) is meant a set $K \subset \mathbb{R}^n$ such that $K + K \subset K$ and $tK \subset K$ for all $t > 0$.

3) $U \subset K_R$ and $V \subset K_R$ for some $R > 0$, where K_R is an R-neighbourhood of the cone K, the cone not containing straight lines.

4) U is an *asymptotically time-like* set, that is, for some (large) $R > 0$ the inequality $|x_1| > \gamma|x'|$ holds for $x \in U$ and $|x| > R$, where $\gamma > 1$ and $x' \equiv (x_2, \ldots, x_n)$, and V is an *asymptotically space-like* set, that is, $|x_1| \leqslant |x'|$ for $x \in V$ and $|x| > R$. (For example, $U = \{(t, x) \in \mathbb{R}^4: x = x(t)\}$, where $|\dot{x}(t)| < 1$, that is, $(t, x(t))$ is a time-like curve, and V is the *light cone* $t^2 = |x|^2$.)

Remark 1.5. If the sets U and V satisfy the condition (1.52′), then their R-neighbourhoods U_R and V_R, and their translates $U + a_1$ and $V + a_2$, also obviously satisfy this condition for all $R > 0$ and all $a_1, a_2 \in \mathbb{R}^n$.

Lemma 1.3 (Schwartz [1966]). *Suppose that the sets $U, V \subset \mathbb{R}^n$ satisfy the condition* (1.52′) *or* (1.52″). *Then for $u, v \in \mathscr{D}'_U \times \mathscr{D}'_V$ the convolution $u * v$ lies in $\mathscr{D}'_{U+V}$ and the map $(u, v) \mapsto u * v$ is continuous in these spaces. For this convolution the formulae* (1.47) *and* (1.48) *hold; the formulae* (1.46) *also hold provided that all the convolutions involved are defined.*

Example 1.20. The formulae (1.54) and (1.47) imply that

$$\delta^{(\alpha)} * u = \partial_x^\alpha(\delta * u) = \partial_x^\alpha u(x), \quad \forall \alpha = (\alpha_1, \ldots, \alpha_n), \tag{1.55}$$

for $u \in \mathscr{D}'(\mathbb{R}^n)$. Therefore the operator $P(\partial_x)$ of (0.1) is a convolution operator with kernel

$$P_c := P\delta = \sum_{|\alpha| \leqslant m} p_\alpha \partial_x^\alpha \delta(x): P_c * u = (P\delta) * u = P(\delta * u) = Pu, \quad u \in \mathscr{D}'(\mathbb{R}^n). \quad (1.56)$$

Definition 1.10 (Vladimirov [1964]). If K is a cone in $\mathbb{R}^n$ not containing straight lines, then $\mathscr{D}'_{(K)}$ denotes the set of distributions $u(x) \in \mathscr{D}'(\mathbb{R}^n)$ with $\operatorname{supp} u \subset K_R$ for some $R > 0$, where K_R is an R-neighbourhood of K. Furthermore, $u_k \overset{\mathscr{D}'_{(K)}}{\to} u$ if $u_k \overset{\mathscr{D}'}{\to} u$ and $\operatorname{supp} u_k \subset K_R$ for all k and some $R > 0$ independent of k.

As we noted above, the convolution is defined for $u, v \in \mathscr{D}'_{K_R}$. Therefore Lemma 1.3 implies

Corollary 1.2. *The spaces $\mathscr{E}'(\mathbb{R}^n)$ and $\mathscr{D}'_{(K)}$ are algebras with identity $\delta(x)$ (see (1.54)) with respect to the operation of forming convolution that are commutative and associative if K is a cone of $\mathbb{R}^n$ not containing straight lines. The convolution is (sequentially) continuous on both $\mathscr{E}' \times \mathscr{E}'$ and $\mathscr{D}'_{(K)} \times \mathscr{D}'_{(K)}$.*

Lemma 1.4. *If $u(x) \in \mathscr{D}'$ (or $\mathscr{E}'$) and $v \in \mathscr{D}$ (or $\mathscr{E}$), then $u * v \in C^\infty(\mathbb{R}^n)$, and*

$$(u * v)(x) = \langle u(x - y), v(y) \rangle = \langle u(y), v(x - y) \rangle. \quad (1.57)$$

This relation follows from (1.45) in view of the continuity of the operation of convolution. The same formula implies easily that $u * v \in C^\infty(\mathbb{R}^n)$.

Let us show that the inclusion

$$\operatorname{sing\,supp} u * v \subset \operatorname{sing\,supp} u + \operatorname{sing\,supp} v \quad (1.58)$$

holds when the convolution is defined, that is, when the condition (1.52) is satisfied.

Indeed, for all $\varepsilon > 0$, the decompositions $u = u_\varepsilon + u'_\varepsilon$ and $v = v_\varepsilon + v'_\varepsilon$ exist such that $\operatorname{supp} u_\varepsilon$ and $\operatorname{supp} v_\varepsilon$ lie in ε-neighbourhoods of $\operatorname{sing\,supp} u$ and $\operatorname{sing\,supp} v$ respectively, and $u'_\varepsilon, v'_\varepsilon \in C^\infty(\mathbb{R}^n)$. Then $u * v = u_\varepsilon * v_\varepsilon + u'_\varepsilon * v_\varepsilon + u_\varepsilon * v'_\varepsilon + u'_\varepsilon * v'_\varepsilon$. All the terms here, except the first, lie in $C^\infty(\mathbb{R}^n)$ by Lemma 1.4. It remains to note that $\operatorname{supp} u_\varepsilon * v_\varepsilon \subset \operatorname{supp} u_\varepsilon + \operatorname{supp} v_\varepsilon$, in view of (1.48), that is, $\operatorname{supp} u_\varepsilon * v_\varepsilon$ lies in the 2ε-neighbourhood of the set $\operatorname{sing\,supp} u + \operatorname{sing\,supp} v$, and $\varepsilon > 0$ is arbitrary.

Note that the sum of the sets on the right-hand sides of (1.48) and (1.58) is empty if one of the summands is empty, and then the set on the left-hand side is also empty.

1.6. Boundary Values of Analytic Functions. Let Ω be a convex open set in $\mathbb{R}^n$, and let $f(x + iy)$ be an analytic function of $z = x + iy$ in a "tubular" domain $T_\Omega \equiv \mathbb{R}^n + i\Omega$. Consider the distributions

$$f_y(x) = f(x + iy) \in \mathscr{D}'(\mathbb{R}^n_x) \quad \text{for all } y \in \Omega. \quad (1.59)$$

Assume that, for each open cone $K \subset \mathbb{R}^n$ such that $K^\delta = \{y \in K: |y| < \delta\} \subset \Omega$ for some $\delta > 0$,

$$|f(x+iy)| \leqslant \frac{c(x)}{|y|^\nu} \quad \text{for } x \in \mathbb{R}^n, \quad y \in K^\delta, \quad \text{where} \quad c(x) \in C(\mathbb{R}^n). \tag{1.60}$$

Here $\nu \in \mathbb{R}$ and $c(x)$ may depend on the cone K and $\delta > 0$.

Lemma 1.5. (Compare with Reed and Simon [1975, 1979], Vladimirov [1964].) *If the condition* (1.60) *holds, the distributions* (1.59) *have a limit as* $y \to 0$ *and* $y \in \Omega$ *in the sense of weak convergence in* $\mathscr{D}'(\mathbb{R}^n)$. *Thus, for all* $y \in \Omega$

$$f_{\varepsilon y}(x) := f(x + i\varepsilon y) \xrightarrow[\varepsilon \to 0+]{} f(x + i0y) \in \mathscr{D}'(\mathbb{R}^n), \tag{1.61}$$

and the limit function (generalized) $f_0(x) \equiv f(x + i0y)$ *is independent of* $y \in \Omega$.

Example 1.21. The function $f(z) = 1/z$, $z \in \mathbb{C}\backslash 0$, has upper and lower limits on $\mathbb{R}$, given by

$$\frac{1}{x + i\varepsilon} \xrightarrow[\varepsilon \to 0\pm]{} \frac{1}{x \pm i0} = \text{p.v.}\,\frac{1}{x} \mp \pi i \delta(x), \tag{1.62}$$

where p.v. $1/x$ denotes the *Cauchy principal value* (see Gel'fand and Shilov [1958a], Shilov [1965] and the formula (2.5') of Chap. 3).

Example 1.22. The function $f(\xi_0, \xi) = \dfrac{1}{-\xi_0^2 + |\xi|^2}$, $\xi_0 \in \mathbb{C}$ and $\xi \in \mathbb{C}^k$, is analytic for $|\text{Im}\,\xi| < |\text{Im}\,\xi_0|$, and

$$\frac{1}{-(\xi_0 \pm i\varepsilon)^2 + |\xi|^2} \xrightarrow[\varepsilon \to 0+]{} \frac{1}{-(\xi_0 \pm i0)^2 + |\xi|^2}, \quad \xi_0 \in \mathbb{R}, \quad \xi \in \mathbb{R}^k. \tag{1.63}$$

Likewise,

$$\begin{aligned} \frac{1}{-i(\xi_0 + i\varepsilon) + |\xi|^2} &\xrightarrow[\varepsilon \to 0+]{} \frac{1}{-i(\xi_0 + i0) + |\xi|^2}, \\ \frac{1}{\xi_0 \pm i\varepsilon + |\xi|^2} &\xrightarrow[\varepsilon \to 0+]{} \frac{1}{\xi_0 \pm i0 + |\xi|^2}. \end{aligned} \tag{1.64}$$

In the one-dimensional case when $n = 1$ and $\Omega = \mathbb{R}^+$, the lemma can be proved as follows. The primitive $f^{(-m)}(z)$ of $f(z)$ of order $m = [\nu] + 1$ is continuous in the half-plane $\text{Im}\,z \geqslant 0$. Therefore $f^{(-m)}(x + i\varepsilon) \xrightarrow[\varepsilon \to 0+]{} f^{(-m)}(x)$ in $C(\mathbb{R})$ and more so in $\mathscr{D}'(\mathbb{R})$. But the operator $\dfrac{d^m}{dx^m}$ is (sequentially) continuous in $\mathscr{D}'(\mathbb{R})$. Therefore

$$\begin{aligned} f(x + i\varepsilon) &= \frac{d^m}{dx^m} f^{(-m)}(x + i\varepsilon) \xrightarrow[\varepsilon \to 0+]{\mathscr{D}'(\mathbb{R})} \frac{d^m}{dx^m} f^{(-m)}(x) \\ &= f(x + i0) \in \mathscr{D}'(\mathbb{R}). \end{aligned} \tag{1.65}$$

In the general case of arbitrary $n \geqslant 1$, we can make a linear change of variables in $\mathbb{R}^n$ such that in the new coordinates $y = (1, 0, \ldots, 0)$ and the domain $x_1 \geqslant 0$, $\ldots$, $x_n \geqslant 0$ is contained in the cone K. Then $f(z)$ is an analytic function of z for

$\operatorname{Im} z_1 > 0, \ldots, \operatorname{Im} z_n > 0$ and $|z| < \delta_1$. To complete the proof, it remains to repeat the argument (1.65) with respect to the variable x_1, instead of x, by keeping the remaining variables $x_2, \ldots, x_n$ fixed. The primitives can be chosen from the condition that $f^{(-m)}(i\delta_{1/2}, x_2, \ldots, x_n) = 0$.

Definition 1.11. Under the hypotheses of Lemma 1.5, the function $f_0(x)$ is called the *boundary value of the analytic function* $f(x+iy)$ in T_Ω and $f(x+iy)$ is called the *analytic continuation of the generalized function* $f(x)$ to T_Ω. We set

$$f(z)|_{\mathbb{R}^n} := f_0(x). \tag{1.66}$$

1.7. The Space of Tempered Distributions. We now introduce new classes of test functions and generalized functions.

Definition 1.12. The *Schwartz space* $S = S(\mathbb{R}^n)$ of rapidly decreasing functions consists of complex-valued functions $\varphi(x) \in C^\infty(\mathbb{R}^n)$ such that

$$|\varphi|_{\alpha,N} := \sup_{x \in \mathbb{R}^n} (1+|x|)^N |\partial_x^\alpha \varphi(x)| < \infty \tag{1.67}$$

for all $\alpha = (\alpha_1, \ldots, \alpha_n)$ and all $N > 0$.

Convergence of a sequence φ_k to φ in the space S is defined by

$$|\varphi_k - \varphi|_{\alpha,N} \to 0 \text{ as } k \to \infty$$

for all α and all N, and it is denoted by $\varphi_k \overset{S}{\to} \varphi$.

As usual, we denote, the space of continuous linear functionals on $S(\mathbb{R}^n)$ by $S'(\mathbb{R}^n)$. The functionals $u \in S'(\mathbb{R}^n)$ are called *temperately growing* or simply *tempered generalized functions* (or *distributions*).

Clearly, $\mathscr{D}(\mathbb{R}^n) \subset S(\mathbb{R}^n)$ is a continuous embedding and $\mathscr{D}(\mathbb{R}^n)$ is everywhere dense in $S(\mathbb{R}^n)$. Therefore the dual map $S'(\mathbb{R}^n) \subset \mathscr{D}'(\mathbb{R}^n)$ is also a continuous embedding (with respect to weak convergence). It can also be shown easily that $\mathscr{E}'(\mathbb{R}^n) \subset S'(\mathbb{R}^n)$. Thus $S' = S'(\mathbb{R}^n)$ is a subset of $\mathscr{D}'(\mathbb{R}^n)$. We can show easily that $S(\mathbb{R}^n)$ is everywhere dense in $S'(\mathbb{R}^n)$.

Clearly, $\delta(x) \in S'$ and $\delta^{(\alpha)}(x) \in S'$ for all α. But $e^x \notin S'$ although $e^x \cos e^x = (\sin e^x)' \in S'$.

Since the operations of differentiation, forming convolution, etc., are defined in $\mathscr{D}'(\mathbb{R}^n)$, the same operations are also defined in $S'(\mathbb{R}^n) \subset \mathscr{D}'(\mathbb{R}^n)$. Under certain conditions, these operations lead to results which remain within $S'(\mathbb{R}^n)$.

For example, $\partial_x^\alpha \colon S'(\mathbb{R}^n) \to S'(\mathbb{R}^n)$ is a continuous operator for all α because it is the adjoint of the continuous operator $(-\partial_x)^\alpha \colon S(\mathbb{R}^n) \to S(\mathbb{R}^n)$.

The operator $u(x) \mapsto g(x)u(x)$ is continuous in $S'(\mathbb{R}^n)$ if

$$|g^{(\alpha)}(x)| \leqslant c_\alpha (1+|x|)^{p_\alpha}, \quad x \in \mathbb{R}^n, \tag{1.68}$$

for all $\alpha = (\alpha_1, \ldots, \alpha_n)$ and certain $c_\alpha, p_\alpha \in \mathbb{R}$. This result follows from the continuity of the operator $\varphi \mapsto g\varphi$ in $S(\mathbb{R}^n)$ when (1.68) holds. It turns out (Volevich and Paneyakh [1965]) that the condition (1.68) is also necessary for the operator of multiplication by $g(x)$ to be continuous in $S'(\mathbb{R}^n)$ (and in $S(\mathbb{R}^n)$).

The condition (1.68) is satisfied by any polynomial $g(x)$ for example. Therefore the operation of multiplication by any polynomial is a continuous operation in both $S'(\mathbb{R}^n)$ and $S(\mathbb{R}^n)$.

For all k and $N = 0, 1, 2, \ldots$, we write

$$\|\varphi\|_{k,N} := \sum_{|\alpha| \leqslant k} |\varphi|_{\alpha,N}. \tag{1.69}$$

As for Lemma 1.2, we can establish

Lemma 1.6. *If* $u(x) \in S'(\mathbb{R}^n)$, *then*

$$|\langle u, \varphi\rangle| \leqslant C\|\varphi\|_{\mu,N}, \quad \varphi \in S(\mathbb{R}^n) \tag{1.70}$$

for some C, μ *and* $N < \infty$.

Example 1.22. 1) If either $u(x) \in C(\mathbb{R}^n)$ or $u(x) \in L_\infty^{\mathrm{loc}}(\mathbb{R}^n)$, and if

$$|u(x)| \leqslant C(1 + |x|)^p, \quad x \in \mathbb{R}^n \tag{1.71}$$

for some C and p, then $u \in S'(\mathbb{R}^n)$; here u is a functional of the form (1.2).

2) If $u(x) \in L_1^{\mathrm{loc}}(\mathbb{R}^n)$ and

$$\int \frac{|u(x)|}{(1 + |x|)^N}\, dx < \infty \tag{1.72}$$

for some $N > 0$, then $u(x) \in S'(\mathbb{R}^n)$.

Remark 1.6. Assume that the function $c(x)$ of (1.60) satisfies an estimate of the form

$$|c(x)| \leqslant C(1 + |x|)^p, \quad x \in \mathbb{R}^n. \tag{1.73}$$

Then it is clear from the proof of Lemma 1.5 that the limit function in (1.61) lies in $S'(\mathbb{R}^n)$ and the convergence in (1.61) is weak convergence in $S'(\mathbb{R}^n)$.

§ 2. Fundamental Solutions of Differential Equations

2.1. The Fundamental Solutions

Definition 2.1. By a *fundamental solution* of the differential operator $P(\partial_x)$, and *of the equation* (0.1), we mean a distribution $\mathscr{E}(x) \in \mathscr{D}'(\mathbb{R}^n)$ such that

$$P(\partial_x)\mathscr{E}(x) = \delta(x), \quad x \in \mathbb{R}^n. \tag{2.1}$$

Examples of fundamental solutions of ordinary differential equations are given in (1.11). Roughly speaking, solutions of the equation (0.1) for an arbitrary right-hand side $f(x)$ are expressed as the convolution of $f(x)$ with suitable fundamental solutions of $P(\partial_x)$ (see (2.2) and (2.21) below).

Proposition 2.1. 1) *If a solution* $u(x)$ *of the equation* (0.1) *lies in* $\mathscr{E}'(\mathbb{R}^n)$, *then also* $f(x) \in \mathscr{E}'(\mathbb{R}^n)$, *and*

$$u(x) = \mathscr{E} * f(x). \tag{2.2}$$

2) *If the right-hand side* $f(x)$ *of the equation* (0.1) *lies in* $\mathscr{E}'(\mathbb{R}^n)$, *then the function* (2.2) *is a (particular) solution of the equation.*

Proof. 1) If $u \in \mathscr{E}'$, then (0.1) implies that

$$\mathscr{E} * Pu = \mathscr{E} * f. \tag{2.3}$$

But $\mathscr{E} * Pu = P\mathscr{E} * u = \delta * u = u$ by (1.47), and then (2.2) follows from (2.3).

2) If $f \in \mathscr{E}'$ and u is given by (2.2) we have, in view of (2.1),

$$Pu = P\mathscr{E} * f = \delta * f = f. \quad \blacksquare \tag{2.4}$$

Proposition 2.1 implies that the operator $\mathscr{E}*: f \mapsto \mathscr{E} * f$ is a left inverse of P on the domain of definition $\mathscr{E}'(\mathbb{R}^n)$ and a right inverse of P on the domain of definition $P^{-1}\mathscr{E}' = \{u \in \mathscr{D}' : Pu \in \mathscr{E}'\}$.

The construction and investigation of properties of various fundamental solutions of differential operators constitutes one of the major problems in the theory of partial differential equations with constant coefficients.

In § 1, Chap. 3, we shall prove

Theorem 2.1 (Ehrenpreis [1954], Gel'fand and Shilov [1958a], Malgrange [1953], Shilov [1965]). *Every differential operator* $P(\partial_x)$ *has a fundamental solution* $\mathscr{E}(x) \in \mathscr{D}'(\mathbb{R}^n)$ *provided that* $P \neq 0$.

A much more difficult result is

Theorem 2.2 (Bernshtejn [1971], Bernshtejn and Gel'fand [1969], Hörmander [1958], Lojasiewicz [1959]). *Every operator* $P(\partial_x) \neq 0$ *has a tempered fundamental solution* $\mathscr{E}(x)$, *that is,* $\mathscr{E}(x) \in S'(\mathbb{R}^n)$.

A fundamental solution is not unique, because any solution of the homogeneous equation can be added to it. This non-uniqueness can sometimes be used for obtaining solutions in the form (2.2) under various conditions on the support of the right-hand side $f(x)$. We have only to choose $\mathscr{E}(x)$ in such a way that the sets $U = \operatorname{supp} \mathscr{E}$ and $V = \operatorname{supp} f$ satisfy (1.52'). Then the convolution $\mathscr{E} * f$ is defined and, as in (2.4), the functon $u = \mathscr{E} * f$ is a solution of the equation (0.1). Thus we have the following generalization of Proposition 2.1.

Proposition 2.2. 1) *Let* $u(x)$ *be a solution of the equation* (0.1) *and let* $\mathscr{E}(x)$ *be a fundamental solution of the operator* $P(\partial_x)$. *Assume that the sets* $U = \operatorname{supp} \mathscr{E}$ *and* $V = \operatorname{supp} u$ *satisfy the condition* (1.52'). *Then the given solution* $u(x)$ *is expressed by the formula* (2.2). 2) *If* $U = \operatorname{supp} \mathscr{E}$ *and* $V = \operatorname{supp} f$ *satisfy* (1.52'), *then* $u = \mathscr{E} * f$ *is a particular solution of* (0.1).

2.2. Examples of Fundamental Solutions

Example 2.1. To solve the equation

$$\left(\frac{d}{dt} + \lambda\right) u(t) = f(t), \quad t \in \mathbb{R}, \tag{2.5}$$

in the form (2.2) when $f(t) = 0$ for either $t < 0$ or $t > 0$, we take the fundamental solution

$$\mathscr{E}^{\pm}(t) = \pm\theta(\pm t)e^{-\lambda t}, \quad t \in \mathbb{R}, \tag{2.5'}$$

where θ is the Heaviside function of Example 1.3. These fundamental solutions are known as the *retarded solution* ($\mathscr{E}^+$) and the *advanced solution* ($\mathscr{E}^-$). This is due to the fact that when $f \in \mathscr{D}(\mathbb{R})$ the convolution $\mathscr{E}^+ * f = u^+$ is expressed, in view of (1.57), by the formula

$$u^+(t) = \mathscr{E}^+ * f(t) = \int_{-\infty}^{t} \mathscr{E}^+(t - \tau)f(\tau)\,d\tau, \quad t \in \mathbb{R}, \tag{2.5''}$$

and it depends only on $f(\tau)$ for $\tau < t$. Similarly

$$u^-(t) = \mathscr{E}^- * f(t) = \int_{t}^{\infty} \mathscr{E}^-(t - \tau)f(\tau)\,d\tau, \quad t \in \mathbb{R}, \tag{2.5'''}$$

depends only on $f(\tau)$ for $\tau > t$. The solutions (2.5″) and (2.5‴) are referred to as the *retarded* and *advanced potentials* respectively.

Example 2.2. Likewise, for the *wave equation*

$$\Box u = \left(\frac{\partial^2}{\partial t^2} - a^2\Delta\right)u(x, t) = f(x, t), \quad t \in \mathbb{R}, \quad x \in \mathbb{R}^k; \quad a > 0, \tag{2.6}$$

in $\mathbb{R}^k$, $k \geqslant 1$, the *retarded* and *advanced fundamental solutions* $\mathscr{E}_k^{\pm}$ are

$$\begin{aligned} \mathscr{E}_1^{\pm} &= \frac{1}{2a}\theta(\pm at - |x|); \quad \mathscr{E}_2^{\pm} = \frac{\theta(\pm at - |x|)}{2\pi a\sqrt{(at)^2 - |x|^2}}; \\ \mathscr{E}_3^{\pm} &= \frac{\theta(\pm t)}{2\pi a}\delta((at)^2 - |x|^2) = \frac{\theta(\pm t)}{4\pi a^2|t|}\delta(a|t| - |x|) \end{aligned} \tag{2.6'}$$

respectively (Bogolyubov and Shirkov [1973], Vladimirov [1981]). These fundamental solutions, named after d'Alembert ($k = 1$), Poisson ($k = 2$) and Kirchhoff ($k = 3$), will be derived in Chap. 4. These formulae enable us to obtain solutions of the equation $\Box u = f(x, t)$ in the form (2.2) for the cases when $f = 0$ for either $t < 0$ or $t > 0$ respectively.

Example 2.3. For the *Klein-Gordon equation*

$$(\Box + m_0^2)u(x, t) = f(x, t), \quad t \in \mathbb{R}, \quad x \in \mathbb{R}^k; \quad m_0 > 0, \tag{2.7}$$

in $\mathbb{R}^k$, the *retarded* and *advanced* fundamental solutions for $k = 1, 2, 3$ (which reduce to (2.6′) as $m_0 \to 0$) are

$$\mathscr{E}_1^{\pm} = \frac{\theta(\pm at - |x|)}{2a}J_0\left(m_0\sqrt{t^2 - \left(\frac{x}{a}\right)^2}\right),$$

$$\mathscr{E}_2^{\pm} = \sqrt{\frac{m_0}{8\pi a^3}}\,\theta(\pm at - |x|)\frac{J_{-1/2}\left(m_0\sqrt{t^2 - \left(\frac{x}{a}\right)^2}\right)}{((at)^2 - x^2)^{1/4}}$$

$$= \frac{\theta(\pm at - |x|)}{2\pi a}\,\frac{\operatorname{Cos} m_0\sqrt{t^2 - \left(\frac{x}{a}\right)^2}}{\sqrt{(at)^2 - x^2}}, \tag{2.7'}$$

$$\mathscr{E}_3^{\pm} = \frac{\theta(\pm t)}{2\pi a}\,\delta((at)^2 - |x|^2) - \frac{m_0\theta(\pm at - |x|)}{4\pi a^2}\,\frac{J_1\left(m_0\sqrt{t^2 - \left(\frac{x}{a}\right)^2}\right)}{\sqrt{(at)^2 - x^2}},$$

respectively (Bogolyubov and Shirkov [1973], Reed and Simon [1975, 1979], Vladimirov [1981]).

Example 2.4. For the *equation of heat conduction* (or *diffusion equation*)

$$\left(\frac{\partial}{\partial t} - a^2\Delta\right)u(x, t) = f(x, t), \quad t \in \mathbb{R}, \quad x \in \mathbb{R}^k; \quad a > 0, \tag{2.8}$$

the *retarded* fundamental solution has the form of the Gaussian distribution

$$\mathscr{E}_k^{+}(x, t) = \frac{\theta(t)}{(2a\sqrt{\pi t})^k}\,e^{-(x^2/4a^2t)} \tag{2.8'}$$

with dispersion $\sim a\sqrt{t}$ at time $t > 0$ (Hörmander [1983, 1985], Vladimirov [1981]).

Example 2.5. The *Schrödinger equation*

$$\left(\frac{\partial}{\partial t} - ia^2\Delta\right)\psi(x, t) = f(x, t), \quad t \in \mathbb{R}, \quad x \in \mathbb{R}^k; \quad a > 0, \tag{2.9}$$

has retarded and advanced fundamental solutions which are analogous to (2.8′) (Hörmander [1983, 1985]):

$$\mathscr{E}_k^{\pm}(x, t) = \frac{\pm\theta(\pm t)}{(2a\sqrt{\pi i t})^k}\,e^{ix^2/4a^2t}; \quad \left(\sqrt{\pm i} = \frac{1 \pm i\sqrt{2}}{2}\right). \tag{2.9'}$$

Example 2.6. The *Laplace equation*

$$\Delta u(x) = f(x), \quad x \in \mathbb{R}^n, \tag{2.10}$$

in $\mathbb{R}^n$ has the fundamental solutions

$$\mathscr{E}_n(x) = \begin{cases} -\dfrac{1}{(n-2)\Omega_n}\dfrac{1}{|x|^{n-2}}, & n \neq 2, \\ -\dfrac{1}{2\pi}\ln\dfrac{1}{|x|}, & n = 2, \end{cases} \tag{2.10'}$$

where Ω_n denotes the surface area of the sphere $|x| = 1$ in $\mathbb{R}^n$ (Vladimirov [1981], Shilov [1965]). In particular, $\mathscr{E}_1(x) = |x|/2$.

Example 2.7. The *Helmholtz equation*

$$(\Delta + \omega^2)u(x) = f(x), \quad x \in \mathrm{R}^k, \quad \omega > 0, \tag{2.11}$$

is obtained from the wave equation (2.6) by substituting $u(x, t) = u(x)e^{-i\omega t}$ and $f(x) = f(x)e^{-i\omega t}$. Its fundamental solutions for $k = 1, 2, 3$ are

$$\mathscr{E}_1^{\pm} = \frac{e^{\pm i\omega|x|}}{\pm 2i\omega}, \quad \mathscr{E}_2^{\pm} = \begin{cases} -\frac{i}{4} H_0^{(1)}(\omega|x|) \\ \frac{i}{4}\overline{H_0^{(1)}(\omega|x|)} \end{cases}, \quad \mathscr{E}_3^{\pm} = -\frac{e^{\pm i\omega|x|}}{4\pi|x|}, \tag{2.11'}$$

respectively, where $H_0^{(1)}$ is the Hankel function (Bateman and Erdélyi [1953], Whittaker and Watson [1927]).

The fundamental solutions enable us to obtain by means of (2.2) the limiting amplitude $u(x)$ of the solution $u(x, t)$ for a given amplitude $f(x)$ of some external source $f(x, t)$ (force, pressure, current etc.). The convolutions with $\mathscr{E}_k^+$ and $\mathscr{E}_k^-$ give the limiting amplitudes $u^+(x)$ and $u^-(x)$ of a diverging wave and a converging wave respectively. This is due to the fact that the solution of (2.11) is not unique and, in order to obtain a unique solution, we must prescribe "boundary conditions" as $|x| \to \infty$. The amplitude $u^+(x)$ satisfies the Sommerfeld condition for lack of radiation (from infinity) and corresponds to the wave which is radiated by a source with density $f(x)$. In contrast, $u^-(x)$ corresponds to the wave which comes from infinity and is absorbed by the medium with an absorbing intensity $f(x)$. For a detailed account of this topic the reader is referred to Grushin [1963a], Tikhonov and Samarskij [1951], Vajnberg [1966].

All the fundamental solutions (2.6′), (2.7′), (2.10′) and (2.11′) will be obtained by a unified method in Chap. 4. For the other fundamental solutions (2.8′) and (2.9′) the reader is referred to Hörmander [1983, 1985] and Vladimirov [1981].

2.3. The Propagation of Waves. By formula (2.2), we can obtain a particular solution of equation (2.6) when $f(x) \in C_0^\infty(\mathbb{R}^{k+1})$. Thus, from (2.6′) and (2.2) we obtain such a solution in the form (1.57):

$$u_k^{\pm}(x, t) = \langle \mathscr{E}_k^{\pm}(x - y, t - \tau), f(y, \tau)\rangle. \tag{2.12}$$

When $k = 1$ this formula becomes

$$u_1^+(x, t) = \frac{1}{2a}\int_{-\infty}^{t}\left(\int_{|y-x|\leqslant a(t-\tau)} f(y, \tau)\, dy\right) d\tau. \tag{2.12'}$$

Similarly, when $k = 2$,

$$u_2^+(x, t) = \frac{1}{2\pi a}\int_{-\infty}^{t}\left(\int_{|y-x|\leqslant a(t-\tau)} \frac{f(y, \tau)\, dy}{\sqrt{a^2(t-\tau)^2 - |x-y|^2}}\right) d\tau \tag{2.12''}$$

and, when $k = 3$,

$$u_3^+(x, t) = \frac{1}{4\pi a^2 t} \int_{-\infty}^{t} \left(\int_{|y-x|=a(t-\tau)} f(y, \tau)\, dy \right) d\tau. \tag{2.12'''}$$

Remark 2.1. 1) We write $\mathfrak{Q}^+ = \{(x, t) \in \mathbb{R}^{k+1}: |x| \leqslant at\}$ and $\mathscr{D} = \operatorname{supp} f(x, t)$. Then, in view of (1.48), it is clear that $\operatorname{supp} u_k^+ \subset \mathscr{D} + \mathfrak{Q}^+$ for the solutions (2.12′) – (2.12‴), because $\operatorname{supp} \mathscr{E}_k^+ \subset \mathfrak{Q}^+$. But $\mathscr{D} + \mathfrak{Q}^+$ intersects each plane $t = t_0$ lying in $\mathbb{R}^{k+1}$ along a set whose diameter does not exceed $d + at_0$, where d is the diameter of $\mathscr{D}$. Hence $u_k^+(x, t_0) = 0$ for $|x| \geqslant \text{const.} + at_0$, that is, the "wave" $u_k^+(x, t)$ has a sharp front edge and is propagated with a speed $\leqslant a$. For rear edges, on the other hand, a difference occurs according to the value of k. For $k = 3$, $\operatorname{supp} \mathscr{E}_k^+ \subset Q^+ \equiv \partial\mathfrak{Q}^+$ and so $\operatorname{supp} u_3^+ \subset \mathscr{D} + Q^+$. But $\mathscr{D} + Q^+$ is an "inflated" cone Q^+ of thickness $\approx d$. Accordingly, $u_3^+(x, t_0) = 0$ for $|x| \leqslant at_0 - \text{const.}$, that is, the wave u_3^+ has also a sharp rear edge. It turns out (see Corollary 4.1, Chap. 4) that for all odd $k \geqslant 3$ the wave equation (2.6) in $\mathbb{R}^k$ similarly has a retarded fundamental solution $\mathscr{E}_k^+$ concentrated on the cone Q^+ and the solution is unique, by Proposition 6.2, Chap. 5. Consequently, the "retarded" solutions $u_k^+ \equiv \mathscr{E}_k * f$ of the equation (2.6), with $f \in \mathscr{D}(\mathbb{R}^{k+1})$, have sharp front and rear edges. In contrast, $\operatorname{supp} \mathscr{E}_k^+$ coincides with $\mathfrak{Q}^+$ for all even $k \geqslant 2$ and for $k = 1$, and the solutions u_k^+ have, in general, no rear edges. This is expressed by saying that "wave diffusion" takes place.

2) For the Klein-Gordon equation (2.7), where $k \geqslant 1$, the retarded fundamental solution $\mathscr{E}_k^+$ is concentrated in $\mathfrak{Q}^+$ but not in Q^+ (see formula (5.12), Chap. 4). Consequently, for all k, the "retarded" solutions $u_k^+ \equiv f * \mathscr{E}_k^+$ of (2.7) have sharp front edge and are propagated with a speed $\leqslant a$ and they have, in general, no rear edge.

3) For both the heat equation (2.8) and the Schrödinger equation (2.9) the supports of the retarded fundamental solutions $\mathscr{E}_k^+$ coincide with the half-space $t \geqslant 0$ for all $k \geqslant 1$. Therefore the retarded solutions $u \equiv \mathscr{E}_k^+ * f$ have neither front nor rear edges, that is, they are propagated with infinite speed.

2.4. The Construction of Fundamental Solutions of Ordinary Differential Equations. When $n = 1$, it is easy to construct the (retarded) fundamental solutions for any operator $P \neq 0$. Thus, with $p_m \neq 0$, let $\mathscr{E}_+(x)$ denote a solution of the following Cauchy problem:

$$\begin{cases} P\mathscr{E}_+(x) = \sum\limits_{k=0}^{m} p_k \mathscr{E}_+^{(k)}(x) = 0, \quad x > 0, \\ \mathscr{E}_+(0+) = 0,, \ldots, \mathscr{E}_+^{(m-2)}(0+) = 0, \mathscr{E}_+^{(m-1)}(0+) = 1/p_m. \end{cases} \tag{2.14}$$

Lemma 2.1. *Let $\mathscr{E}_+(x)$ be extended to $x < 0$ by setting it equal to zero and let*

$$\mathscr{E}(x) = \begin{cases} \mathscr{E}_+(x), & x > 0, \\ 0, & x < 0. \end{cases} \tag{2.15}$$

Then $\mathscr{E}(x)$ is a fundamental solution of the operator

$$P = \sum_{k=0}^{m} p_k \frac{d^k}{dx^k}.$$

Proof. By formula (1.10), we obtain

$$\mathscr{E}^{(k)}(x) = \{\mathscr{E}^{(k)}(x)\}, \quad x \in \mathbb{R} \tag{2.16}$$

for $k = 0, \ldots, m-1$ in succession because, by (2.14) and (2.15), $h_k = \mathscr{E}^{(k-1)}(0+) - \mathscr{E}^{(k-1)}(0-) = 0$ for $k = 1, \ldots, m-1$ and, for $k = m$,

$$\mathscr{E}^{(m)}(x) = \{\mathscr{E}^{(m)}(x)\} + \frac{1}{p_m}\delta(x). \tag{2.17}$$

On multiplying (2.16) and (2.17) by p_k and p_m and summing, we obtain (2.1) because for $x \neq 0$ $\mathscr{E}(x)$ satisfies the homogeneous equation

$$P\mathscr{E}(x) = 0. \quad \blacksquare$$

2.5. A Mean Value Theorem. Basic for the investigation of the smoothness of solutions of equations (0.1) is the following *mean value theorem* (Gel'fand and Shilov [1958a], Shilov [1965]).

Take a point $x_0 \in \mathbb{R}^n$ and let $\varphi(x), \hat{\varphi}(x), \hat{\hat{\varphi}}(x) \in \mathscr{D}(\mathbb{R}^n)$. Suppose that $x_0 \in \omega$, where ω is an open domain of $\mathbb{R}^n$. Further, assume that

$$\varphi(x) = 1 \quad \text{for } x \in \omega, \quad \hat{\varphi}(x) = 1 \quad \text{for } x \in \operatorname{supp} \varphi, \quad \hat{\hat{\varphi}}(x) = 1 \quad \text{for } x \in \operatorname{supp} \hat{\varphi}. \tag{2.18}$$

Let $u(x)$ be any solution of (0.1) in a domain $\Omega \supset \operatorname{supp} \hat{\hat{\varphi}} \supset \omega$. We define

$$w(x) = \hat{\hat{\varphi}}(x)(1 - \varphi(x))u(x). \tag{2.19}$$

Then clearly

$$\operatorname{supp} w \subset \operatorname{supp} \hat{\hat{\varphi}} \backslash \omega. \tag{2.20}$$

Lemma 2.2. *Let $\mathscr{E}(x) \in \mathscr{D}'(\mathbb{R}^n)$ be an arbitrary fundamental solution of $P(\partial_x)$. Then, for $x \in \omega$, $u(x)$ is expressed in terms of f and w as follows:*

$$u(x) = \mathscr{E} * (\hat{\varphi} f - \hat{\varphi} P w)(x), \quad x \in \omega. \tag{2.21}$$

Proof. It follows from (0.1) that

$$\begin{aligned} P(\varphi u) &= \hat{\varphi} P(\varphi u) = \hat{\varphi}(f - P((1-\varphi)u)) \\ &= \hat{\varphi} f - \hat{\varphi} P(\hat{\hat{\varphi}}(1-\varphi)u). \end{aligned} \tag{2.22}$$

Then, by Part 1 of Proposition 2.1,

$$\varphi u = \mathscr{E} * (\hat{\varphi} f - \hat{\varphi} P w) \tag{2.23}$$

since $\varphi u \in \mathscr{E}'(\mathbb{R}^n)$. $\blacksquare$

We note that (2.21) expresses the solution $u(x)$, with $x \in \omega$, in terms of the right-hand side of equation (0.1) and the "values" of the solution $u(x)$ in the domain supp $\hat{\hat{\varphi}}\backslash\omega$. Thus (2.21) is an analogue of the Poisson formula for harmonic functions.

Corollary 2.1. *If* $f(x) \in C^\infty(\Omega)$ *and the solution* $u(x)$ *of the equation* (0.1) *is smooth in a neighbourhood of* $\partial\Omega$, *then* $u(x) \in C^\infty(\Omega)$.

Proof. We take a domain $\omega \subset \Omega$ so that $u \in C^\infty(\Omega\backslash\omega)$, and we take functions φ, $\hat{\varphi}$, $\hat{\hat{\varphi}} \in C_0^\infty(\Omega)$ satisfying (2.18). Then $\hat{\varphi} f \in C_0^\infty(\Omega)$ and $w \in C_0^\infty(\Omega)$. Therefore (2.21) implies that $u \in C^\infty(\omega)$. ■

The formula (2.21) also enables us to obtain sharper results on the smoothness of solutions of equations of the form (0.1) (see Grushin [1963b]).

Chapter 2
Fourier Transformation of Generalized Functions

In this chapter we construct decompositions of the form (0.2) for any generalized functions and investigate the properties of these decompositions.

§ 1. Fourier Transformation of Test Functions

1.1. Fourier Transformation of Rapidly Decreasing Functions. Any function $\varphi(x) \in S(\mathbb{R}^n)$ and, in particular, $\varphi \in \mathscr{D}(\mathbb{R}^n)$, can be expressed as a *Fourier integral*

$$\varphi(x) = (2\pi)^{-n} \int_{\mathbb{R}^n} e^{-ix\xi}\tilde{\varphi}(\xi)\,d\xi := F^{-1}\varphi; \quad x\xi = x_1\xi_1 + \cdots + x_n\xi_n, \tag{1.1}$$

where the "Fourier coefficients" $\tilde{\varphi}(\xi)$ are given by the formula

$$\tilde{\varphi}(\xi) = \int_{\mathbb{R}^n} e^{ix\xi}\varphi(x)\,dx := F\varphi, \quad \xi \in \mathbb{R}^n. \tag{1.2}$$

Here we denote by F the transformation operator $\varphi \mapsto \tilde{\varphi}$ and by F^{-1} the inverse operator $\tilde{\varphi} \mapsto \varphi$.

1.2. Properties of the Fourier Transformation

1) *The differentiation formulae.* It follows from (1.1) and (1.2) that $\partial_x^\alpha\varphi(x) = (2\pi)^{-n}\int(-i\xi)^\alpha e^{-ix\xi}\tilde{\varphi}(\xi)\,d\xi$ and $\partial_\xi^\alpha\tilde{\varphi}(\xi) = \int(ix)^\alpha e^{ix\xi}\varphi(x)\,dx$ for all $\alpha = (\alpha_1, \ldots, \alpha_n)$. Therefore, for $\varphi \in S$,

$$F\partial_x^\alpha\varphi = (-i\xi)^\alpha\tilde{\varphi}(\xi), \quad F(x^\alpha\varphi(x)) = (-i\partial_\xi)^\alpha\tilde{\varphi}(\xi). \tag{1.3}$$

2) *The translation formulae.* The formula (1.1) implies that

$$F(\varphi(x+a)) = e^{-ia\xi}\tilde{\varphi}(\xi), \quad F(e^{iax}\varphi(x)) = \tilde{\varphi}(\xi + a) \tag{1.4}$$

for $\varphi \in S(\mathbb{R}^n)$ and all $a \in \mathbb{R}^n$.

3) **Proposition 1.1** (Gel'fand and Shilov [1958a], Shilov [1965]). *F is an isomorphism of $S(\mathbb{R}^n)$ onto $S(\mathbb{R}^n)$.*

4) For $\varphi, \psi \in S(\mathbb{R}^n)$, we define the convolution $\varphi * \psi$ by the formula (1.45), Chap. 1. Then, for $\xi \in \mathbb{R}^n$, by the Fubini theorem we have

$$F(\varphi * \psi)(\xi) = \int e^{ix\xi}\left(\int \varphi(x-y)\psi(y)\,dy\right)dx$$

$$= \int e^{i\xi y}\left(\int e^{i\xi(x-y)}\varphi(x-y)\,dx\right)\psi(y)\,dy = \tilde{\varphi}(\xi)\tilde{\psi}(\xi),$$

$$F(\varphi\psi)(\xi) = \int e^{ix\xi}\varphi(x)\left((2\pi)^{-n}\int e^{-ix\eta}\tilde{\psi}(\eta)\,d\eta\right)dx = (2\pi)^{-n}(\tilde{\varphi} * \tilde{\psi})(\xi).$$

Thus, for $\varphi, \psi \in S(\mathbb{R}^n)$, we obtain

$$F(\varphi * \psi)(\xi) = \tilde{\varphi}(\xi)\tilde{\psi}(\xi), \quad F(\varphi\psi)(\xi) = (2\pi)^{-n}(\tilde{\varphi} * \tilde{\psi})(\xi), \quad \xi \in \mathbb{R}^n. \tag{1.5}$$

1.3. Fourier Transformation of Functions with Compact Support. If $\varphi(x) \in \mathcal{D}(\mathbb{R}^n)$, the formula (1.2) shows that $\tilde{\varphi}(\xi)$ is an entire function on $\mathbb{C}^n$, and

$$|\tilde{\varphi}(\xi)| \leqslant Ce^{A|\operatorname{Im}\xi|}, \quad \xi \in \mathbb{C}^n, \tag{1.6}$$

provided that $\varphi(x) \equiv 0$ for $|x| > A$. We write $\mathcal{D}_A \equiv \{\varphi \in \mathcal{D}(\mathbb{R}^n): \varphi(x) \equiv 0$ for $|x| > A\}$. If $\varphi \in \mathcal{D}_A$, then $x^\alpha\varphi(x) \in \mathcal{D}_A$ and $\partial_x^\alpha\varphi(x) \in \mathcal{D}_A$. In view of (1.3), the functions $\partial_\xi^\alpha\tilde{\varphi}(\xi)$ and $\xi^\alpha\tilde{\varphi}(\xi)$ also satisfy estimates of the form (1.6).

Theorem 1.1 (Paley-Wiener; see Gel'fand and Shilov [1958a], Hörmander [1963], Vladimirov [1964]). *If $\varphi \in \mathcal{D}(\mathbb{R}^n)$, then $\tilde{\varphi}(\xi)$ is an entire function on $\mathbb{C}^n$ and, for all α and $N > 0$,*

$$(1 + |\xi|)^N |\partial_\xi^\alpha \tilde{\varphi}(\xi)| \leqslant C_{\alpha,N}e^{A|\operatorname{Im}\xi|}, \quad \xi \in \mathbb{C}^n, \tag{1.7}$$

for some A and $C_{\alpha,N} \geqslant 0$ which depend on φ. Conversely, if $\tilde{\varphi}(\xi)$ is an entire function on $\mathbb{C}^n$ which satisfies estimates of the form (1.57) *for all α and N, then $\varphi(x) \equiv F^{-1}\tilde{\varphi} \in \mathcal{D}_A(\mathbb{R}^n)$.*

Definition 1.1. We denote by $Z = Z(\mathbb{C}^n)$ the space of entire functions $\tilde{\varphi}$ in $\mathbb{C}^n$ which satisfy estimates of the form (1.57) for all α and N. By definition, a sequence $\psi_k \in Z$ converges to $0 \in Z$ if

$$|\psi_k|_{\alpha,N} := \sup_{\xi \in \mathbb{C}^n} e^{-A|\operatorname{Im}\xi|}(1 + |\xi|)^N |\partial_\xi^\alpha \psi_k(\xi)| \underset{k\to\infty}{\to} 0 \ \forall \alpha, N \tag{1.8}$$

with some $A > 0$, and we then write $\psi_k \overset{Z}{\to} 0$.

Theorem 1.1′ (Gel'fand and Shilov [1958a, 1958b]). *The operator $F: \mathcal{D} \to Z$ is continuous and so is its inverse $F^{-1}: Z \to \mathcal{D}$.*

For $\varphi \in \mathscr{D}$ all the properties (1.3) and (1.4) of the Fourier transformation hold, and the second relation of (1.4) is valid for all $a \in \mathbb{C}^n$.

§ 2. Fourier Transformation of Tempered Generalized Functions

2.1. Closure of the Fourier Transformation with Respect to Continuity. The Fourier transformation $F: S \to S$ can be extended by continuity to $S'(\mathbb{R}^n)$. To do this, we observe that for any $u \in S(\mathbb{R}^n)$ the *Parseval identity*

$$\langle \tilde{u}(\xi), \varphi(\xi) \rangle = \langle u(x), \tilde{\varphi}(x) \rangle \quad \forall \varphi \in S(\mathbb{R}^n), \tag{2.1}$$

holds, as a direct consequence of (1.2). By taking the closure with respect to continuity, we have

Definition 2.1. For $u(x) \in S'(\mathbb{R}^n)$ the functional $Fu = \tilde{u}(\xi)$ is defined by the identity (2.1).

Proposition 2.1. *The operator $F: u \mapsto \tilde{u}$ acts continuously from $S'(\mathbb{R}^n)$ into $S'(\mathbb{R}^n)$. Similarly, $F^{-1}: \tilde{u} \mapsto u$ is also a continuous map from $S'(\mathbb{R}^n)$ onto $S'(\mathbb{R}^n)$.*

Example 2.1. 1) On taking $u(x) = \delta(x)$ and using (1.2), we have

$$\langle \tilde{\delta}, \varphi \rangle = \langle \delta, \tilde{\varphi} \rangle = \tilde{\varphi}(0) = \int_{\xi \in \mathbb{R}^n} \varphi(\xi)\, d\xi = \langle 1, \varphi \rangle \Rightarrow \tilde{\delta}(\xi) \equiv 1. \tag{2.2}$$

2) Conversely, for $u(x) \equiv 1$,

$$\langle \tilde{1}, \varphi \rangle = \langle 1, \tilde{\varphi} \rangle = \int \tilde{\varphi}(x)\, dx = (2\pi)^n \varphi(0) \Rightarrow \tilde{1} = (2\pi)^n \delta(x), \tag{2.3}$$

by (1.1).

2.2. Properties of the Fourier Transformation

1) By taking closure with respect to continuity, we find from (1.3) that, for all $u \in S'(\mathbb{R}^n)$,

$$F(\partial_x^\alpha u(x)) = (-i\xi)^\alpha \tilde{u}(\xi), \quad F(x^\alpha u(x)) = (-i\partial_\xi)^\alpha \tilde{u}(\xi) \quad \forall \alpha. \tag{2.4}$$

Corollary 2.1. *For any operator $P(\partial_x)$ of the form* (0.1)

$$F(P(\partial_x)u(x)) = P(-i\xi)\tilde{u}(\xi) = \tilde{P}(\xi)\tilde{u}(\xi), \quad u \in S'(\mathbb{R}^n), \tag{2.5}$$

where $P(-i\xi) = \tilde{P}(\xi)$ is the polynomial

$$P(-i\xi) = \tilde{P}(\xi) = \sum_{|\alpha| \leqslant m} p_\alpha (-i\xi)^\alpha, \quad \xi \in \mathbb{R}^n. \tag{2.6}$$

Definition 2.2. The polynomial $\tilde{P}(\xi) \equiv P(-i\xi)$ is called the *symbol* of the operator $P(\partial_x)$.

The formulae (2.2) and (2.3), together with (2.4), imply that

$$F(\delta^{(\alpha)}(x)) = (-i\xi)^\alpha, \quad Fx^\alpha = (2\pi)^n(-i)^{|\alpha|}\delta^{(\alpha)}(\xi) \quad \forall\alpha. \tag{2.7}$$

2) By taking the closure with respect to continuity, we find from (1.4) that

$$F(u(x+a)) = e^{-ia\xi}\tilde{u}(\xi), \quad F(e^{iax}u(x)) = \tilde{u}(\xi + a), \quad a \in \mathbb{R}^n, \quad u \in S'(\mathbb{R}^n). \tag{2.8}$$

On applying (2.2) and (2.3), we obtain from these results, for example, the formulae

$$F(\delta(x+a)) = e^{-ia\xi}, \quad F(e^{iax}) = (2\pi)^n\delta(\xi + a), \quad a \in \mathbb{R}^n. \tag{2.9}$$

3) By taking the closure with respect to continuity, we can deduce easily from (1.5) that

$$F(u * v) = \tilde{u}(\xi)\tilde{v}(\xi) \quad \text{for } (u, v) \in S \times S' \text{ (or } S' \times S). \tag{2.10}$$

Similarly, we can also deduce from (1.5) that

$$F(uv) = (2\pi)^{-n}\tilde{u} * \tilde{v} \quad \text{for } u \in S'(\mathbb{R}^n) \text{ and } v \in S(\mathbb{R}^n). \tag{2.11}$$

Here the convolution $\tilde{u} * \tilde{v}$ can be defined, for example, by the formula (1.57) in Chap. 1.

4) *The Parseval theory.* Note that $F\colon S'(\mathbb{R}^n) \to S'(\mathbb{R}^n)$ is an isomorphism, by Proposition 2.1. In particular, $F\colon L_2(\mathbb{R}^n) \to S'(\mathbb{R}^n)$.

Theorem 2.1 (Parseval; see Gel'fand and Shilov [1958a], Shilov [1965], Titchmarsh [1937]). *The operator F is an isomorphism from $L_2(\mathbb{R}^n)$ onto $L_2(\mathbb{R}^n)$, and*

$$\int |\tilde{u}(\xi)|^2\,d\xi = (2\pi)^n \int |u(x)|^2\,dx. \tag{2.12}$$

5) *Linear change of variables.* With $C \in GL_{\mathbb{R}}(n)$ and $u(x) \in S'(\mathbb{R}^n)$, we define $u_C(y) = u(Cy)$ for $y \in \mathbb{R}^n$ (see formula (1.33), Chap. 1). Then

$$\tilde{u}_C(\eta) = \frac{1}{|\det C|}\tilde{u}((C^{-1})^t\eta), \quad \eta \in \mathbb{R}^n. \tag{2.13}$$

In particular,

$$\tilde{u}_C(\eta) = \tilde{u}(C\eta) = (\tilde{u})_C(\eta), \quad \eta \in \mathbb{R}^n, \tag{2.14}$$

if $C \in O(n)$.

When $u(x) \in S$ the formula (2.13) follows from (1.2) since

$$\begin{aligned} \tilde{u}_C(\eta) &= \int e^{iy\eta}u(Cy)\,dy \\ &= \int e^{i(C^{-1}x)\eta}u(x)\frac{dx}{|\det C|} = \frac{1}{|\det C|}\tilde{u}((C^{-1})^t\eta), \quad \eta \in \mathbb{R}^n, \end{aligned}$$

while for $u \in S'$ (2.13) follows by taking the closure with respect to continuity.

Corollary 2.2. *The formula* (2.14) *implies that under an orthogonal change of variables in* $\mathbb{R}^n$ *the Fourier transform* $\tilde{u}(\xi)$ *of a generalized function* $u(x) \in S'(\mathbb{R}^n)$ *transforms as an invariantly defined generalized function on* $\mathbb{R}^n$. *Therefore, by* (2.5), *the symbol* $\tilde{P}(\xi)$ *of the differential operator* $P(\partial_x)$ *is also transformed under an orthogonal change of variables as a function which is invariantly defined on* $\mathbb{R}^n$.

On the other hand, under a general change of variables $C \in GL_{\mathbb{R}}(n)$ we find from (2.13) and (2.5) that the symbol $\tilde{P}_C$ of the operator $P(\partial_x)$ takes the form

$$\tilde{P}_C(\eta) = \tilde{P}((C^{-1})^t\eta), \quad \eta \in \mathbb{R}^n, \tag{2.15}$$

in the coordinates $y = C^{-1}x$.

Corollary 2.3. *The symbol* $\tilde{P}$ *of the operator* P *is an invariantly defined function on the space* $\mathbb{R}^{n*}$, *the dual space of* $\mathbb{R}^n$ *which consists of* (*continuous*) *linear functionals on* $\mathbb{R}^n$.

Proof. Let $l: \mathbb{R}^n \to \mathbb{R}$ be a linear functional defined on the coordinates $(x_1, \dots, x_n)$ by means of the row vector $\xi = (\xi_1, \dots, \xi_n) \in \mathbb{R}^n$, that is, $l(x) = \xi x = \xi_1 x_1 + \cdots + \xi_n x_n$. Then, in terms of the coordinates $y = C^{-1}x$, the same functional l has the form

$$l_C(y) = l(x) = \xi \cdot Cy = C^t\xi \cdot y = \eta \cdot y. \tag{2.16}$$

Thus l_C is defined by the row vector $\eta = C^t\xi$. But then (2.15) implies that

$$\tilde{P}_C(\eta) = \tilde{P}_C(C^t\xi) = \tilde{P}(\xi), \quad \xi \in \mathbb{R}^n. \tag{2.17}$$

This means that the values of $\tilde{P}_C$ and P coincide on the same functional $l \in \mathbb{R}^{n*}$. ■

2.3. Methods for Computing Fourier Transforms. The formula (1.2) can be used for obtaining Fourier transforms of a wide class of generalized functions, not only for $\varphi \in S(\mathbb{R}^n)$.

Lemma 2.1. *If* $u(x) \in L_1(\mathbb{R}^n)$, *then* $u(x) \in S'(\mathbb{R}^n)$ *and* $\tilde{u}(\xi) \in C_b(\mathbb{R}^n)$. *Furthermore* (*compare with* (1.2)),

$$\begin{gathered}(Fu)(\xi) = \tilde{u}(\xi) = \int e^{ix\xi}u(x)\,dx, \quad \xi \in \mathbb{R}^n; \\ \max_{\xi \in \mathbb{R}^n} |\tilde{u}(\xi)| \leqslant \int |u(x)|\,dx.\end{gathered} \tag{2.18}$$

Proof. Using Definition 2.1, the formula (1.2) and Fubini's theorem, we have

$$\langle \tilde{u}, \varphi \rangle = \int u(x)\left(\int e^{ix\xi}\varphi(\xi)\,d\xi\right)dx = \int\left(\int e^{ix\xi}u(x)\,dx\right)\varphi(\xi)\,d\xi$$

for all $\varphi \in S(\mathbb{R}^n)$. This gives (2.18), and consequently $\tilde{u} \in C_b(\mathbb{R}^n)$. ■

Corollary 2.4. *If $u(x) \in L_1^{\mathrm{loc}}(\mathbb{R}^n)$ and*

$$\int \frac{|u(x)|}{(1+|x|^2)^N}\,dx < \infty \tag{2.18$'$}$$

for some $N = 1, 2, \ldots$, then $u(x) \in S'(\mathbb{R}^n)$, and

$$Fu = \tilde{u}(\xi) = (1 - \Delta_\xi)^N \int e^{ix\xi} \frac{u(x)}{(1+|x|^2)^N}\,dx, \tag{2.18$''$}$$

This last result follows from the second formula in (2.4); *here the derivatives involved are in the sense of generalized functions.*

Example 2.2. If $u(x) \in L_2(\mathbb{R})$, then $\dfrac{u(x)}{x+i} \in L_1(\mathbb{R})$ by the Cauchy-Bunyakovskij inequality. Therefore, as for (2.18″),

$$\tilde{u}(\xi) = \left(-i\frac{d}{d\xi} + i\right) \int e^{ix\xi} \frac{u(x)}{x+i}\,dx, \quad \xi \in \mathbb{R}. \tag{2.18$'''$}$$

(Compare with Titchmarsh [1937].)

Lemma 2.2 (Hörmander [1983, 1985]). *If $u(x) \in \mathscr{E}'(\mathbb{R}^n)$, then $\tilde{u}(\xi) \in C^\infty(\mathbb{R}^n)$, and*

$$\tilde{u}(\xi) = \langle u(x), e^{ix\xi} \rangle, \quad \xi \in \mathbb{R}^n, \tag{2.19}$$

where the scalar product is taken in the sense of the formula (1.40) *in Chap.* 1.

Example 2.3.

$$\begin{aligned} F(\delta^{(\alpha)}(x))(\xi) &= \langle \delta^{(\alpha)}(x), e^{ix\xi} \rangle = (-i\xi)^\alpha; \\ F(\delta(x-h))(\xi) &= \langle \delta(x-h), e^{ix\xi} \rangle = e^{ih\xi}. \end{aligned} \tag{2.19$'$}$$

2.4. Examples of the Computation of Fourier Transforms

1) For the one-dimensional Gaussian distribution with dispersion $\sigma > 0$

$$F\frac{e^{-(x^2/2\sigma)}}{\sqrt{2\pi\sigma}} = \int \frac{e^{-(x^2/2\sigma)+ix\xi}}{\sqrt{2\pi\sigma}}\,dx = \int \frac{e^{-((x-i\sigma\xi)^2/2\sigma)-(\sigma\xi^2/2)}}{\sqrt{2\pi\sigma}}\,dx = e^{-(\sigma\xi^2/2)}. \tag{2.20}$$

2) If we close the contour of integration from above (in the domain $\operatorname{Im} x > 0$) when $\xi > 0$ and from below (in the domain $\operatorname{Im} x < 0$) when $\xi < 0$, we obtain

$$F\frac{1}{x-a} = \begin{cases} 2\pi i\theta(\xi)e^{ia\xi} & \text{if } \operatorname{Im} a > 0, \\ -2\pi i\theta(-\xi)e^{ia\xi} & \text{if } \operatorname{Im} a < 0 \end{cases} \tag{2.21}$$

for $a \in \mathbb{C}\backslash\mathbb{R}$, by using the Cauchy residue theorem, as in the case of (2.18″).

3) For the Cauchy distribution $\dfrac{1}{1+x^2}$, the last formula gives

$$\begin{aligned} F\frac{1}{1+x^2} &= \frac{1}{2i}F\left(\frac{1}{x-i} - \frac{1}{x+i}\right) \\ &= \pi\theta(\xi)e^{-\xi} + \pi\theta(-\xi)e^{\xi} = \pi e^{-|\xi|}. \end{aligned} \tag{2.22}$$

4) For Re $\lambda > 0$, we have

$$F(\theta(x)e^{-\lambda x}) = \int_0^\infty e^{-(\lambda - i\xi)x}\,dx = i\frac{1}{\xi + i\lambda} \tag{2.23}$$

and, likewise,

$$F(\theta(-x)e^{\lambda x}) = -i\frac{1}{\xi - i\lambda}. \tag{2.23'}$$

5) Since the Fourier transformation is continuous, the previous results imply that, as $\lambda \to 0+$,

$$F(\theta(x))(\xi) = i\frac{1}{\xi + i0}$$

and, similarly,

$$F(\theta(-x))(\xi) = -i\frac{1}{\xi - i0}. \tag{2.24}$$

6) Adding these two relations, we obtain the *Sokhotskij-Plemelj "jump" formula*

$$F(1)(\xi) = 2\pi\delta(\xi) = i\left(\frac{1}{\xi + i0} - \frac{1}{\xi - i0}\right) \tag{2.25}$$

(see (1.62) in Chap. 1).

7) In view of (2.4), the formula (2.23) yields

$$F(\theta(x)x^k e^{-\lambda x})(\xi) = \left(-i\frac{d}{d\xi}\right)^k i\frac{1}{\xi + i\lambda} = \frac{i^{k+1}k!}{(\xi + i\lambda)^{k+1}} \tag{2.26}$$

for Re $\lambda > 0$ and $k = 0, 1, \ldots$, and, similarly, (2.23′) yields

$$F(\theta(-x)x^k e^{\lambda x})(\xi) = -\frac{i^{k+1}k!}{(\xi - i\lambda)^{k+1}}. \tag{2.27}$$

§ 3. The Sobolev Function Spaces

Definition 3.1 (Agranovich and Vishik [1964], Hörmander [1983, 1985], Sobolev [1936], Volevich and Paneyakh [1965]). For $s \in \mathbb{R}$, we denote by $H_s = H_s(\mathbb{R}^n)$ the space of generalized functions $u(x) \in S'(\mathbb{R}^n)$ for which $\tilde{u}(\xi) \in L_1^{\text{loc}}(\mathbb{R}^n)$ and

$$\|u\|_s^2 = \int (1 + |\xi|)^{2s} |\tilde{u}(\xi)|^2\,d\xi < \infty. \tag{3.1}$$

Example 3.1. $\delta(x) \in H_s(\mathbb{R}^n)$ for $s < -n/2$, because $\tilde{\delta}(\xi) \equiv 1$.

We note that H_s is a complete Hilbert space because the map $u(x) \mapsto (1+|\xi|)^s \tilde{u}(\xi)$ is an isometry from $H_s(\mathbb{R}^n)$ onto $L_2(\mathbb{R}^n)$. Clearly, $H_{s_1}(\mathbb{R}^n) \subset H_{s_2}(\mathbb{R}^n)$ is a continuous embedding if $s_1 \geqslant s_2$. If s is a non-negative integer l, then $u(x) \in H_s$ if and only if $\partial_x^\alpha u(x) \in L_2(\mathbb{R}^n)$ for $|\alpha| \leqslant l$, and then the metric (3.1) is equivalent to the metric

$$|||u|||_l^2 = \sum_{|\alpha| \leqslant l} \int_{\mathbb{R}^n} |\partial_x^\alpha u(x)|^2 \, dx \tag{3.2}$$

(see Agranovich and Vishik [1964], Volevich and Paneyakh [1965]). This result is a consequence of Theorem 2.1 and (2.4). It can be shown easily that $\mathscr{D}(\mathbb{R}^n)$ is everywhere dense in $H_s(\mathbb{R}^n)$ for $s \in \mathbb{R}$ (see Hörmander [1983, 1985]).

Proposition 3.1. *For $s \in \mathbb{R}$, the operator P of the form* (0.1) *acts continuously from $H_s(\mathbb{R}^n)$ into $H_{s-m}(\mathbb{R}^n)$.*

Proof. For any $u \in H_s(\mathbb{R}^n)$, we have, by (2.5),

$$\begin{aligned} \|Pu\|_{s-m}^2 &= \int (1+|\xi|)^{2(s-m)} |\widetilde{Pu}(\xi)|^2 \, d\xi \\ &= \int (1+|\xi|)^{2(s-m)} |\tilde{P}(\xi)|^2 |\tilde{u}(\xi)|^2 \, d\xi \\ &\leqslant C \int (1+|\xi|)^{2s} |\tilde{u}(\xi)|^2 = C \|u\|_s^2. \quad \blacksquare \end{aligned}$$

§4. Fourier Transformation of Rapidly Growing Generalized Functions

4.1. Functionals on the Space $Z(\mathbb{C}^n)$

Definition 4.1. We denote by $Z' = Z'(\mathbb{C}^n)$ the space of continuous linear functionals on $Z(\mathbb{C}^n)$ (see Definition 1.1).

For example, if $w(\xi) \in S'(\mathbb{R}^n)$, then the scalar product $\langle w(\xi), \varphi|_{\mathbb{R}^n} \rangle$ is defined for $\varphi \in Z(\mathbb{C}^n)$ since $\varphi|_{\mathbb{R}^n} \in S(\mathbb{R}^n)$, by (1.7). We write

$$\langle v(z), \varphi(z) \rangle = \langle w(\xi), \varphi|_{\mathbb{R}^n} \rangle, \quad \varphi \in Z(\mathbb{C}^n). \tag{4.1}$$

Then $v(z) \in Z'(\mathbb{C}^n)$ because the map $\varphi \mapsto \varphi|_{\mathbb{R}^n}$ is a continuous embedding $Z(\mathbb{C}^n) \subset S(\mathbb{R}^n)$. Also, $Z(\mathbb{C}^n)$ is everywhere dense in $S(\mathbb{R}^n)$, since $\mathscr{D}(\mathbb{R}^n)$ is everywhere dense in $S(\mathbb{R}^n)$ and $F\colon S(\mathbb{R}^n) \to S(\mathbb{R}^n)$ is an isomorphism, by Proposition 1.1 and, further, $F\colon \mathscr{D}(\mathbb{R}^n) \to Z(\mathbb{C}^n)$ is an isomorphism, by Theorem 1.1′. Consequently, the dual map $S'(\mathbb{R}^n) \to Z'(\mathbb{C}^n)$ which maps w into v by formula (4.1) is also an embedding. Thus $S'(\mathbb{R}^n)$ is a subset of $Z'(\mathbb{C}^n)$.

Let us construct more general functionals belonging to $Z'(\mathbb{C}^n)$. We take an element $w(\xi, \eta) \in S'(\mathbb{R}^{2n})$, where $\mathbb{R}^{2n}$ denotes the usual formulation of the complex space $\mathbb{C}^n$ as a real space. We assume that

$$\operatorname{supp} w \subset T_B$$

for some $B \geqslant 0$ where

$$T_B = \{z = \xi + i\eta \in \mathbb{C}^n : |\eta| \leqslant B, \xi, \eta \in \mathbb{R}^n\}. \tag{4.2}$$

Next we take a cut-off function $\zeta_B(\eta)$ with the following properties:

$$\zeta_B(\eta) \in \mathbb{C}_0^\infty(\mathbb{R}^n), \quad \zeta_B(\eta) = 1 \quad \text{if } |\eta| \leqslant B, \quad \zeta_B(\eta) = 0 \quad \text{if } |\eta| \geqslant B + 1. \tag{4.3}$$

Then the estimates (1.7) imply that $\zeta_B(\eta)\varphi(\xi + i\eta) \in S(\mathbb{R}^{2n})$ for $\varphi \in Z(\mathbb{C}^n)$. We define the scalar product

$$\langle w(\xi, \eta), \varphi(\xi + i\eta)\rangle = \langle w(\xi, \eta), \zeta_B(\eta)\varphi(\xi + i\eta)\rangle, \quad \varphi \in Z(\mathbb{C}^n), \tag{4.4}$$

where $\langle \cdot, \cdot \rangle$ on the right-hand side denotes the duality between $S'(\mathbb{R}^{2n})$ and $S(\mathbb{R}^{2n})$.

The functional (4.4) does not depend on the choice of ζ_B with the required properties (4.3) and, by Definition 1.1, it is continuous on the space $Z(\mathbb{C}^n)$.

Definition 4.2. A functional $v(z) \in Z'(\mathbb{C}^n)$ has a local density $w(\xi, \eta)$ on the set T_B if it can be represented as

$$\langle v(z), \varphi(z)\rangle = \langle w(\xi, \eta), \varphi(\xi + i\eta)\rangle, \quad \varphi \in Z(\mathbb{C}^n), \tag{4.5}$$

where $w(\xi, \eta) \in S'(\mathbb{R}^{2n})$ and $\operatorname{supp} w \subset T_B$.

Example 4.1. $\langle \delta_{2n}(\xi, \eta), \varphi(\xi + i\eta)\rangle = \varphi(0)$, $\langle \delta_n(\eta), \varphi(\xi + i\eta)\rangle = \int_{\mathbb{R}^n} \varphi(\xi)\, d\xi$.
Note that the functional (4.1) has also the local density $w(\xi, \eta) = w(\xi) \times \delta_n(\eta)$.

4.2. Fourier Transformation on the Space $\mathscr{D}'(\mathbb{R}^n)$

Proposition 4.1. *The operator* $F\colon S'(\mathbb{R}^n) \to S'(\mathbb{R}^n)$ *defined by the identity* (2.1) *can be extended to a continuous map* $F\colon \mathscr{D}'(\mathbb{R}^n) \to Z'(\mathbb{C}^n)$ *in such a way that the diagram*

$$\begin{array}{ccc} S'(\mathbb{R}^n) & \xrightarrow{F} & S'(\mathbb{R}^n) \\ \cap & & \cap \\ \mathscr{D}'(\mathbb{R}^n) & \xrightarrow{F} & Z'(\mathbb{R}^n) \end{array}$$

becomes commutative. Here the embedding $S' \subset Z'$ *is defined by* (4.1).

Proof. Definition 2.1 implies that, for $u \in S'$,

$$\langle \tilde{u}(\xi), \varphi(\xi)\rangle = \langle u(x), \tilde{\varphi}(x)\rangle, \quad \varphi \in Z(\mathbb{C}^n). \tag{4.6}$$

It remains to note that the map $F\colon \varphi \mapsto \tilde{\varphi}$ acts continuously from $Z(\mathbb{C}^n)$ into $\mathscr{D}(\mathbb{C}^n)$ as does the operator F^{-1} (see Theorem 1.1′). ■

If we take the closure with respect to continuity in (4.6) we are led to the next definition.

Definition 4.3. For $u \in \mathscr{D}'$ define u by the identity (4.6).

The identity (4.6) and Theorem 1.1′ together lead to

Proposition 4.1′. *The operators* $F\colon \mathscr{D}'(\mathbb{R}^n) \to Z'(\mathbb{C}^n)$ *and* $F^{-1}\colon Z'(\mathbb{C}^n) \to \mathscr{D}'(\mathbb{R}^n)$ *are continuous.*

4.3. Operations on the Space $Z'(\mathbb{C}^n)$. Since $\mathscr{D}(\mathbb{R}^n)$ is dense in $\mathscr{D}'(\mathbb{R}^n)$ and $F\colon \mathscr{D}' \to Z'$ is an isomorphism, it follows that $Z = F(\mathscr{D})$ is dense in Z'. On Z there are defined the operations of differentiation, multiplication by a polynomial and by e^{iax}, where $a \in \mathbb{R}^n$, linear real coordinate transformations and the operation of translation by a complex vector. By continuity all these operations can be extended to $v \in Z'$ by means of the following identities, in which $\varphi \in Z(\mathbb{C}^n)$:

$$\begin{aligned}
\langle \partial_z^\alpha v(z), \varphi(z)\rangle &= \langle v(z), (-\partial_z)^\alpha \varphi(z)\rangle \quad \forall \alpha = (\alpha_1, \ldots, \alpha_n),\\
\langle z^\alpha v(z), \varphi(z)\rangle &= \langle v(z), z^\alpha \varphi(z)\rangle,\\
\langle e^{iaz} v(z), \varphi(z)\rangle &= \langle v(z), e^{iaz}\varphi(z)\rangle, \quad a \in \mathbb{R}^n,\\
\langle v(z+a), \varphi(z)\rangle &= \langle v(\zeta), \varphi(\zeta - a)\rangle, \quad a \in \mathbb{C}^n,\\
\langle v(Cz), \varphi(z)\rangle &= \left\langle v(\zeta), \frac{1}{|\det C|}\varphi(C^{-1}\zeta)\right\rangle, \quad C \in GL_{\mathbb{R}}(n).
\end{aligned} \tag{4.7}$$

4.4. Properties of the Fourier Transformation. By taking the closure with respect to continuity, we obtain from (2.4), (2.5) and (2.8) the following identities for $u \in \mathscr{D}'(\mathbb{R}^n)$:

$$\begin{aligned}
F(\partial_x^\alpha u) = (-iz)^\alpha \tilde{u}(z), \quad F(x^\alpha u) &= (-i\partial_z)^\alpha \tilde{u}(z) \forall \alpha,\\
F(P(\partial_x)u) &= \tilde{P}(z)\tilde{u}(z),\\
F(u(x+a)) &= e^{-iaz}\tilde{u}(z), \quad a \in \mathbb{R}^n,\\
F(e^{iax}u(x)) &= \tilde{u}(z+a), \quad a \in \mathbb{C}^n.
\end{aligned} \tag{4.8}$$

We note that the last identity is valid for all $a \in \mathbb{C}^n$ and $u \in \mathscr{D}'(\mathbb{R}^n)$ (as well as for $u \in \mathscr{D}(\mathbb{R}^n)$) in contrast to the second formula of (2.8).

4.5. Analytic Functionals. By this term we mean functionals $v(z) \in Z'$ having analytic local density (Gel'fand and Shilov [1958a]). More precisely, let $f(z)$ be a holomorphic function of $z \in T_\Omega$, where Ω is a connected domain in $\mathbb{R}^n$ and

$$T_\Omega := \{z = \xi + i\eta \in \mathbb{C}^n\colon \eta = \operatorname{Im} z \in \Omega\} \tag{4.9}$$

is a tubular domain of $\mathbb{C}^n$ with base Ω. We assume that

$$|f(z)| \leqslant C(1+|z|)^{\mu} \quad \text{for } z \in T_{\Omega}, |\operatorname{Im} z| \leqslant B \tag{4.10}$$

for all $B > 0$ and some $C = C_B$ and $\mu = \mu_B$. We can then define an *analytic functional* by the formula (see Gel'fand and Shilov [1958a])

$$\langle f|_{T_{\Omega}}, \varphi \rangle = \int_{\operatorname{Im} z = \eta} f(z)\varphi(z)\, dz, \quad \varphi \in Z; \quad \eta \in \Omega. \tag{4.11}$$

This functional does not depend on $\eta \in \Omega$, as can be established easily using the Cauchy theorem together with the estimates (4.10) and (1.7) and the fact that Ω is a connected domain. Note that Ω and T_{Ω} can be assumed to be convex, by the Bochner theorem (Gunning and Rossi [1965], Hörmander [1973], Vladimirov [1964]).

Example 4.2. Let $n = 1$, $f(z) = 1/z$ and $T_{\Omega} = \{z \in \mathbb{C}\colon \operatorname{Im} z > 0\}$. Then

$$\left\langle \frac{1}{z}\bigg|_{\operatorname{Im} z > 0}, \varphi \right\rangle = \int_{\operatorname{Im} z = \varepsilon} \frac{\varphi(z)}{z}\, dz, \quad \varepsilon > 0. \tag{4.11'}$$

In the domain $\operatorname{Im} z < 0$ the function $1/z$ defines another functional and, by the Cauchy theorem,

$$\frac{1}{z}\bigg|_{\operatorname{Im} z < 0} - \frac{1}{z}\bigg|_{\operatorname{Im} z > 0} = 2\pi i \delta(\xi), \quad \xi \in \mathbb{R}. \tag{4.11''}$$

(Compare with (2.25).)

Proposition 4.2. *If $v \in Z'(\mathbb{C}^n)$ is an analytic functional, so that $v = f|_{T_{\Omega}}$, then, for $z \in T_{\Omega}$, the function $f(z)$ is uniquely determined by v.*

The result follows because the function space $\{\tilde{\varphi}(\xi + i\eta)\colon \tilde{\varphi} \in Z\}$, with $\eta \in \mathbb{R}^n$, is dense in $S(\mathbb{R}^n)$.

Proposition 4.3. *Any functional of the form* (4.5) (*that is, any functional with a local density*) *can be represented as a finite sum of* (*not more than* 2^n) *analytic functionals of the form* (4.11).

Proof. In the one-dimensional case, for any continuous function $w(\xi)$ such that $|w(\xi)| \leqslant C(1+|\xi|)^{\mu}$, with $\xi \in \mathbb{R}$, we claim that

$$\begin{aligned} &\langle w(\xi) \times \delta(\eta), \varphi(\xi + i\eta)\rangle \\ &= \int_{\mathbb{R}} w(\xi)\varphi(\xi)\, d\xi \\ &= \frac{1}{2\pi i}\left(\int_{\operatorname{Im} z = -1} f(z)\varphi(z)\, dz - \int_{\operatorname{Im} z = 1} f(z)\varphi(z)\, dz\right), \end{aligned} \tag{4.12}$$

where

$$f(z) = \int_{\mathbb{R}} \frac{w(\xi)(2i)^{\mu+1}\, d\xi}{(z-\xi)(z-\xi+2i)^{\mu+1}}. \tag{4.13}$$

Indeed, (4.12) and (4.13) follow from the representation

$$\varphi(\xi) = \frac{1}{2\pi i}\left(\int_{\operatorname{Im} z=-1} \frac{\varphi(z)(2i)^{\mu+1}\,dz}{(z-\xi)(z-\xi+2i)^{\mu+1}} - \int_{\operatorname{Im} z=1} \frac{\varphi(z)(2i)^{\mu+1}\,dz}{(z-\xi)(z-\xi+2i)^{\mu+1}}\right). \tag{4.14}$$

A similar construction can be carried out for the general n-dimensional case and for any functional of the form (4.5).

§ 5. The Paley-Wiener Theory

5.1. Fourier Transform of Generalized Functions with Compact Supports. We note that the Fourier transform of the generalized functions $\delta(x)$ and $\delta^{(\alpha)}(x)$ (with compact supports) belonging to $\mathscr{E}'$ are polynomials (see (2.2) and (2.19′)). It turns out that, similarly, the Fourier transform $\tilde{u}(\xi)$ of any distribution $u(x) \in \mathscr{E}'$ is an entire function on $\mathbb{C}^n$. The following theorem sharpens Lemma 2.2.

Theorem 5.1 (Hörmander [1983, 1985], Schwartz [1966], Vladimirov [1964]). *If $u(x) \in \mathscr{E}'(\mathbb{R}^n)$, then $\tilde{u}(\xi) \in C^\infty(\mathbb{R}^n)$ and $\tilde{u}$ can be extended from $\mathbb{R}^n$ as an entire function of $z \in \mathbb{C}^n$ given by*

$$\tilde{u}(z) = \langle u(x), e^{ixz}\rangle, \quad z \in \mathbb{C}^n \tag{5.1}$$

(*compare with* (2.19)). *For some C, A and $\mu \geqslant 0$*

$$|\tilde{u}(z)| \leqslant Ce^{A|\operatorname{Im} z|}(1+|z|)^\mu, \quad z \in \mathbb{C}^n. \tag{5.1′}$$

Conversely, if $\tilde{u}(z)$ is an entire function in $\mathbb{C}^n$ satisfying (5.1′) *for some C, A and μ, then $\tilde{u}(z)|_{\mathbb{R}^n} = \tilde{u}(\xi) \in S'(\mathbb{R}^n)$ and $u \equiv F^{-1}\tilde{u} \in \mathscr{E}'(\mathbb{R}^n)$, and furthermore, $u(x) = 0$ for $|x| > A$.*

Corollary 5.1. $\mathscr{E}'(\mathbb{R}^n) \subset H_{-\infty}(\mathbb{R}^n) = \bigcup_{s\in\mathbb{R}} H_s(\mathbb{R}^n)$.

The result follows because for $\operatorname{Im} z = 0$ (5.1′) implies that $|\tilde{u}(\xi)| \leqslant C(1+|\xi|)^\mu$, $\xi \in \mathbb{R}^n$. Thus $u(x) \in H_s(\mathbb{R}^n)$ for $s < -\mu - n/2$.

5.2. Tempered Distributions with Support in a Cone. Let K be a closed cone in $\mathbb{R}^n$ with vertex at the origin and not containing straight lines. By definition, $K + K \subset K$ and $tK \subset K$ for $t > 0$; thus K is a convex set. An example is $K = \overline{\mathbb{R}}_+ = \{x \in \mathbb{R}: x \geqslant 0\}$. Let K^* be the *dual cone* defined by

$$K^* = \{\eta \in \mathbb{R}^n: \eta x > 0 \text{ for } x \in K\backslash 0\}. \tag{5.2}$$

Evidently, K^* is an open set and $K^* \neq \phi$ if and only if K does not contain straight lines.

Let us take a cut-off function $\psi(x) \in C^\infty(\mathbb{R}^n)$ such that

$$\psi(x) = \begin{cases} 1 & \text{if } x \in K, \\ 0 & \text{if } \rho(x, K) \geqslant 1; \end{cases} \quad \sup_{x\in\mathbb{R}^n} |\psi^{(\alpha)}(x)| < \infty \quad \forall\alpha. \tag{5.3}$$

Clearly,

$$\psi(x)e^{ixz} \in S(\mathbb{R}^n) \quad \text{for } \operatorname{Im} z \in K^*. \tag{5.4}$$

For $u \in S'_K \equiv \{u \in S'(\mathbb{R}^n); \operatorname{supp} u \subset K\}$, we define the scalar product

$$\langle u, e^{ixz} \rangle = \langle u, \psi(x)e^{ixz} \rangle, \quad \operatorname{Im} z \in K^*. \tag{5.5}$$

This definition of the scalar product is independent of the function $\psi(x)$ chosen with the required properties (5.3).

Theorem 5.2 (Hörmander [1983, 1985], Komech [1974], Merzon [1973], Paley and Wiener [1934], Reed and Simon [1975, 1979], Vladimirov [1964]). *If $u \in S'_K$, then $\tilde{u}(\xi) \in S'(\mathbb{R}^n)$ can be extended in the sense of Definition* 1.11, *Chap.* 1, *as an analytic function $\tilde{u}(z)$ in the tubular domain $T_{K^*} = \{z \in \mathbb{C}^n\colon \operatorname{Im} z \in K^*\}$ (see* (4.9)), *and*

$$\tilde{u}(z) = \langle u(x), e^{ixz} \rangle, \quad z \in T_{K^*}, \tag{5.6}$$

where the scalar product is understood as in (5.5). *Furthermore, the estimate*

$$|\tilde{u}(z)| \leqslant C\frac{(1+|z|)^\mu}{\rho^\nu}, \quad z \in T_{K^*}, \tag{5.6'}$$

holds for some C, μ and ν where $\rho = \rho(z, \partial T_{K^*}) = \rho(\operatorname{Im} z, \partial K^*)$ (compare with the formulae (1.60) *and* (1.73) *of Chap.* 1). *Conversely, if $\tilde{u}(z)$ is a holomorphic function in T_{K^*} satisfying* (5.6′), *then $\tilde{u}|_{\mathbb{R}^n} \in S'$ and $\tilde{u}$ has a unique representation of the form* (5.6), *where $u = F^{-1}(\tilde{u}|_{\mathbb{R}^n}) \in S'_K$.*

Definition 5.1 (Hörmander [1983, 1985], Vladimirov [1964]; compare with Definition 1.10). We denote by $S'_{(K)}$ the space of functions $u(x) \in S'(\mathbb{R}^n)$ for which $\operatorname{supp} u(x+a) \subset K$ for some $a \in \mathbb{R}^n$.

In view of the first formula of (2.8), Theorem 5.2 evidently leads to

Theorem 5.2′. *If $u(x) \in S'_{(K)}$, then $\tilde{u}(\xi)$ has an analytic extension to the domain T_{K^*}. If $\operatorname{supp} u(x+a) \subset K$, where $a \in \mathbb{R}^n$, then*

$$\tilde{u}(z) = \langle u(x), \psi(x-a)e^{ixz} \rangle, \quad z \in T_{K^*}. \tag{5.7}$$

Moreover, the estimate

$$|\tilde{u}(z)| \leqslant C\frac{(1+|z|)^\mu}{\rho^\nu} e^{A|\operatorname{Im} z|}, \quad z \in T_{K^*}, \tag{5.7'}$$

holds for some C, μ, ν and $A > 0$. Conversely, if $\tilde{u}(z)$ is any holomorphic function in T_{K^*} satisfying (5.7′), *then $\tilde{u}(z)|_{\mathbb{R}^n} \in S'$ and $\tilde{u}(z)$ has a unique representation of the form* (5.7), *where $u = F^{-1}(\tilde{u}(z)|_{\mathbb{R}^n}) \in S'_{(K)}$.*

5.3. Exponentially Growing Distributions Having Support in a Cone. As above, let K be a closed convex cone in $\mathbb{R}^n$ not containing straight lines.

A distribution $u(x) \in \mathscr{D}'(\mathbb{R}^n)$ is said to be an *exponentially growing distribution* if

$$e^{-B \cdot M(x)} u(x) \in S'(\mathbb{R}^n)$$

for some $B > 0$, where

$$M(x) = \sqrt{1 + |x|^2}. \tag{5.8}$$

Definition 5.2. $\mathscr{D}'E(\mathbb{R}^n)$ denotes the set of all exponentially growing distributions $u(x) \in \mathscr{D}'(\mathbb{R}^n)$, and $\mathscr{D}'E_K = \mathscr{D}'E(\mathbb{R}^n) \cap \mathscr{D}'_K$.

With $R > 0$, we write

$$K_R^* = \{\eta \in K^*: \rho(\eta, \partial K^*) > R\}. \tag{5.9}$$

Clearly,

$$\psi(x) e^{ixz} e^{B \cdot M(x)} \in S(\mathbb{R}^n), \tag{5.9'}$$

with $\operatorname{Im} z \in K_R^*$, for a sufficiently large $R = R(B) > 0$. Therefore for such an R the scalar product

$$\begin{aligned} f(z) &= \langle u(x), \psi(x) e^{ixz} \rangle \\ &:= \langle e^{-B \cdot M(x)} u(x), \psi(x) e^{ixz} e^{B \cdot M(x)} \rangle \end{aligned} \tag{5.10}$$

is defined for $\operatorname{Im} z \in K_R^*$.

Theorem 5.3. *If an exponentially growing distribution u lies in $\mathscr{D}'_K$, then $\tilde{u} = Fu \in Z'$ is an analytic functional of the form* (4.11) *in the domain $\Omega = T_{K_R^*}$ for a sufficiently large $R > 0$; here the function $f(z)$ is holomorphic in $T_{K_R^*}$ and is defined by the formula* (5.10). *An estimate of the form* (5.6′) *holds: for some C and μ*

$$|f(z)| \leqslant C(1 + |z|)^{\mu}, \quad \operatorname{Im} z \in K_R^*. \tag{5.11}$$

Conversely, if $f(z)$ is any holomorphic function in $T_{K_R^*}$ satisfying (5.11) *for some C and μ, then the analytic functional $f|_{T_{K_R^*}}$ is the Fourier transform of an exponentially growing distribution $u \in \mathscr{D}'E_K$.*

The proof of this theorem is almost word for word the same as that of Theorem 5.2. Indeed, the theorem can be deduced from Theorem 5.2 because, by (5.9′), $u_\eta(x) \equiv e^{-x\eta} u(x) \in S'_K$ for $\eta \in K_R^*$, and $\tilde{u}_\eta(z) = \tilde{f}(z + i\eta)$ is a holomorphic function in T_{K^*} which is continuous in $\overline{T}_{K^*}$.

Corollary 5.2. *The Fourier transform of any exponentially growing distribution has a local density in the sense of Definition* 4.2. *The converse statement is also true.*

The direct statement is obtained from Theorem 5.3 by decomposing an exponentially growing distribution into a sum whose terms have supports in convex cones not containing straight lines. The converse statement is obvious for analytic functionals of the form (4.11), because they are the Fourier transforms of the product of exponentials of the form $e^{\eta x}$ and tempered distributions belonging to $S'(\mathbb{R}^n)$. It remains to note that any functional having a local density is, by Proposition 4.3, equal to a finite sum of analytic functionals of the form (4.11).

Definition 5.3. We denote by $\mathscr{D}'E_{(K)}$ the space of functions $u(x) \in \mathscr{D}'E(\mathbb{R}^n)$ for which $\operatorname{supp} u(x + a) \subset K$ for some $a \in \mathbb{R}^n$. A sequence $u_k \xrightarrow{\mathscr{D}'E_{(K)}} u$ if $e^{-B \cdot M(x)} u_k(x - a) \xrightarrow{S'_K} e^{-B \cdot M(x)} u(x - a)$ for some $B > 0$ and $a \in \mathbb{R}^n$ which are independent of k (by definition, convergence in S'_K coincides with weak convergence in $S'(\mathbb{R}^n)$).

In view of the first formula in (2.8), Theorem 5.3 evidently leads to

Theorem 5.3′. *If $u(x) \in \mathscr{D}'E_{(K)}$, then $\tilde{u} \equiv Fu$ is an analytic functional in $T_{K_R^*}$ for a sufficiently large $R > 0$ with a local density of the form* (5.10) *which satisfies the estimate*

$$|f(z)| \leqslant C(1 + |z|)^\mu e^{A|\operatorname{Im} z|}, \quad z \in T_{K_R^*}, \tag{5.12}$$

for some C, μ and $A > 0$. The converse also holds.

For example, when $u(x) = \theta(x + a)e^{3x}$ the local density is, by (5.10),

$$f(z) = \int_{-a}^{\infty} e^{izx} e^{3x}\, dx = -\frac{e^{-a(iz+3)}}{iz + 3} \quad \text{for } \operatorname{Im} z > 3. \tag{5.13}$$

§6. Convolution and Fourier Transform

According to (1.5),

$$F(u * v)(\xi) = \tilde{u}(\xi)\tilde{v}(\xi), \quad \xi \in \mathbb{R}^n, \tag{6.1}$$

for $u, v \in \mathscr{D}(\mathbb{R}^n)$.

Theorem 6.1. *If $u(x) \in \mathscr{E}'(\mathbb{R}^n)$ and $v(x) \in \mathscr{D}'(\mathbb{R}^n)$, then*

$$F(u * v) = \tilde{u}(z)\tilde{v}, \tag{6.2}$$

where $\tilde{u}(z)$ is the holomorphic function (5.1) *and $\tilde{v} \equiv Fv \in Z'(\mathbb{C}^n)$.*

To prove this result, we approximate u and v by functions belonging to $\mathscr{D}(\mathbb{R}^n)$ and use (6.1) as well as continuity of all the operations involved in (6.2). These operations are convolution on $\mathscr{E}' \times \mathscr{D}'$, Fourier transformation and multiplication on the right-hand side.

As above, let K be a closed convex cone in $\mathbb{R}^n$ not containing straight lines.

Theorem 6.2. *If $u, v \in \mathscr{D}'E_{(K)}$, then*

$$F(u * v)(z) = \tilde{u}(z)\tilde{v}(z), \quad z \in T_{K_R^*}, \tag{6.3}$$

for a sufficiently large $R > 0$. Here $\tilde{u}(z)$ and $\tilde{v}(z)$ are analytic functions in $T_{K_R^*}$ which define the functionals $\tilde{u}, \tilde{v} \in Z'$ by a formula of the form (4.11), *and the left-hand side of* (6.3) *is an analytic local density of the functional $F(u * v) \in Z'$ in $T_{K_R^*}$.*

Proof. We approximate u and v by functions belonging to $\mathscr{E}'$ which have the form $u_\varepsilon(x) = \psi(\varepsilon x)u(x)$ and $v_\varepsilon(x) = \psi(\varepsilon x)v(x)$, where $\psi(x) \in C_0^\infty(\mathbb{R}^n)$ and $\psi(x) = 1$ for $|x| < 1$. Then, clearly,

$$\left(u_\varepsilon \xrightarrow{\mathscr{D}'E_{(K)}} u,\ v_\varepsilon \xrightarrow{\mathscr{D}'E_{(K)}} v\right) \Rightarrow \left(u_\varepsilon * v_\varepsilon \xrightarrow{\mathscr{D}'_{(K)}} u * v\right) \tag{6.4}$$

as $\varepsilon \to 0$. Moreover, it is clear from (5.10) that, as $\varepsilon \to 0$,

$$\tilde{u}_\varepsilon(z) \to \tilde{u}(z), \quad \tilde{v}_\varepsilon(z) \to \tilde{v}(z) \tag{6.5}$$

for $z \in T_{K_R^*}$, with a sufficiently large $R > 0$. Finally, the functions $\tilde{u}_\varepsilon(z)$ and $\tilde{v}_\varepsilon(z)$ satisfy estimates of the form (5.12) uniformly in ε. Therefore, as $\varepsilon \to 0$, we obtain (6.3) by applying (6.2) to u_ε and v_ε. ■

Chapter 3
Existence and Uniqueness of Solutions of Differential Equations

§ 1. The Problem of Division

1.1. The Problem of Division in Classes of Rapidly Growing Distributions. We consider the equation (0.1) with $f(x) \in \mathscr{D}'(\mathbb{R}^n)$. The solution $u(x)$ of this equation is also sought in $\mathscr{D}'(\mathbb{R}^n)$. Let us applying the Fourier transformation to both sides of (0.1). This results, in view of formula (4.8) in Chap. 2, in the "algebraic" equation

$$\tilde{P}(z)\tilde{u}(z) = \tilde{f}(z), \tag{1.1}$$

where the equality is understood as the equality of functionals belonging to $Z'(\mathbb{C}^n)$.

Theorem 1.1 (Ehrenpreis [1954], Malgrange [1953]). *The equation* (1.1) *has a solution* $\tilde{u} \in Z'(\mathbb{C}^n)$ *for all* $\tilde{f} \in Z'(\mathbb{C}^n)$ *provided that* $\tilde{P}(\xi) \not\equiv 0$.

Corollary 1.1. *The equation* (1.1) *has a solution* $u(x) \in \mathscr{D}'(\mathbb{R}^n)$ *for all* $f \in \mathscr{D}'(\mathbb{R}^n)$ *if* $P \neq 0$.

We note that Theorem 2.1 in Chap. 1 follows from this result when $f = \delta(x)$.

To prove Theorem 1.1 we apply the Hahn-Banach theorem on the extension of continuous linear functionals in locally convex spaces (Yosida [1965]). First we note that (1.1) is equivalent to

$$\langle \tilde{u}(\xi), \tilde{P}(\xi)\psi(\xi) \rangle = \langle \tilde{f}(\xi), \psi(\xi) \rangle, \tag{1.2}$$

where $\psi \in Z(\mathbb{C}^n)$. Thus the desired functional $\tilde{u}$ is known in the space of test functions of the form $\tilde{P}(\xi)\psi(\xi)$ with $\psi \in Z(\mathbb{C}^n)$. Let us denote this space by $\tilde{P}Z$.

To complete the proof of Theorem 1.1 it remains to verify that $\tilde{u}$ can be extended from $\tilde{P}Z$ as a continuous linear functional on Z. This fact follows from the Hahn-Banach theorem once we have established that the functional $\tilde{u}$ is continuous on $\tilde{P}Z$ endowed with the topology induced by the embedding $\tilde{P}Z \subset Z$. To do this, we require the following Lemma.

Lemma 1.1. *If* $\varphi(z) = \tilde{P}(z)\psi(z)$, *where* $\psi \in Z(\mathbb{C}^n)$, *and*

$$|\varphi(z)| \leqslant \frac{C_N(\varphi)}{(1+|z|)^N} e^{A|\operatorname{Im} z|} \quad \text{for } z \in \mathbb{C}^n, \tag{1.3}$$

then also

$$|\psi(z)| \leqslant \frac{\tilde{C}_N(\psi)}{(1+|z|)^N} e^{A|\operatorname{Im} z|} \quad \text{for } z \in \mathbb{C}^n, \tag{1.4}$$

and

$$\tilde{C}_N(\psi) \leqslant B_N C_N(\varphi), \tag{1.5}$$

where B_N *is independent of* φ *(and* A*).*

We establish the lemma by following the method of Ehrenpreis [1954], Merzon [1973] and Vladimirov [1964]. First, by the Cauchy theorem,

$$\psi(z) = \frac{1}{2\pi i} \int_{\gamma_z} \frac{\psi(z+\lambda e)\, d\lambda}{\lambda} \tag{1.6}$$

where γ_z is a positively oriented contour in the complex plane enclosing the point $\lambda = 0$ and $e \in \mathbb{C}^n \backslash 0$. Next we choose the vector $e \in \mathbb{R}^n$ such that

$$a \equiv \tilde{P}_m(e) \neq 0, \quad \text{where} \quad \tilde{P}_m(z) \equiv \sum_{|\alpha|=m} p_\alpha(-iz)^\alpha. \tag{1.7}$$

Then $\tilde{P}(z+\lambda e) = a\lambda^m + Q_{m-1}(z,\lambda)$, where Q_{m-1} is a polynomial in λ of degree $\leqslant m-1$. Therefore $\tilde{P}(z+\lambda e) = a(\lambda - \lambda_1(z))\dots(\lambda - \lambda_m(z))$. This shows that the contour γ_z can be chosen for all $z \in \mathbb{C}^n$ such that γ_z depends piecewise continuously on z, and

$$\frac{1}{2} \leqslant \sup_{\lambda \in \gamma_z} |\lambda| \leqslant 1, \quad |\gamma_z| \leqslant c_1(m), \quad \inf_{\lambda \in \gamma_z} |\tilde{P}(z+\lambda e)| \geqslant c(m) \cdot |a|, \tag{1.8}$$

where $c(m) > 0$ is independent of z and $|\gamma_z|$ denotes the length of γ_z. Substituting $\varphi/\tilde{P}$ for ψ in (1.6), we immediately obtain (1.4) and (1.5) from (1.3) and (1.8).

Corollary 1.2. *Lemma* 1.1 *implies that the map* $\varphi(\xi) = \tilde{P}\psi \mapsto \psi(\xi) = \varphi/\tilde{P}$ *is continuous as a map from* $\tilde{P}Z$ *into* $Z(\mathbb{C}^n)$.

Now since $\tilde{f}$ is a continuous functional on Z, it follows that $\tilde{u}$, defined on $\tilde{P}Z$ by (1.2), is continuous on $\tilde{P}Z$. But then, by the Hahn-Banach theorem, $\tilde{u}$ can be extended as a continuous functional on $Z(\mathbb{C}^n)$, completing the proof of Theorem 1.1. ■

1.2. The Problem of Division in Classes of Exponentially Growing Generalized Functions. The Hörmander Staircase

Theorem 1.2. *Let* $f \in \mathscr{D}'E(\mathbb{R}^n)$. *Then the equation* (0.1) *has a solution* $u \in \mathscr{D}'E(\mathbb{R}^n)$ *provided that* $P \neq 0$.

Proof. According to Proposition 4.3 in Chap. 2, we can assume that $\tilde{f}$ is an analytic functional of the form (4.11) of Chap. 2, where Ω is an open connected domain of $\mathbb{R}^n$. Thus, for all $\psi \in Z$,

$$\langle \tilde{f}(z), \psi(z) \rangle = \int_{\operatorname{Im} z = \eta} \tilde{f}(z)\psi(z)\, dz \quad \forall \eta \in \Omega. \tag{1.9}$$

The required functional $\tilde{u}$ acts on the teest functions $\varphi = \tilde{P}\psi \in \tilde{P}Z$ according to the formula

$$\langle \tilde{u}, \varphi \rangle = \left\langle \tilde{f}, \frac{\varphi}{\tilde{P}} \right\rangle = \int_{\operatorname{Im} z = \eta} \tilde{f}(z) \frac{\varphi(z)}{\tilde{P}(z)}\, dz \quad \forall \varphi \in \tilde{P}Z. \tag{1.10}$$

In this integral we can replace $\psi = \varphi/\tilde{P}$ by the expression (1.6) and obtain an extension of $\tilde{u}$ to $\varphi \in Z$, because (1.8) holds for all $z \in \mathbb{C}^n$. There is, however, another vivid construction, known as the *Hörmander staircase*. To carry out this construction, we make a linear coordinate transformation ξ in $\mathbb{R}^n$ so that the vector e of (1.7) becomes the first basis vector. To keep the discussion simple, we assume that the coordinates $\xi_1, \ldots, \xi_n$ themselves have this property, that is, $e = (1, 0, \ldots, 0)$. Then, by the Cauchy theorem, the formula (1.10) for $\varphi \in \tilde{P}Z$ can be transformed into

$$\langle \tilde{u}, \varphi \rangle = \int_{\operatorname{Im} z' = \eta'} \left(\int_{\operatorname{Im} z_1 = h(z')} \frac{f(z)\varphi(z)}{\tilde{P}(z)}\, dz_1 \right) dz', \tag{1.11}$$

where $z' = (z_2, \ldots, z_n)$, $\eta' = (\eta_2, \ldots, \eta_n)$ and $(h, \eta') \in \Omega$. Note that, by Bochner's theorem (Hörmander [1973], Vladimirov [1964]), Ω can be assumed to be convex. Thus $(h, \eta') \in \Omega$ for $h \in \left]\alpha(\eta'), \beta(\eta')\right[$, where $\alpha(\eta') < \beta(\eta')$. It remains to choose $h(z')$, for all z', such that, on the straight line $\operatorname{Im} z_1 = h(z')$ of the complex plane, $P(z_1, z')$ has no roots as a polynomial in z_1. This is possible because $\tilde{P}(z) = a(z_1 - \lambda_1(z'))\ldots(z_1 - \lambda_m(z'))$ and the roots $\lambda_1(z'), \ldots, \lambda_m(z')$ depend continuously on z' since, by (1.7), $a = \tilde{P}_m(1, 0, \ldots, 0) \neq 0$. Therefore we can choose $h(z')$, depending piecewise continuously on z' for $\operatorname{Im} z' = \eta'$, so that

$$h(z') \in \left]\alpha(\eta'), \beta(\eta')\right[\text{ and } |\tilde{P}(z)| \geqslant \delta > 0 \quad \text{for } \operatorname{Im} z_1 = h(z'), \quad \operatorname{Im} z' = \eta'. \tag{1.12}$$

The set $H \equiv \{z \in \mathbb{C}^n \colon \operatorname{Im} z' = \eta', \operatorname{Im} z_1 = h(z')\}$ is known as the *Hörmander staircase*. In view of the estimate (4.10) in Chap. 2, it follows from (1.12) that the functional $\tilde{u}$ of (1.11) is continuous on Z. ∎

Corollary 1.3. *The equation* (2.1) *in Chap.* 1 *has a solution* $\mathscr{E}(x) \in \mathscr{D}'E(\mathbb{R}^n)$ *provided that* $P(\partial_x) \neq 0$.

The result follows because $\delta(x) \in \mathscr{D}'E(\mathbb{R}^n)$.

1.3. The Problem of Division in Classes of Tempered Distributions. We consider the equation (0.1) with $f(x) \in S'(\mathbb{R}^n)$, where the solution $u(x)$ is also sought in $S'(\mathbb{R}^n)$. We apply the Fourier transformation to both sides of (0.1). Then, by formula (2.5) in Chap. 2, we obtain the "algebraic" equation

$$\tilde{P}(\xi)\tilde{u}(\xi) = \tilde{f}(\xi), \quad \xi \in \mathbb{R}^n, \tag{1.13}$$

which, in contrast to (1.1), is an equation in $S'(\mathbb{R}^n)$. This division problem was solved in the works of Bernshtejn [1971], Bernshtejn and Gel'fand [1969], Hörmander [1958], Lojasiewicz [1959].

Theorem 1.3. *The equation* (1.13) *has a solution* $\tilde{u}(\xi) \in S'(\mathbb{R}^n)$ *for any right-hand side* $\tilde{f}(\xi) \in S'(\mathbb{R}^n)$ *provided that* $\tilde{P}(\xi) \not\equiv 0$.

Corollary 1.4. *If* $P \neq 0$, *the equation* (0.1) *has a solution* $u \in S'$ *for all* $f \in S'$.

When $f = \delta(x)$, this result leads to Theorem 2.2 in Chap. 1.

The equation (1.1) can be solved easily when

$$|\tilde{P}(\xi)| \geqslant c(1 + |\xi|)^m \tag{1.14}$$

for $\xi \in \mathbb{R}^n$ and some $c > 0$. Such operators P are called *strictly elliptic*. For example, the operator $K \equiv \Delta - m_0^2$, with $m_0 \in \mathbb{R}\backslash 0$, is strictly elliptic, because its symbol $\tilde{K}$ is

$$\tilde{K}(\xi) = -|\xi|^2 - m_0^2 \Rightarrow |\tilde{K}(\xi)| \geqslant c(1 + |\xi|)^2, \quad \xi \in \mathbb{R}^n.$$

In contrast, the *Helmholtz operator* $H \equiv \Delta + \omega^2$, with $\omega \in \mathbb{R}\backslash 0$, is not strictly elliptic, because its symbol

$$\tilde{H}(\xi) = -|\xi|^2 + \omega^2 \Rightarrow \tilde{H}(\xi) = 0 \text{ on the sphere } |\xi| = |\omega|.$$

If P is strictly elliptic, the equation (1.13) is easily solved:

$$\tilde{u}(\xi) = \frac{\tilde{f}(\xi)}{\tilde{P}(\xi)} = \frac{1}{\tilde{P}(\xi)} \cdot \tilde{f}(\xi), \quad \xi \in \mathbb{R}^n. \tag{1.15}$$

Notice that the operator of multiplication by $1/\tilde{P}(\xi)$ is continuous in $S'(\mathbb{R}^n)$ because, by (1.14), $g(\xi) \equiv 1/\tilde{P}(\xi)$ satisfies (1.68) in Chap. 1.

For example, on applying the Fourier transformation the equation $\left(\frac{d^2}{dx^2} - m_0^2\right)\mathscr{E}(x) = \delta(x)$, $x \in \mathbb{R}$, becomes $(-|\xi|^2 - m_0^2)\tilde{\mathscr{E}}(x) = 1$, whence $\mathscr{E}(x) = \frac{1}{2\pi}\int e^{-ix\xi}\frac{1}{-\xi^2 - m_0^2}\,d\xi = -\frac{e^{-m_0|x|}}{2m_0}$ (compare with (2.22) in Chap. 2 and (1.11) in Chap. 1).

It is almost as simple to solve the division problem for the case when $\tilde{P}(\xi) \neq 0$ for $\xi \in \mathbb{R}^n$. Indeed, in this case $\tilde{u}(\xi)$ can be defined by (1.15) as an element of $\mathscr{D}'(\mathbb{R}^n)$. However, we must show further that $\tilde{u}(\xi) \in S'(\mathbb{R}^n)$, but this assertion follows from an estimate of the type (1.14). Thus we have

Lemma 1.2. *If $\tilde{P}(\xi) \neq 0$ for $\xi \in \mathbb{R}^n$, then*

$$|\tilde{P}(\xi)| \geqslant C(1 + |\xi|)^p, \quad \xi \in \mathbb{R}^n, \tag{1.16}$$

where p can be positive as well as negative.

To prove this lemma we invoke the Seidenberg-Tarski principle (Gorin [1961], Hörmander [1983, 1985], Trèves [1961]). It can be deduced easily from (1.16) that, as before, $g(\xi) \equiv 1/\tilde{P}(\xi)$ satisfies (1.68) in Chap. 1. Hence (1.15) implies that $\tilde{u}(\xi) \in S'$, since $\tilde{f} \in S'(\mathbb{R}^n)$.

In the general case, the symbol $\tilde{P}(\xi)$ may vanish at points $\xi \in \mathbb{R}^n$. We write

$$\mathscr{S} = \{\xi \in \mathbb{R}^n : \tilde{P}(\xi) = 0\}, \quad \mathscr{R} = \mathbb{R}^n \backslash \mathscr{S}. \tag{1.17}$$

Then (1.13) defines a unique restriction $\tilde{u}(\xi)$ to $\mathscr{R}$ (see Definition 1.5, Chap. 1) given by

$$\begin{aligned} \tilde{u}(\xi)|_{\mathscr{R}} &= \frac{1}{\tilde{P}(\xi)}(\tilde{f}(\xi)|_{\mathscr{R}}) \Leftrightarrow \langle \tilde{u}(\xi), \varphi(\xi) \rangle \\ &= \left\langle \tilde{f}(\xi), \frac{\varphi(\xi)}{\tilde{P}(\xi)} \right\rangle, \quad \varphi \in \mathscr{D}(\mathscr{R}). \end{aligned} \tag{1.18}$$

Here, by definition, $\dfrac{\varphi(\xi)}{\tilde{P}(\xi)} = 0$ for $\xi \notin \operatorname{supp} \varphi$ and hence for all $\xi \in \mathscr{S}$.

Thus in order to solve the division problem, we extend the functional (1.18), defined in $\mathscr{R}$, to a distribution $\tilde{u}(\xi) \in S'(\mathbb{R}^n)$. This problem, in turn, is referred to as the *problem of regularization* and, by Theorem 1.3, it always has a solution provided that $\tilde{P}(\xi) \not\equiv 0$.

Remark 1.1. If $\mathscr{S} \neq \phi$, the solution of the regularization problem is never unique. For example, we can always add $c\delta(\xi - \xi_0)$ to $\tilde{u}(\xi)$, where $\xi_0 \in \mathscr{S}$, because $\tilde{P}(\xi)\delta(\xi - \xi_0) = 0$, $\xi \in \mathbb{R}^n$. If grad $\tilde{P}(\xi) \neq 0$ for $\xi \in \mathscr{S}$ then also $\tilde{P}(\xi)\delta(\tilde{P}(\xi)) = 0$ and thus we can add $c\delta(\tilde{P}(\xi))$ to $\tilde{u}$ if $\delta(\tilde{P}(\xi)) \in S'(\mathbb{R}^n)$. For example, condition (1.24) of Chap. 1 is satisfied for the symbol $-|\xi|^2 + \omega^2$ of the Helmholtz operator $H = \Delta + \omega^2$, with $\omega \in \mathbb{R}\backslash 0$, and $\delta(-|\xi|^2 + \omega^2) \in S'(\mathbb{R}^n)$ (see (1.23), Chap. 1).

§2. Regularization. The Methods of "Subtraction" and Exit to the Complex Domain and the Riesz Power Method

It is sufficient to solve the regularization problem for the case when

$$\tilde{f}(\xi) \in C(\mathbb{R}^n) \text{ and } |\tilde{f}(\xi)| \leqslant c(1 + |\xi|)^N, \quad \xi \in \mathbb{R}^n, \tag{2.1}$$

because any distribution $\tilde{f} \in S'(\mathbb{R}^n)$ can be expresssed as a finite sum of derivatives of functions having the properties (2.1). Thus we assume that $\tilde{f}(\xi)$ satisfies the conditions (2.1), and we write $R(\xi) = \dfrac{\tilde{f}(\xi)}{\tilde{P}(\xi)}$, $\xi \in \mathscr{R}$. In general, $R(\xi) \notin L_1^{\mathrm{loc}}(\mathbb{R}^n)$, and therefore for this $R(\xi)$ we cannot in general construct a generalized function $\hat{R}$ using formula (1.2) in Chap. 1. Therefore $R(\xi)$ is referred to as a *formal function*.

Definition 2.1. A generalized function $\hat{R}(\xi) \in S'(\mathbb{R}^n)$ is known as a *regularization of a formal function* $R(\xi) \in C(\mathbb{R}^n \backslash \mathscr{S})$ if

$$\langle \hat{R}, \varphi \rangle = \langle R, \varphi \rangle = \int_{\mathscr{R}} R(\xi)\varphi(\xi)\, d\xi \quad \forall \varphi \in \mathscr{D}(\mathbb{R}^n \backslash \mathscr{S}). \tag{2.2}$$

If $\hat{R}^1$ and $\hat{R}^2$ are two distinct regularizations of $R(\xi)$, then (2.2) implies that $\operatorname{supp}(\hat{R}^1 - \hat{R}^2) \subset \mathscr{S}$.

The complexity of the regularization problem depends on the construction of the set $\mathscr{S}$ and on the nature of growth of $R(\xi)$ near $\mathscr{S}$. In the proof of Theorem 1.3, the following two factors play a decisive role.

1) The algebraic manifold $\mathscr{S}$ of (1.17) is the union of a finite number of sets of analytic submanifolds of $\mathbb{R}^n$ having different dimension:

$$\mathscr{S} = \bigcup_{k=0}^{n-1} \mathscr{S}^k, \quad \dim \mathscr{S}^k = k, \tag{2.3}$$

where $\mathscr{S}^k$ is an analytic manifold of $\mathbb{R}^n$.

2) $1/\tilde{P}(\xi)$ grows near $\mathscr{S}$ not faster than some (negative) power of distance from $\mathscr{S}$ (the *Lojasiewicz inequality*; see Lojasiewicz [1959]). Thus, for some N and ν,

$$\frac{1}{|\tilde{P}(\xi)|} \leqslant \frac{c(1 + |\xi|)^N}{\rho^\nu(\xi)}, \quad \rho(\xi) := \rho(\xi, \mathscr{S}), \quad \xi \in \mathbb{R}^n \backslash \mathscr{S}. \tag{2.4}$$

To establish (2.4), we use, as in (1.16), the Seidenberg-Tarski principle (Gorin [1961], Hörmander [1983, 1985]).

2.1. The Method of Subtraction (Ehrenpreis [1954]). To keep the exposition simple, we confine ourselves to the case when $\mathscr{S}_k$ is a smooth submanifold in $\mathbb{R}^n$, that is, when (2.3) contains just one member $\mathscr{S}^k$, where $0 \leqslant k \leqslant n-1$.

We first examine the case $k = 0$ so that $\mathscr{S} = \mathscr{S}^0$ is a discrete point set (if $\mathscr{S}^0$ is not discrete, (2.4) does not hold). Thus $\mathscr{S} = \{a_j \in \mathbb{R}^n, j = 1, \ldots\}$ and the ε_j-neighbourhood ($\varepsilon_j > 0$) of a_j does not contain other points a_i. Then a regularization can be constructed by subtracting from $\varphi(x)$ a truncated Taylor series:

$$\begin{aligned}\langle \hat{R}, \varphi \rangle := \sum_j \int_{|\xi - a_j| < \varepsilon_j/2} & R(\xi)\Bigg[\varphi(\xi) - \varphi(a_j) - \cdots \\ & - \sum_{|\alpha| = |\nu| - n} \frac{\varphi^{(\alpha)}(a_j)}{\alpha!}(\xi - a_j)^\alpha\Bigg] d\xi + \int_{\mathscr{R}_{(\varepsilon)}} R(\xi)\varphi(\xi)\, d\xi,\end{aligned} \tag{2.5}$$

where

$$\mathscr{R}_{(\varepsilon)} := \{\xi \in \mathscr{R}: |\xi - a_j| > \varepsilon_j/2\ \forall j\}, \quad \varphi \in \mathscr{D}(\mathbb{R}^n).$$

Clearly, $\hat{R} \in \mathscr{D}'(\mathbb{R}^n)$ and $\langle \hat{R}, \varphi \rangle = \int R(\xi)\varphi(\xi)\, d\xi$ if supp $\varphi \subset \mathscr{R}$. Thus $\hat{R} \in \mathscr{D}'$ is a regularization of $R(\xi)$ in the sense of (2.2). However, we must show that an $\hat{R} \in S'$ does exist.

Example 2.1. For $n = 1$ and $R(\xi) = 1/\xi$, the regularization $\hat{R}$ can be defined as the Cauchy principal value. Namely, for $\varphi(x) \in S(\mathbb{R})$,

$$\langle \hat{R}, \varphi \rangle = \left\langle \text{p.v.}\,\frac{1}{\xi}, \varphi(\xi) \right\rangle := \lim_{\varepsilon \to 0+} \int_{|\xi| > \varepsilon} \frac{\varphi(\xi)}{\xi}\, d\xi$$

$$= \int_{-1}^{1} \frac{\varphi(\xi) - \varphi(0)}{\xi}\, d\xi + \int_{|\xi|>1} \frac{\varphi(\xi)}{\xi}\, d\xi. \qquad (2.5')$$

Example 2.2. For the Helmholtz operator $H = \Delta + \omega^2$, $\omega \in \mathbb{R}\backslash 0$, in $\mathbb{R}^n$ the symbol $\tilde{H}(\xi)$ is $-|\xi|^2 + \omega^2$ and grad $\tilde{H}(\xi) = -2\xi \neq 0$ if $\tilde{H}(\xi) = 0$. Therefore the regularization of $R \equiv 1/\tilde{H}(\xi)$ can be defined by the formula $\hat{R}(\xi) = \text{p.v.}\,\dfrac{1}{-|\xi|^2 + \omega^2}$.

We now consider the more general case when $\mathscr{S} = \mathscr{S}^k$ for $1 \leqslant k \leqslant n - 1$. For simplicity, assume that $\mathscr{S}^k$ is a compact smooth manifold. With $\varepsilon > 0$, we denote by $\mathscr{S}^k_\varepsilon$ an ε-neighbourhood of $\mathscr{S}^k$. Then for some $\varepsilon > 0$ there is a diffeomorphism $h: \mathscr{S}^k_\varepsilon \to \mathscr{S}^k \times D^{n-k}$, where D^{n-k} is a ball in $\mathbb{R}^{n-k}$ of radius 1 (or ε, the value being immaterial), that maps $\xi \in \mathscr{S}^k_\varepsilon$ into $h(\xi) = (\xi^s, \xi^D)$, where $\xi^s \in \mathscr{S}^k$, $\xi^D \in D^{n-k}$ and $h(\mathscr{S}^k) = \mathscr{S}^k \times \{0\}$. Then the regularization $\hat{R}$ can be constructed in a form that is analogous to (2.5). Namely, for $\varphi \in S(\mathbb{R}^n)$,

$$\langle \hat{R}, \varphi \rangle = \int_{\mathscr{S}^k_\varepsilon} R(\xi) \Bigg[\varphi_h(\xi^s, \xi^D) - \varphi_h(\xi^s, 0) - \cdots$$

$$- \sum_{|\alpha| = |\nu| - (n-k)} \frac{\partial^\alpha_{\xi^D} \varphi_h(\xi^s, 0)(\xi^D)^\alpha}{\alpha!} \Bigg] d\xi + \int_{\mathscr{R}_{(\varepsilon)}} R(\xi)\varphi(\xi)\, d\xi,$$

where

$$\mathscr{R}_{(\varepsilon)} = \mathbb{R}^n \backslash \mathscr{S}_\varepsilon, \quad \varphi_h(\xi^s, \xi^D) = \varphi(h^{-1}(\xi^s, \xi^D)) = \varphi(\xi).$$

Here $\partial^\alpha_{\xi^D} \varphi_h(\xi^s, 0)$ denotes the derivative of $\varphi_h(\xi^s, \xi^D)$ at the point $\xi^D = 0$, and $(\xi^s, 0) \in h(\mathscr{S}^k)$.

Since $\mathscr{S}^k$ is a compact manifold, it follows that $\hat{R} \in S'(\mathbb{R}^n)$.

In the general case when $\mathscr{S}$ is of the form (2.3), the regularization $\hat{R} \in S'(\mathbb{R}^n)$ is constructed in a similar fashion by using (2.4). In this method, suitable coordinates are constructed in the neighbourhood of each "layer" $\mathscr{S}^k$ of $\mathscr{S}$ by means of the so-called *σ-process* (or otherwise known as the method of *resolution of singularity*). (See Hörmander [1958], Lojasiewicz [1959].)

This is how we solve the problem of regularization, and we can also establish Theorem 1.3 along the same lines (Hörmander [1958], Lojasiewicz [1959]).

2.2. The Method of Exit to the Complex Domain. Such a method of regularization is applicable when, with $\xi \in \mathscr{R}$, $R(\xi)$ is, in the sense of Definition 1.11 in Chap. 1, a restriction to $\mathscr{R}$ of a function $R(z)$ that is holomorphic in a tubular domain of $\mathbb{C}^n$. The existence of a regularization follows from Lemma 1.5 in Chap. 1.

Without going into the general formulation, we confine ourselves to concrete examples.

1) For the function $R(\xi) = 1/\xi$, $\xi \in \mathbb{R}\backslash 0$, we can define a regularization by the formulae $\hat{R}_+(\xi) = \dfrac{1}{\xi + i0}$ or $\hat{R}_-(\xi) = \dfrac{1}{\xi - i0}$. Furthermore, $F^{-1}\hat{R}_\pm(\xi) = \mp i\theta(\pm x) = 0$ for $x \lessgtr 0$. This result follows from Theorem 5.2 in Chap. 2, because $\hat{R}_\pm(\xi)$ has an analytic extension to the set $\operatorname{Im} \xi \gtrless 0$.

Corollary 2.1. *For the Helmholtz operator of Example 2.2, the regularization of* $R(\xi) \equiv \dfrac{1}{\tilde{H}(\xi)}$ *can be defined as the superposition of* $\dfrac{1}{\xi + i0}$ *and* $\zeta \equiv \tilde{H}(\xi) = -|\xi|^2 + \omega^2$ *because* $\operatorname{grad} \tilde{H}(\xi) = -2\xi \neq 0$ *for* $\tilde{H}(\xi) = 0$. *Thus* $\hat{R}_\pm(\xi) = \dfrac{1}{-|\xi|^2 + \omega^2 \pm i0}$.

Remark 2.1. The regularization $\hat{R}_+(\xi)$ is used for finding the limiting amplitude of diverging waves. This amplitude corresponds to the limiting absorption and satisfies for $\omega > 0$ the Sommerfeld condition for the absence of radiation (see Grushin [1963a], Tikhonov and Samarskij [1951], Vajnberg [1966]).

2) For the *wave operator*

$$\Box = \frac{\partial^2}{\partial t^2} - a^2 \Delta, \quad a > 0, \tag{2.6}$$

the symbol is $\tilde{\Box}(\tau, \xi) = -\tau^2 + a^2|\xi|^2$, where $\tau \in \mathbb{R}$ and $\xi \in \mathbb{R}^k$. It can be verified easily that $\tilde{\Box}(\tau, \xi) \neq 0$ in the tubular domain T_{K^*} with base $K^* = \{(s, \eta) \in \mathbb{R} \times \mathbb{R}^k : |s| > a|\eta|\}$. This domain has two connected domains $K^*_\pm$ in which $s > 0$ and $s < 0$ respectively. Therefore we can define two different regularizations

$$\left(\frac{1}{\tilde{\Box}(\tau, \xi)}\right)_\pm = \frac{1}{-(\tau \pm i0)^2 + a^2|\xi|^2}, \quad (\tau, \xi) \in \mathbb{R} \times \mathbb{R}^k.$$

It follows from Theorem 5.2 in Chap. 2 that

$$\mathscr{E}_\pm(t, x) = F^{-1}\left(\left(\frac{1}{\tilde{\Box}(\tau, \xi)}\right)_\pm\right) = 0 \quad \text{for } \pm at < |x|.$$

To see this, we write $K_\pm = \{(t, x) \in \mathbb{R} \times \mathbb{R}^k : \pm at > |x|\}$. Then the cones $K^*_\pm$ are dual to $K_\pm$ and $\tilde{\mathscr{E}}_\pm(\tau, \xi)$ have analytic extensions to $T_{K^*_\pm}$.

3) Similarly, for the *Klein-Gordon operator* $\mathscr{K} = \Box + m^2$, $m > 0$, the symbol $\tilde{\mathscr{K}}(\tau, \xi) = \tilde{\Box}(\tau, \xi) + m^2 \neq 0$ in T_{K^*}. Therefore there are two distinct regularizations

$$\left(\frac{1}{\tilde{\mathscr{K}}(\tau, \xi)}\right)_{\pm} = \frac{1}{-(\tau \pm i0)^2 + a^2|\xi|^2 + m^2}, \quad (\tau, \xi) \in \mathbb{R} \times \mathbb{R}^k,$$

and Theorem 5.2 in Chap. 2 implies that

$$\mathscr{E}_{\pm}(t, x) = F^{-1}\left(\frac{1}{\tilde{\mathscr{K}}(\tau, \xi)}\right)_{\pm} = 0 \quad \text{for } \pm at < |x|.$$

2.3. The Riesz Method of Complex Powers. Roughly speaking, the idea of the *Riesz method* is that $\tilde{P}^{\lambda}(\xi)$ is an analytic function of $\lambda \in \mathbb{C}$ with values in $S'(\mathbb{R}^n)$ for $\operatorname{Re}\lambda > 0$, and therefore we can conjecture first that $\tilde{P}^{\lambda}(\xi)$ has an analytic extension to the domain $\operatorname{Re}\lambda \leqslant 0$, and second that $\tilde{P}^{-1}(\xi)$ (the value of the extension of $\tilde{P}^{\lambda}$ for $\lambda = -1$) is a regularization of the formal function $1/\tilde{P}(\xi)$.

The conjecture turns out to be valid with certain refinements. Namely,

1) $\tilde{P}^{\lambda}(\xi)$ has an extension as a meromorphic function of $\lambda \in \mathbb{C}$ with values in $S'(\mathbb{R}^n)$, and

2) the regularization of $1/\tilde{P}(\xi)$ is the regular part of the meromorphic function $\tilde{P}^{\lambda}$ for $\lambda = -1$.

We now formulate the precise results in this direction (Bernshtejn [1971], Bernshtejn and Gel'fand [1969], Gel'fand and Shilov [1958a]).

Let $\tilde{P}(\xi)$ be a real polynomial in $\xi \in \mathbb{R}^n$. For $\operatorname{Re}\lambda > 0$, we define

$$\tilde{P}^{\lambda}(\xi) = \begin{cases} \tilde{P}^{\lambda}(\xi) & \text{if } \tilde{P}(\xi) \geqslant 0, \\ (-1)^{\lambda}|\tilde{P}(\xi)|^{\lambda} & \text{if } \tilde{P}(\xi) < 0, \end{cases} \tag{2.7}$$

where $(-1)^{\lambda}$ denotes an arbitrary branch; for example, $(-1)^{\lambda} = e^{\lambda\pi i}$. It is understood, of course, that $a^{\lambda} = e^{\lambda \ln a}$ for $a > 0$. Note that, for $\operatorname{Re}\lambda > 0$ and $k = 0, 1, 2, 3, \ldots,$

$$\tilde{P}^{\lambda+k}(\xi) = \tilde{P}^{k}(\xi) \cdot \tilde{P}^{\lambda}(\xi), \quad \xi \in \mathbb{R}^n. \tag{2.8}$$

For $\operatorname{Re}\lambda \geqslant 0$, the function $\tilde{P}^{\lambda}(\xi)$ is continuous and, with this function, we can construct a distribution belonging to $S'(\mathbb{R}^n)$. We note that

$$\tilde{P}^{\lambda}(\xi) \to 1 \text{ as } \lambda \to 0, \quad \operatorname{Re}\lambda > 0. \tag{2.9}$$

Let $\tilde{P}_+$ denote the function

$$\tilde{P}_+(\xi) = \begin{cases} \tilde{P}(\xi), & \tilde{P}(\xi) \geqslant 0, \\ 0, & \tilde{P}(\xi) < 0. \end{cases} \tag{2.10}$$

Then the following result has been established in Bernshtejn [1971], Bernshtejn and Gel'fand [1969].

Theorem 2.1. 1) *The function* $\tilde{P}_+^{\lambda}$ *(and* $|\tilde{P}|^{\lambda}$*) can be extended from the domain* $\operatorname{Re}\lambda \geqslant 0$ *as a meromorphic function of* $\lambda \in \mathbb{C}$ *with values in* $S'(\mathbb{R}^n)$.

2) *The set* Π *of poles of the function* $\tilde{P}_+^{\lambda}$ *(and* $|\tilde{P}|^{\lambda}$*) of* λ *is the union of a finite number of sets that form an arithmetic progression.*

The proof of this theorem makes use of the σ-process for resolving the singularities of the algebraic manifold $\mathcal{S}$ (see (1.17)). There is a simplified proof in Bernshtejn [1971] that does not use the σ-process.

Theorem 2.1 also implies the analyticity of the extension of $\tilde{P}^\lambda$ because

$$\tilde{P}^\lambda = \tilde{P}_+^\lambda + (-1)^\lambda(-\tilde{P})_+^\lambda \quad \text{for } \operatorname{Re}\lambda \geqslant 0.$$

Thus the analytic extension of $\tilde{P}^\lambda$ from the domain $\operatorname{Re}\lambda \geqslant 0$ either has an isolated pole of order N at the point $\lambda = -1$ or is regular (then $N = 0$). Hence

$$\tilde{P}^\lambda(\xi) = \sum_{k=1}^{N} \frac{r_k(\xi)}{(\lambda+1)^k} + \hat{R}_\lambda(\xi), \tag{2.11}$$

where $r_k \in S'(\mathbb{R}^n)$ and $\hat{R}_\lambda(\xi)$ is an analytic function of λ in a neighbourhood of the point $\lambda = -1$ with values in $S'(\mathbb{R}^n)$.

Proposition 5.1. *The function $\hat{R}_{-1}(\xi) \in S'(\mathbb{R}^n)$ is a regularization of the formal function $1/\tilde{P}(\xi)$ and*

$$\tilde{P}(\xi)\cdot\hat{R}_{-1}(\xi) = 1, \quad \xi \in \mathbb{R}^n, \tag{2.12}$$

where the equality is understood in the sense of generalized functions.

Proof. It follows from (2.9) and (2.8) that

$$\tilde{P}(\xi)\cdot\tilde{P}^\lambda(\xi) = \tilde{P}^{\lambda+1}(\xi) \to 1 \quad \text{as } \lambda \to -1, \quad \operatorname{Re}\lambda > -1. \tag{2.13}$$

We note that the equality holds in the sense of $S'(\mathbb{R}^n)$ for all $\lambda \in \mathbb{C}$ except $\lambda \in \Pi$ because of the uniqueness of the analytic extension. But (2.13) and (2.11) imply that

$$\sum_{k=1}^{N} \frac{\tilde{P}(\xi)\cdot r_k(\xi)}{(\lambda+1)^k} + \tilde{P}(\xi)\tilde{R}_\lambda(\xi) \to 1 \quad \text{as } \lambda \to -1, \quad \operatorname{Re}\lambda > -1.$$

This is possible only if

$$\tilde{P}(\xi)\cdot r_k(\xi) = 0, \quad \xi \in \mathbb{R}^n \ \forall k = 1,\ldots,N \quad \text{and} \quad \tilde{P}(\xi)\cdot\hat{R}_{-1}(\xi) = 1. \quad \blacksquare$$

§ 3. Equations in a Convex Cone. An Operational Calculus

3.1. Equations in a Cone. Let K be a closed (convex) cone of $\mathbb{R}^n$, that is, $K + K \subset K$ and $tK \subset K$ for all $t \geqslant 0$. Suppose that K does not contain straight lines. We consider the differential equation (0.1) in K. More precisely, we consider the equation in the class of functions having support in K:

$$P(\partial_x)u(x) = f(x), \quad x \in \mathbb{R}^n, \tag{3.1}$$

where

$$f \in S'_K(\mathbb{R}^n) := \{f \in S' : \operatorname{supp} f \subset K\}.$$

The solution u is also sought in S'_K. By Theorem 5.2 in Chap. 2, the functions $\tilde{u}(\xi)$ and $\tilde{f}(\xi)$ have analytic extensions to T_{K*} and the identity (1.13) also holds in T_{K*}:

$$\tilde{P}(z)\tilde{u}(z) = \tilde{f}(z), \quad z \in T_{K*}. \tag{3.2}$$

Let us formulate necessary and sufficient conditions for the equation (3.1) to have a solution in the class $S'_K(\mathbb{R}^n)$. We define

$$V := \{z \in \mathbb{C}^n : \tilde{P}(z) = 0\} \quad \text{and} \quad V_{K*} := V \cap T_{K*}. \tag{3.3}$$

Further, $\mathscr{Y}_{K*}(\tilde{P})$ denotes the ideal generated by the polynomial $\tilde{P}(z)$ in the ring of holomorphic functions in T_{K*}.

The principal observation is that

$$\tilde{f}(z) = 0 \quad \text{for } z \in V_{K*}, \tag{3.4}$$

in view of (3.2). Moreover, (3.2) implies that

$$\tilde{f}(z) \in \mathscr{Y}_{K*}(\tilde{P}). \tag{3.5}$$

Using the method of proof of Lemma 1.1 as well as Theorem 5.2 in Chap. 2, we can establish

Theorem 3.1 (Merzon [1973], Vladimirov [1964]). 1) *For the equation* (3.1) *to have a solution* $u \in S'_K$ *it is necessary and sufficient that* (3.5) *holds, that is,* $\tilde{f}(z)/\tilde{P}(z)$ *is a holomorphic function in* T_{K*}.

2) *If* grad $\tilde{P}(z) \neq 0$ *for all* $z \in V_{K*}$ *except for an analytic submanifold in* T_{K*} *of codimension* $\geqslant 2$, *then the condition* (3.5) *is equivalent to* (3.4).

Example 3.1. Let $n = 2$ and $K = \{x \in \mathbb{R}^2 : x_1 \geqslant 0, x_2 \geqslant 0\}$ be the first quadrant of the plane. Then for the equation

$$\Delta u(x) - u(x) = f(x), \quad x \in \mathbb{R}^2, \tag{3.6}$$

to have a solution u in the class S'_K, with $f \in S'_K$, it is necessary and sufficient that

$$\tilde{f}(z_1, z_2) = 0 \quad \text{for } z_1^2 + z_2^2 + 1 = 0, \quad \operatorname{Im} z_1 > 0, \quad \operatorname{Im} z_2 > 0. \tag{3.7}$$

We next examine (3.1) in the classes of exponentially growing distributions. Thus u and f now lie in $\mathscr{D}'E_K = \mathscr{D}'E(\mathbb{R}^n) \cap \mathscr{D}'_K$ (see Definition 5.2, Chap. 2). Then, by Theorem 5.3 in Chap. 2, the functionals $\tilde{u}, \tilde{f} \in Z'$ are holomorphic in the domain $T_{K_R^*}$ for a sufficiently large $R > 0$ (see formula (5.9), Chap. 2). Denoting holomorphic local densities of the functionals $\tilde{u}$ and $\tilde{f}$ also by $\tilde{u}(z)$ and $\tilde{f}(z)$ and using Proposition 4.2, Chap. 2, we obtain from (3.1) the identity (3.2) holding in $T_{K_R^*}$ instead of T_{K*}. Thus

$$\tilde{P}(z)\tilde{u}(z) = \tilde{f}(z), \quad z \in T_{K_R^*}. \tag{3.8}$$

Hence the conditions (3.4) and (3.5) are also satisfied when K^* is replaced by K_R^* with a sufficiently large $R > 0$.

As for Theorem 3.1, we can use the method of proof of Lemma 1.1, as well as Theorem 5.3 in Chap. 2, to establish

Theorem 3.1′. 1) *For the equation* (3.1), *with* $f \in \mathscr{D}'E_K$, *to have a solution* $u \in \mathscr{D}'E_K$ *it is necessary and sufficient that*

$$\tilde{f}(z) \in \mathscr{Y}_{K_R^*}(\tilde{P}) \quad \text{for some } R > 0. \tag{3.9}$$

2) *If* grad $\tilde{P}(z) \neq 0$ *for all* $z \in V_{K_R^*} := V \cap T_{K_R^*}$ *except for a submanifold in* $T_{K_R^*}$ *of codimension* $\geqslant 2$, *then* (3.9) *is equivalent to the condition that*

$$\tilde{f}(z) = 0 \quad \text{for } z \in V_{K_R^*}. \tag{3.10}$$

Here $\mathscr{Y}_{K_R^*}(\tilde{P})$ *is an ideal generated by the polynomial* $\tilde{P}(z)$ *in the ring of holomorphic functions in* $T_{K_R^*}$.

3.2. An Operational Calculus. Let $n = 1$ and $K = \overline{\mathbb{R}}_+ := \{x \in \mathbb{R}: x \geqslant 0\}$. We consider the ordinary differential equation (0.1)

$$P\left(\frac{d}{dx}\right)u(x) = \sum_{k=0}^{m} p_k \left(\frac{d}{dx}\right)^k u(x) = f(x) \tag{3.11}$$

on the semi-axis $x > 0$ with $f(x) \in \mathscr{D}'E(\overline{\mathbb{R}}_+)$. To simplify the discussion, we assume that $f(x) \in C(\overline{\mathbb{R}}_+)$ and $u(x) \in C^m(\overline{\mathbb{R}}_+)$. In order to find a unique solution $u(x)$, we prescribe the initial conditions

$$u(0+) = u^0, \ldots, u^{(m-1)}(0+) = u^{m-1}. \tag{3.12}$$

We solve the Cauchy problem (3.11), (3.12) by means of the Fourier transformation. For functions having support in $\overline{\mathbb{R}}_+$ the Fourier transformation is sometimes called the *Fourier-Laplace transformation*. We first extend (3.11) to the whole real axis $\mathbb{R}$ in the form

$$P\left(\frac{d}{dx}\right)u_0(x) = lf(x) \equiv f_0(x) + \sum_{k=0}^{m-1} b_k(U_{(0)})\delta^{(k)}(x), \quad x \in \mathbb{R}, \tag{3.13}$$

where we use the notation

$$u_0(x) := \begin{cases} u(x), & x > 0, \\ 0, & x < 0; \end{cases} \quad f_0(x) := \begin{cases} f(x), & x > 0, \\ 0, & x < 0; \end{cases}$$
$$U_{(0)} := (u^0, \ldots, u^{m-1}), \tag{3.14}$$

and $b_k(U_{(0)})$ are constants. The formula (1.10) in Chap. 1 implies that

$$b_{m-1}(U_{(0)}) = p_m u^0, b_{m-2}(U_{(0)}) = p_m u^1 + p_{m-1}u^0, \ldots.$$

The equation (3.13) has the form (3.1). We assume that the growth of $u(x)$ and $f(x)$ as $x \to +\infty$ is at most exponential, that is, we assume that

$$|u(x)| \leqslant C_1 e^{Bx}, \quad |f(x)| \leqslant C_2 e^{Bx}, \quad x > 0. \tag{3.15}$$

Then, by Theorem 5.3 in Chap. 2 (formula (5.10), Chap. 2),

$$\tilde{u}_0(z) = \int_0^\infty e^{izx} u(x)\, dx,$$
$$\tilde{f}_0(z) = \int_0^\infty e^{izx} f(x)\, dx \quad \text{for } \operatorname{Im} z > B. \tag{3.16}$$

We note that, for the case in question, $K^* = \mathbb{R}_+ = \{s \in \mathbb{R}: s > 0\}$ and so $T_{K^*} = \{z \in \mathbb{C}: \operatorname{Im} z > 0\}$ and $K_R^* = \{s \in \mathbb{R}: s > R\}$, while $T_{K_R^*} = \{z \in \mathbb{C}: \operatorname{Im} z > R\}$. By (3.8) and (3.13), we obtain

$$\tilde{P}(z)\tilde{u}_0(z) = \widetilde{lf}(z) := \tilde{f}_0(z) + \sum_{k=0}^{m-1} b_k(-iz)^k \quad \text{for } \operatorname{Im} z > B. \tag{3.17}$$

This gives the local density $\tilde{u}_0(z)$ of the functional $\tilde{u}_0 := Fu_0$ to be

$$\tilde{u}_0(z) = \frac{\widetilde{lf}(z)}{\tilde{P}(z)} \quad \text{for } \operatorname{Im} z > B. \tag{3.18}$$

But, by Theorem 3.1′, the functional $\tilde{u}_0 \in Z'(\mathbb{C})$ defined by this density in the domain $\operatorname{Im} z > R$ is the Fourier transform of a distribution $u \in \mathscr{D}'E_{\overline{\mathbb{R}}_+}$ if and only if $\tilde{u}_0(z)$ is holomorphic for $\operatorname{Im} z > R$, that is, if and only if (3.9) holds.

The polynomial $\tilde{P}(z)$ has m zeros (taking multiplicity into account):

$$\tilde{P}(z_k) = 0, \quad k = 1, \ldots, m, \quad z_k \in \mathbb{C}. \tag{3.19}$$

Therefore $\tilde{P}(z) \neq 0$ for $\operatorname{Im} z > M := \max_k \operatorname{Im} z_k$, and thus (3.9) holds for $R > \max(M, B)$.

Thus, by Theorem 3.1′, the equation (3.13) has a solution $u_0(x) \in \mathscr{D}'E_{\overline{\mathbb{R}}_+}$ given by

$$u_0 = F^{-1}(\tilde{u}_0(z)|_{\operatorname{Im} z > \max(M,B)}), \tag{3.20}$$

where $\tilde{u}_0(z)$ is found from (3.18), and $\tilde{u}_0(z)$ is unique under the conditions (3.15). The solution u_0 has the simplest form when $f(x) \equiv 0$ and the roots z_j are simple, that is, $z_i \neq z_j$ if $i \neq j$. Then the formula (3.18) can be transformed into

$$\tilde{u}_0(z) = i \sum_{k=1}^m \frac{d_k(U_{(0)})}{z - z_k}, \tag{3.21}$$

whence we obtain, as for (5.13) in Chap. 2,

$$u_0(x) = \sum_{k=1}^m d_k(U_{(0)})\theta(x)e^{-iz_k x} \Rightarrow u(x) = \sum_{k=1}^m d_k(U_{(0)})e^{-iz_k x}, \quad x > 0. \tag{3.22}$$

In the general case $u_0(x)$ can be found from (3.18) by means of the tables for *Laplace transforms* (Ditkin and Prudnikov [1966], Doetsch [1967]) which differ from Fourier-Laplace transforms in that z is replaced by $s = -iz$.

We note that if we require a solution u for which $|u(x)| \leqslant Ce^{Dx}$, $x > 0$, the density $\tilde{u}_0(z)$ must be analytic for $\operatorname{Im} z > D$. Therefore (3.17) implies the necessity of the condition that

$$\widetilde{lf}(z_k) = 0 \quad \text{for } \operatorname{Im} z_k > D. \tag{3.23}$$

This condition is also sufficient if the roots z_k are simple, with $\operatorname{Im} z_k \neq D$ for all k, and $f = 0$. If, however, the root z_k has multiplicity ν_k, then (3.23) is replaced by the condition that

$$\widetilde{lf}^{(j)}(z_k) = 0 \quad \text{for } \operatorname{Im} z_k > D, \quad j = 0, \ldots, \nu_k - 1.$$

This condition on $u^0, \ldots, u^{m-1}$ when $f = 0$ is necessary and sufficient for the existence of a solution $u(x)$ such that $|u(x)| \leqslant ce^{Dx}$ for $x > 0$ provided that $\operatorname{Im} z_k \neq D$ for all k.

Example 3.2. Consider the problem

$$u''(x) - 3u'(x) = 0, \quad x > 0; \quad u(0+) = 1, \quad u'(0+) = 2.$$

Then (3.17) becomes

$$(-z^2 + 3iz)\tilde{u}_0(z) = -iz - 1, \quad \operatorname{Im} z > 3,$$

and hence

$$\tilde{u}_0(z) = \frac{iz + 1}{z^2 - 3iz} = \frac{A}{z} + \frac{B}{z - 3i}, \quad \operatorname{Im} z > 3,$$

where $A = \frac{i}{3}$ and $B = \frac{2i}{3}$. Then, by (3.22),

$$u_0(x) = \tfrac{1}{3}\theta(x) + \tfrac{2}{3}\theta(x)e^{3x} \Rightarrow u(x) = \tfrac{1}{3} + \tfrac{2}{3}e^{3x} \quad \text{for } x > 0.$$

3.3. Differential-difference Equations on a Semi-axis (Ditkin and Prudnikov [1966], Doetsch [1967]). We consider an *equation with a retarded argument* on the semi-axis $x > 0$. Thus, we consider the equation

$$\sum_{k=0}^{m} p_k u^{(k)}(x - h_k) - f(x), \quad x > 0, \tag{3.24}$$

where $p_m \neq 0$ and $h_k \geqslant h_m = 0$ for $0 \leqslant k \leqslant m - 1$. We assume that $f(x)$ satisfies the condition (3.15). For the initial conditions we can require that

$$u(x) = 0 \quad \text{for } x < 0, \quad u \in \mathscr{D}'E_{\overline{\mathbb{R}}_+} \cap C^m(\mathbb{R}).$$

We note that (3.24) is a *convolution equation*

$$(P * u)(x) = f_0(x), \quad x \in \mathbb{R}, \tag{3.25}$$

where

$$P(x) := \sum_{k=0}^{m} p_k \delta^{(k)}(x - h_k).$$

Therefore, by Theorem 6.2 in Chap. 2, it follows from (3.15) that

$$\tilde{P}(z) \cdot \tilde{u}(z) = \tilde{f}(z), \quad \operatorname{Im} z > B,$$

as for (3.8). Here

$$\widetilde{P}(z) = \sum_{k=0}^{m} p_k(-iz)^k e^{ih_k z}, \quad z \in \mathbb{C},$$

is the *symbol of the convolution operator* (3.25) which is equal to the Fourier transform of the "convolution kernel" $P(x)$. Since $p_m \neq 0$, $h_k \geqslant h_m = 0$ for $1 \leqslant k \leqslant m-1$ and $|e^{ih_k z}| = e^{-h_k \operatorname{Im} z}$, it follows that

$$|\widetilde{P}(z)| \geqslant c|z|^m \quad \text{for } \operatorname{Im} z > M$$

with some $c > 0$ and a sufficiently large $M > 0$. This implies that the quotient

$$\tilde{u}(z) = \frac{\tilde{f}(z)}{\widetilde{P}(z)},$$

where $\operatorname{Im} z > R = \max(M, B)$, satisfies estimates of the form (5.11) in Chap. 2. Accordingly, by Theorem 5.3 in Chap. 2, the analytic functional $\tilde{u}(z)|_{\operatorname{Im} z > R}$ is the Fourier transform of a distribution $u \in \mathscr{D}'E_{\overline{\mathbb{R}}_+}$ satisfying the equations (3.24) and (3.25).

§4. Propagation of Singularities and Smoothness of Solutions

4.1. Characteristics of Differential Equations. We denote by P_m the *principal homogeneous part* of the operator P in (0.1):

$$P_m(\partial_x) := \sum_{|\alpha|=m} p_\alpha \partial_x^\alpha. \tag{4.1}$$

With such a homogeneous operator P_m, we first examine (0.1) in the case $f = 0$:

$$P_m(\partial_x)u(x) = 0, \quad x \in \mathbb{R}^n. \tag{4.2}$$

We seek the solutions of this equation in the form of plane waves, that is,

$$u(x) = w(\xi \cdot x), \quad x \in \mathbb{R}^n; \quad \xi \in \mathbb{R}^{n*}\backslash 0; \tag{4.3}$$

where $\xi \cdot x := \xi_1 x_1 + \cdots + \xi_n x_n$ and $w \in \mathscr{D}'(\mathbb{R})$. To do this we substitute (4.3) into (4.2) and obtain

$$P_m(\partial_x)w(\xi x) = w^{(m)}(\xi x) \sum_{|\alpha|=m} p_\alpha \xi^\alpha = 0. \tag{4.4}$$

We note that $P_m(\xi) = \sum_{|\alpha|=m} p_\alpha \xi^\alpha = i^m \widetilde{P}_m(\xi)$. From (4.4) there follows

Proposition 4.1. *The function* (4.3) *satisfies* (4.2) *for all* $w \in \mathscr{D}'(\mathbb{R})$ *if the covector* $\xi \in \mathbb{R}^{n*}\backslash 0$ *satisfies the* ***characteristic equation***

$$P_m(\xi) := \sum_{|\alpha|=m} p_\alpha \xi^\alpha = 0 \Leftrightarrow \widetilde{P}_m(\xi) = 0. \tag{4.5}$$

Definition 4.1. 1) The covector $\xi \in \mathbb{R}^{n*}\backslash 0$ satisfying (4.5) is known as the *characteristic (co)normal of equation* (0.1) (or *of the operator* $P(\partial_x)$ in (0.1)).

2) The set K of all characteristic normals $\xi \in \mathbb{R}^{n*}\backslash 0$ is called the *characteristic cone of the operator P*:

$$K = \{\xi \in \mathbb{R}^{n*}\backslash 0: P_m(\xi) = 0\}. \tag{4.6}$$

3) The hyperplane $\xi_{x_0}^{\perp} \equiv \{v \in T_{x_0}\mathbb{R}^n: \langle \xi, v \rangle = 0\}$ in $T_{x_0}\mathbb{R}^n$ is called the *characteristic plane* of (0.1) at a point x_0 if the covector $\xi \in \mathbb{R}^{n*}\backslash 0$ is a characteristic conormal of (0.1).

4) A hypersurface $\mathscr{S}$ of class C^1 in $\mathbb{R}^n$ is called a *characteristic hypersurface* at a point x_0 if the tangent plane at this point is characteristic, and it is called a *characteristic* of (0.1) if $\mathscr{S}$ is a characteristic surface at each of its points.

Example 4.1. The plane $\{x \in \mathbb{R}^n: (x - x_0)\cdot\xi = 0\}$ of $\mathbb{R}^n$ is a characteristic if $\xi \in K$.

Corollary 4.1. *For any characteristic of* (4.2) *which has the form* $\{x \in \mathbb{R}^n: (x - x_0)\cdot\xi = 0\}$ *the equation* (4.2) *has solutions* $u(x)$ *that are discontinuous on this plane.*

For example, $u(x) = \theta((x - x_0)\cdot\xi)$ is a function of the form (4.3) and hence satisfies (4.2).

Remark 4.1. We make a linear coordinate transformation $x = Cy$ in $\mathbb{R}^n_x$. Let the plane $y_n = 0$ be a characteristic of (4.2), that is, let the covector $\xi = (C^{-1})^t(0, \ldots, 0, 1)$ be characteristic. Then, by Theorem 2.15 in Chap. 2, the symbol of P_m in the new y-coordinates is $(\tilde{P}_m)_C(\eta) = \tilde{P}_m((C^{-1})^t\eta)$, where $\eta \in \mathbb{R}^n$. Therefore

$$(\tilde{P}_m)_C(0, \ldots, 0, 1) = \tilde{P}_m(\xi) = 0. \tag{4.7}$$

But $(\tilde{P}_m)_C(0, \ldots, 0, 1)$ is the coefficient of $\partial_{y_n}^m u(y)$ in the operator $(P_m)_C(\partial_y)$. Therefore (4.7) implies that in the new y-coordinates (4.2) takes the form

$$\sum_{\substack{|\alpha|=m \\ \alpha_n<m}} q_\alpha \partial_y^\alpha u_C(y) = 0, \quad y \in \mathbb{R}^n. \tag{4.8}$$

This form of the equation clearly explains why the solutions of (4.2) may have discontinuities along the characteristic hyperplanes $y_n = \text{const}$. Indeed, (4.8) does not contain the leading derivative $\partial_{y_n}^m$ in a direction transverse to these planes, but each term in (4.8) contains at least one derivative with respect to the contact variables $y_1, \ldots, y_{n-1}$. Therefore any function which is constant on the hyperplanes $y_n = \text{const.}$ satisfies (4.8) and, in particular, so does any discontinuous function.

Example 4.2. For the wave operator (2.6) the characteristic equation (4.5) is

$$\tau^2 = a^2|\xi|^2, \quad (\tau, \xi) \in \mathbb{R} \times \mathbb{R}^k. \tag{4.9}$$

This equation defines a conical surfaces K in $\mathbb{R}^{k+1}$. The plane wave

$$u(t, x) = \theta(at - x_1) = \theta((a, -1, 0, \ldots, 0)\cdot(t, x)), \tag{4.10}$$

where

$$(a, -1, 0, \ldots, 0) \in K,$$

runs in the direction of the x_1-axis with speed a and its front lies on the plane

$$F := \{x_1 = at\} = (a, -1, 0, \ldots, 0)_0^{\perp}. \tag{4.10'}$$

Definition 4.2. The operator $P(\partial_x)$ is said to be *Petrovskij strictly hyperbolic* in the variable x_n if the characteristic equation (4.5) has m distinct real roots $\xi_n = \tau_j(\xi')$, for all $\xi' = (\xi_1, \ldots, \xi_{n-1}) \in \mathbb{R}^{n-1}\backslash 0$, that is, if

$$P_m(-i\xi) = p_{(0,\ldots,0,m)}(-i)^m(\xi_n - \tau_1(\xi'))\ldots(\xi_n - \tau_m(\xi')), \quad \xi \in \mathbb{R}^n. \tag{4.11}$$

For example, the wave equation (2.6) and the equation (2.7) in Chap. 1 are hyperbolic in t in this sense.

4.2. Wave Fronts, Bicharacteristics and Propagation of Singularities

Definition 4.3. Char P denotes the set of all characteristic conormals of an operator P that "emanate" from all possible points $x \in \mathbb{R}^n$. That is,

$$\text{Char } P = \{(x, \xi) \in T^*\mathbb{R}^n\backslash O^*: x \in \mathbb{R}^n, \tilde{P}_m(\xi) = 0\}. \tag{4.12}$$

Here $T^*\mathbb{R}^n$ denotes the cotangent bundle of $\mathbb{R}^n$ and 0^* its zero section. Evidently, Char $P \approx \mathbb{R}^n \times K$ (see (4.6)). For example, for the wave operator (2.6)

$$\text{Char } \Box = R^{k+1} \times K, \quad \text{where } K = \{(\tau, \xi) \in R \times R^k: \tau^2 = a^2|\xi|^2\}. \tag{4.12'}$$

In the modern treatment of singularities of solutions to differential equations a central role is played by the notion of the wave front of a generalized function. For $u(x) \in \mathscr{D}'(\mathbb{R}^n)$, the *wave front* $WF(u)$ at a point $x \in \mathbb{R}^n$ is, roughly speaking, the set of all directions $\xi \in T_x^*\mathbb{R}^n$ along which $u(x)$ fails to be smooth.

Definition 4.4. A point $(x_0, \xi_0) \in T^*\mathbb{R}^n\backslash 0^*$ does not lie in $WF(u)$ if there exists a function $\chi(x) \in C_0^\infty(\mathbb{R}^n)$ such that $\chi(x_0) \neq 0$ and for all N

$$|\widetilde{\chi u}(\xi)| \leqslant \frac{C_N}{(1 + |\xi|)^N} \tag{4.13}$$

for $\xi \in \Gamma_N(\xi_0)$, where $\Gamma_N(\xi_0)$ is a conical neighbourhood of the point ξ_0 in $\mathbb{R}^n$. Accordingly, $WF(u)$ consists of the remaining points $(x_0, \xi_0) \in T^*\mathbb{R}^n\backslash 0^*$ for which such χ and $\Gamma_N(\xi_0)$ do not exist.

Clearly, $WF(u)$ is a closed set in $T^*\mathbb{R}^n\backslash 0^*$, and

$$\text{sing supp } u = \pi WF(u), \tag{4.14}$$

where π is the projection of $T^*\mathbb{R}^n$ onto $\mathbb{R}^n$.

The following lemma exposes the significance of the notion of the wave front. Let $u \in S'$ be a real generalized function, that is, let $\langle u, \overline{\varphi}\rangle = \langle u, \varphi\rangle$ for all $\varphi \in \mathscr{D}(\mathbb{R}^n)$. Then we have

Lemma 4.1. *If* $(x_0; (1, 0, \ldots, 0)) \notin WF(u)$, *then there exists a function* $\chi(x) \in C_0^\infty(\mathbb{R}^n)$ *for which* $\chi(x_0) \neq 0$ *and*

$$\chi(x)u(x) = v(x_1; x') \in C^\infty(\mathbb{R}_{x_1}; \mathscr{D}'(\mathbb{R}_{x'}^{n-1})), \quad x' \equiv (x_2, \ldots, x_n). \tag{4.15}$$

That is, $v(x_1; \cdot)$ *is a smooth function of the parameter* $x_1 \in \mathbb{R}$ *with values in* $\mathscr{D}'(\mathbb{R}^{n-1})$, *and*

$$\langle \chi(x)u(x), \varphi(x) \rangle = \int_{-\infty}^{+\infty} \langle v(x_1; \cdot), \varphi(x_1, \cdot) \rangle \, dx_1 \quad \forall \varphi \in \mathscr{D}(\mathbb{R}^n). \tag{4.16}$$

Example 4.3. The wave u of (4.10) is discontinuous on the front F (see (4.10′)), and $WF(u)$ coincides with the bundle NF of normal to F; that is,

$$WF(u) = NF := \{((x, t), (\xi, \tau)) \in \mathbb{R}^{k+1} \times \mathbb{R}^{k+1} : (x, t) \in F, (\xi, \tau) \perp F\}. \tag{4.17}$$

We note that, clearly, $WF(u) \subset NF$, by definition 4.3, while the inclusion $NF \subset WF(u)$ is a consequence of Lemma 4.1.

Example 4.4. Since $\chi(x)\delta(x - x_0) = \chi(x_0) \cdot \delta(x - x_0)$ and $F(\delta(x - x_0)) = e^{ix_0\xi}$, it follows that

$$WF(\delta(x - x_0)) = x_0 \times (\mathbb{R}^n \backslash 0). \tag{4.18}$$

Theorem 4.1 (Hörmander [1983, 1985], Shubin [1978]). *If* $u \in \mathscr{D}'(\mathbb{R}^n)$ *is a solution of* (2.1), *then*

$$WF(u) \subset (\operatorname{Char} P \cup WF(f)). \tag{4.19}$$

Roughly speaking, a solution $u(x)$ fails to be smooth at a point x in the direction $\xi \in T_x^*\mathbb{R}^n$ only if either $f(x)$ is not smooth in the direction ξ or P is a "characteristic" operator in this direction.

Example 4.5. For the wave (4.10)

$$WF(u) = NF \subset \operatorname{Char} \Box = \mathbb{R}^{k+1} \times K. \tag{4.20}$$

The inclusion occurs here because the normal to F can be computed by the formula $(\tau, \xi) = \operatorname{grad}(at - x_1) = (1, 0, \ldots, 0, -a)$, that is, $\tau = -a$ and $\xi = (1, 0, \ldots, 0)$, implying (4.9). Thus the inclusion (4.19) for the wave (4.10) does occur. Notice also that $WF(f) = \phi$ here because $f = 0$.

For a more precise characterization of the singularities of a solution u of (0.1) we introduce the notion of a *bicharacteristic* of (0.1). We assume that the leading part P_m of (0.1) has real coefficients. Consider the *Hamiltonian system*

$$\begin{cases} \dot{x}(s) = H'_\xi = \operatorname{grad} P_m(\xi), \\ \dot{\xi}(s) = -H'_x = 0, \quad s \in \mathbb{R}, \end{cases} \tag{4.21}$$

in $T^*\mathbb{R}^n$ with the Hamiltonian $H(x, \xi) := P_m(\xi) = \tilde{P}_m(i\xi)$. Clearly, the solutions of this system are the straight lines

$$\begin{pmatrix} x(s) \\ \xi(s) \end{pmatrix} = \Phi_s \begin{pmatrix} x_0 \\ \xi_0 \end{pmatrix} = \begin{pmatrix} x_0 + \operatorname{grad} P_m(\xi_0) \cdot s \\ \xi_0 \end{pmatrix}, \quad s \in \mathbb{R}, \tag{4.22}$$

where $x_0 = x(0)$ and $\xi_0 = \xi(0)$. It is well known (and also clear from (4.22)) that the Hamiltonian $H(x, \xi) = P_m(\xi)$ is preserved along the trajectories (4.22). In particular, the set Char P is invariant with respect to the Hamiltonian flow Φ_s, $s \in \mathbb{R}$, of (4.21).

Definition 4.4. The *bicharacteristics* (or *bicharacteristic strips*; see Courant [1962], Gårding, Kotaké and Leray [1964]) of (0.1) are curves in $T^*\mathbb{R}^n$ which are trajectories of the Hamiltonian system (4.21) lying in Char P (that is, on which the symbol $\tilde{P}_m(\xi)$ vanishes). The *rays* of the equation (0.1) are projections of its bicharacteristics under the map π: $T^*\mathbb{R}^n \to \mathbb{R}^n$.

Theorem 4.2 (Hörmander [1983, 1985], Shubin [1978]). *Let the leading part P_m of* (0.1) *have real coefficients and let P_m be an* ***operator of principal type***, *that is, let*

$$\operatorname{grad} P_m(\xi) \neq 0 \quad \text{for } P_m(\xi) = 0, \quad \xi \in \mathbb{R}^n \backslash 0. \tag{4.23}$$

Then, for $f(x) \in C^\infty(\mathbb{R}^n)$, $WF(u)$ is a subset of Char P *and is invariant under the Hamiltonian flow Φ_s, $s \in \mathbb{R}$, of the system* (4.21). *In other words, if $(x_0, \xi_0) \in WF(u)$ then $(x_0, \xi_0) \in$* Char P *and the whole bicharacteristic* (4.22) *also lies in $WF(u)$.*

Roughly speaking, the singularities of the solution $u(x)$ of (0.1) are propagated along the bicharacteristics. We note that the hypotheses of Theorem 4.2 are satisfied by, for example, the equations (2.6), (2.7), (2.10) and (2.11) in Chap. 1.

Remark 4.2. Under condition (4.23), the projections of the bicharacteristics (4.22) of (0.1) onto $\mathbb{R}^n_x$, that is, the rays of (0.1), are roughly speaking intersections of all "infinitely close" characteristic planes $\xi^\perp_{x_0}$ with the characteristic normals ξ which are close to ξ_0. Indeed, if $P_m(\xi_0) = 0$ and $\xi_0 \neq 0$, then the plane $(\xi_0)^\perp_{x_0}$ in $\mathbb{R}^n_x$ passing through x_0 and orthogonal to ξ_0 is, by Definition 4.1, a characteristic plane for (0.1). By (4.23), the surface $K := \{\xi \in \mathbb{R}^n : P_m(\xi) = 0\}$ is smooth at the point ξ_0 and the vector grad $P_m(\xi_0)$ is orthogonal to $T_{\xi_0}K$. But K is a conical surface, because $P_m(\xi)$ is a homogeneous polynomial. Consequently, $\xi_0 \in T_{\xi_0}K$, and so grad $P_m(\xi_0)$ is orthogonal to ξ_0. The same result also follows from the Euler formula: $\xi_0 \cdot \operatorname{grad} P_m(\xi_0) = mP_m(\xi_0) = 0$. Hence

$$\begin{gathered}\xi \cdot \operatorname{grad} P_m(\xi_0) = (\xi - \xi_0) \cdot \operatorname{grad} P_m(\xi_0) = O(|\xi - \xi_0|^2) \\ \text{for } \xi \in K \quad \text{and} \quad \xi \to \xi_0.\end{gathered} \tag{4.24}$$

This implies that

$$\widehat{(\xi^\perp_{x_0}, \operatorname{grad} P_m(\xi_0))} = O(|\xi - \xi_0|^2) \text{ for } \xi \in K, \quad \xi \to \xi_0, \tag{4.24$'$}$$

where the left-hand side denotes the angle between the characteristic hyperplane $\xi^\perp_{x_0}$, having normal ξ close to ξ_0, and the vector grad $P_m(\xi_0)$. It remains to note that grad $P_m(\xi_0)$ is the direction vector of the projection of the bicharacteristic (4.22) onto $\mathbb{R}^n_x$ and (4.24′) signifies that the vector grad $P_m(\xi_0)$ "lies in $\xi^\perp_{x_0}$ to within $O(|\xi - \xi_0|^2)$" for $\xi \in K$ and $\xi \to \xi_0$.

Definition 4.5. The conical surface

$$K_{x_0} = \operatorname{Char} P \cap T^*_{x_0}\mathbb{R}^n \tag{4.25}$$

in $T^*_{x_0}\mathbb{R}^n$ is called a *characteristic cone of* (0.1) at the point x_0. The conical surface Q_{x_0} in $T_{x_0}\mathbb{R}^n$ defined by the formula

$$Q_{x_0} := \{v \in T_{x_0}\mathbb{R}^n : v = s \operatorname{grad} P_m(\xi),\ \xi \in K_{x_0},\ s \in \mathbb{R}\} \tag{4.26}$$

is called a *characteristic conoid of* (0.1) at the point x_0 (Courant [1962], Gårding, Kotaké and Leray [1964]).

Remark 4.3. Under condition (4.23), the surface Q_{x_0} consists of normals to the planes which are tangent to the surface K_{x_0}. The conoid Q_{x_0} is, roughly speaking, an enveloping surface for the family of characteristic hyperplanes passing through x_0. Indeed, it follows from (4.24) that $\xi\cdot(\operatorname{grad} P_m(\xi_0) - \operatorname{grad} P_m(\xi)) = O(|\xi-\xi_0|^2)$ for $\xi_0, \xi \in K$ and $\xi_0 \to \xi$, that is, $\xi^{\perp}_{x_0}$ touches Q_{x_0} along the line $\{s \operatorname{grad} P_m(\xi), s \in \mathbb{R}\} \subset Q_{x_0}$ if $\det \dfrac{\partial^2 P_m(\xi)}{\partial\xi_i \partial\xi_j} \neq 0$.

Theorem 4.2 can be extended to the case of non-smooth functions $f(x)$.

Theorem 4.2′ (Hörmander [1983, 1985], Shubin [1978]). *If P satisfies the hypothesis of Theorem* 4.2, *then*

$$(WF(u)\backslash WF(f)) \subset \operatorname{Char} P. \tag{4.27}$$

Furthermore, if $(x_0, \xi_0) \in WF(u)\backslash WF(f)$, then $WF(u)$ also contains the intersection of the bicharacteristic (4.22) *with the connected component of the open set $T^*\mathbb{R}^n\backslash WF(f)$ which contains the point (x_0, ξ_0).*

For example, $WF(\delta) = 0 \times (\mathbb{R}^{n*}\backslash 0)$, by (4.18). Therefore, with $f(x) = \delta(x)$, Theorem 4.2′ leads to

Corollary 4.2. *Let P satisfy the hypothesis of Theorem* 4.2 *and let $\mathscr{E}(x)$ be any fundamental solution of P. If $(x_0, \xi_0) \in WF(\mathscr{E})\backslash[0 \times (\mathbb{R}^{n*}\backslash 0)]$ and the line $\{x_0 + \operatorname{grad} P(\xi_0)\cdot s, s \in \mathbb{R}\}$ in $\mathbb{R}^n$ does not pass through the point $x = 0$, then it lies completely in* sing supp $\mathscr{E}$. *If, however, the line passes through $x = 0$, then* sing supp $\mathscr{E}$ *contains, at least, the ray through the point x_0.*

Thus sing supp $\mathscr{E}$ is a projection onto $\mathbb{R}^n$ of a family of bicharacteristics and semi-bicharacteristics of P.

§ 5. Smoothness of Solutions of Elliptic Equations. Hypoellipticity

5.1. Smoothness of Generalized Solutions of Elliptic Equations

Definition 5.1. The operator $P(\partial_x)$ is called *elliptic* if

$$\tilde{P}_m(\xi) \neq 0 \quad \text{for } \xi \in \mathbb{R}^n\backslash 0; \tag{5.1}$$

that is, if $P(\partial_x)$ does not have (real) characteristics:

$$\text{Char}\, P = \phi. \tag{5.1'}$$

Example 5.1. The *Laplace operator* in $\mathbb{R}^n$ is elliptic because $\tilde{\Delta}(\xi) = -|\xi|^2 \neq 0$ for $\xi \in \mathbb{R}^n \backslash 0$. Similarly, the Helmholtz operator $H = \Delta + \omega^2$ is elliptic because $\tilde{H}_2(\xi) = -|\xi|^2 \neq 0$ for $\xi \in \mathbb{R}^n \backslash 0$.

The *Cauchy-Riemann operator* $\dfrac{\partial}{\partial \bar{z}} = \dfrac{1}{2}\left(\dfrac{\partial}{\partial x} + i\dfrac{\partial}{\partial y}\right)$ on $\mathbb{R}^2$ is elliptic because its symbol $\frac{1}{2}(-i\xi + \eta) \neq 0$ for $(\xi, \eta) \in \mathbb{R}^2 \backslash 0$.

Proposition 5.1 (Hörmander [1983, 1985], Vishik and Ehskin [1965]). *If $n \geqslant 3$, the order m of the elliptic operator $P(\partial_x)$ is an even integer.*

Note that this statement is obviously true for all $n \geqslant 2$ if the operator $P(\partial_x)$ has real coefficients.

Remark 5.1. When $n = 1$, all equations of the form (0.1) are elliptic provided that $P(\partial_x) \neq 0$.

Theorem 5.1 (the *Weyl lemma*; Shubin [1978], Weyl [1940]). *If the operator $P(\partial_x)$ of the equation* (0.1) *is elliptic and $f(x) \in C^\infty(\Omega)$, where Ω is a domain of $\mathbb{R}^n$, then also $u(x) \in C^\infty(\Omega)$.*

We note that sing supp $f =$ sing supp $Pu \subset$ sing supp u, while the reverse inclusion is a consequence of Theorem 5.1. Therefore this theorem is equivalent to the equality

$$\text{sing supp}\, u = \text{sing supp}\, f \tag{5.2}$$

if the operator $P(\partial_x)$ is elliptic.

Remark 5.2. If the operator P is not elliptic but is, for example, homogeneous, that is, if $P = P_m(\partial_x)$, then (5.2) does not hold in general. Indeed, if $P_m(\xi_0) = 0$ for some $\xi_0 \in \mathbb{R}^n \backslash 0$, then the homogeneous equation $P_m(\partial_x)u(x) = 0$, $x \in \mathbb{R}^n$, has discontinuous solutions, by Corollary 4.1.

Remark 5.3. Theorem 5.1 is a particular case of Theorem 4.1. In fact, if P is elliptic, then Char $P = \phi$, and so (4.19) implies that

$$WF(u) = WF(f). \tag{5.3}$$

This result, together with (4.14), leads to (5.2).

5.2. Hypoelliptic Operators. It turns out that (5.2) holds not only for elliptic operators.

Definition 5.2. The operator $P(\partial_x)$ is called *hypoelliptic* if (5.2) holds for any solution $u(x)$ of (0.1). In other words, the smoothness of $f(x)$ in a domain $\Omega \subset \mathbb{R}^n$ implies the smoothness of the solution $u(x)$ of (0.1) in the same domain.

It is known that the hypoellipticity of $P(\partial_x)$ is related to the location of the complex zeros of the symbol $\tilde{P}(z)$ and, in particular, to their distance from the real space. We denote

$$K_{\mathbb{C}}(P) := \{z \in \mathbb{C}^n : \tilde{P}(z) = 0\}. \tag{5.4}$$

Theorem 5.2 (Shilov [1965], Trèves [1961]). *The operator $P(\partial_x)$ is hypoelliptic if and only if*

$$|\mathrm{Im}\, z| \geqslant A|\mathrm{Re}\, z|^\kappa - B \text{ for some } A > 0, B, \text{ and } \kappa > 0, \tag{5.5}$$

on the algebraic manifold (5.4) *or, equivalently,*

$$|\mathrm{Im}\, z| \to \infty \quad \text{for } z \in K_{\mathbb{C}}(P) \text{ and } |\mathrm{Re}\, z| \to \infty. \tag{5.5'}$$

As it turns out, it is also possible to characterize the hypoelliptic operators in terms of the behaviour of their symbols in the real domain.

Theorem 5.3 (Shilov [1965], Trèves [1961]). *The hypoellipticity of the operator $P(\partial_x)$ is equivalent to each of the following two conditions:*

$$\lim_{|\xi|\to\infty} \frac{\tilde{P}(\xi + \theta)}{\tilde{P}(\xi)} = 1 \quad \text{for } \forall\theta \in \mathbb{R}^n; \tag{5.6}$$

$$\lim_{|\xi|\to\infty} \frac{\partial_{\xi_j}\tilde{P}(\xi)}{\tilde{P}(\xi)} = 0 \quad \text{for } \forall j = 1, \ldots, n. \tag{5.6'}$$

Hypoellipticity is also related in a simple way to the smoothness of fundamental solutions.

If P is a hypoelliptic operator, then, by Definition 5.2, every fundamental solution $\mathscr{E}(x)$ of P is smooth for $x \neq 0$ because $\delta(x)$ is smooth for $x \neq 0$, where we recall that $\delta(x) = 0$ when $x \neq 0$.

Lemma 2.2 of Chap. 1 shows that the converse of this statement is also true.

Theorem 5.4. *If the equation* (0.1) *has at least one fundamental solution $\mathscr{E}(x)$ which is smooth for $x \neq 0$, then the operator $P(\partial_x)$ is hypoelliptic.*

Proof. Let $f(x) \in \mathscr{D}'(\mathbb{R}^n)$ and let $f|_\Omega \in C^\infty(\Omega)$, where Ω is a domain in $\mathbb{R}^n$. Take any point $x_0 \in \Omega$ and functions φ, ϕ and $\hat{\phi}$ as in (2.18) of Chap. 1. Let $\omega \subset \operatorname{supp} \hat{\phi} \subset \Omega$. Then, by (2.21) in Chap. 1,

$$u(x) = (\mathscr{E} * \hat{\phi} f)(x) - (\mathscr{E} * \hat{\phi} P w)(x) \quad \text{for } x \in \omega.$$

This implies that $u(x) \in C^\infty$ for $x \in \omega$. In fact, $\hat{\phi} f \in \mathscr{D}$ and then $\mathscr{E} * \hat{\phi} f \in C^\infty(\mathbb{R}^n)$ by Lemma 1.4 of Chap. 1. On the other hand, by (1.58) in Chap. 1,

$$\operatorname{sing\,supp} \mathscr{E} * \hat{\phi} P w \subset \{0\} + \operatorname{supp} w = \operatorname{supp} w,$$

but supp w does not intersect ω. ∎

Corollary 5.1. *When $n = 1$, all the operators $P \neq 0$ of the form* (0.1) *are hypoelliptic.*

This result is a consequence of the fact that all the fundamental solutions (2.15) of Chap. 1 are smooth for $x \neq 0$.

Chapter 4
The Function P_+^λ for Polynomials of Second-degree and its Application in the Construction of Fundamental Solutions

In this chapter we first study in § 1 and § 2 the analytic continuation of the function P_+^λ for polynomials $P(x)$ of the second (and first) degree. Then using this analytic continuation we contruct in § 3–§ 5 fundamental solutions for second-order equations with non-degenerate quadratic form having arbitrary signature and arbitrary number of independent variables.

We shall obtain, in particular, all the fundamental solutions (2.6′), (2.7′), (2.10′) and (2.11′) of Chap. 1. The fundamental solutions constructed in § 3–§ 5 are invariant under the group of transformations which preserve the principal part of the equation. The solutions are functions of "radius" in a (generally speaking) indefinite metric connected with the equation. In particular, the fundamental solutions of the Klein-Gordon equation in (2.7) and (2.7′) of Chapter 1 turn out to be the analytic continuation of the Bessel functions of the Lorentz interval with respect to the order of the Bessel function.

§ 1. The Function P_+^λ for the Case when P is a Real Linear Function

1.1. Analytic Continuation with Respect to λ. If $P(x) = p_1x_1 + \cdots + p_nx_n + p_0$ and $P(x) \not\equiv$ const., an affine transformation of variables enables us to assume that $P(x) = x_1$. Therefore it remains only to consider the one-dimensional case $n = 1$, and

$$\langle P_+^\lambda(x), \varphi(x)\rangle = \langle x_+^\lambda, \varphi(x)\rangle$$
$$= \int_0^\infty x^\lambda \varphi(x)\,dx \ \forall \varphi \in \mathscr{D}(R), \quad \operatorname{Re}\lambda > -1 \qquad (1.1)$$

(see formula (2.10), Chap. 1). The integral converges and depends analytically on λ if $\operatorname{Re}\lambda > -1$.

Theorem 1.1. *The function $\lambda \to x_+^\lambda$, which is holomorphic as a map of the domain $\operatorname{Re}\lambda > -1$ into $\mathscr{D}'(\mathbb{R})$, can continued as a meromorphic function of $\lambda \in \mathbb{C}$ in the sense of Gel'fand and Shilov [1958a] with values in $\mathscr{D}'(\mathbb{R})$. It has simple poles at the points $\lambda = -1, -2, \ldots$ and the residues are given by*

$$\operatorname*{res}_{\lambda=-k} x_+^\lambda = \frac{(-1)^{k-1}\delta^{(k-1)}(x)}{(k-1)!}, \quad k = 1, 2, \ldots. \qquad (1.2)$$

Proof. Subtracting from $\varphi(x)$ the Taylor series about $x = 0$, we obtain

$$\begin{aligned}\langle x_+^\lambda, \varphi(x)\rangle &= \int_0^\infty x^\lambda \varphi(x)\,dx \\ &= \int_0^1 x^\lambda\left[\varphi(x) - \varphi(0) - \cdots - \frac{\varphi^{(k-1)}(0)}{(k-1)!}x^{k-1}\right]dx \\ &\quad + \int_1^\infty x^\lambda \varphi(x)\,dx + \varphi(0)\int_0^1 x^\lambda\,dx + \cdots \\ &\quad + \varphi^{(k-1)}(0)\int_0^1 \frac{x^{\lambda+k-1}}{(k-1)!}dx \end{aligned} \tag{1.3}$$

for all $\varphi \in \mathscr{D}(\mathbb{R})$ and $\operatorname{Re}\lambda > -1$. The integrals $\int_0^1 \ldots$ can be easily computed:

$$\int_0^1 \frac{x^{\lambda+k-1}}{(k-1)!}dx = \frac{1}{(\lambda+k)(k-1)!}, \quad \operatorname{Re}\lambda > -k. \tag{1.4}$$

It is clear that the first integral on the right-hand side of (1.3) is holomorphic for $\operatorname{Re}\lambda > -k-1$, while the second integral is holomorphic for all $\lambda \in \mathbb{C}$. The remaining integrals have smple poles at the points $\lambda = -k$ with residues $\varphi^{(k-1)}(0)/(k-1)!$ $(k = 1, 2, \ldots)$. ■

Alternative Proof. For $\operatorname{Re}\lambda > -1$,

$$\begin{aligned}\langle x_+^\lambda, \varphi(x)\rangle &= \int_0^\infty x^\lambda \varphi(x)\,dx = -\int_0^\infty \frac{x^{\lambda+1}}{\lambda+1}\varphi'(x)\,dx = \cdots \\ &= (-1)^k \int_0^\infty \frac{x^{\lambda+k}}{(\lambda+1)\ldots(\lambda+k)}\varphi^{(k)}(x)\,dx.\end{aligned} \tag{1.5}$$

Consequently, the function

$$\langle(\lambda+1)\ldots(\lambda+k)x_+^\lambda, \varphi(x)\rangle = (-1)^k \int_0^\infty x^{\lambda+k}\varphi^{(k)}(x)\,dx \tag{1.6}$$

is holomorphic in λ for $\operatorname{Re}\lambda > -k-1$, and

$$\operatorname*{res}_{\lambda=-k}\langle x_+^\lambda, \varphi(x)\rangle = \frac{(-k)^k \int_0^\infty \varphi^{(k)}(x)\,dx}{(-k+1)\ldots(-1)} = \frac{\varphi^{(k-1)}(0)}{(k-1)!}. \quad ■ \tag{1.7}$$

In x_+^λ, we use the substitution $x = h(y) = |y|^2 - \omega^2$, where $y \in \mathbb{R}^n$ and $\omega \in \mathbb{R}\backslash 0$. Since $\operatorname{sing\,supp} x_+^\lambda = \{0\}$ and $\operatorname{grad} h(y) \neq 0$ for $|y|^2 = \omega^2$, it follows that $(h(y))_+^\lambda = (|y|^2 - \omega^2)_+^\lambda \in \mathscr{D}'(\mathbb{R}^n)$, in view of Remark 1.4, Chap. 1. Moreover, the map $u(x) \mapsto u(h(y))$ is continuous in u in the class of functions $u(x)$ having $\operatorname{sing\,supp} u \not\ni 0$ with the appropriate convergence.

Corollary 1.1. *The function* $(|y|^2 - \omega^2)_+^\lambda$ *is a meromorphic function of* $\lambda \in \mathbb{C}$ *with simple poles at* $\lambda = -1, -2, \ldots$ *and residues*

$$\operatorname*{res}_{\lambda=-k} (|y|^2 - \omega^2)_+^\lambda = \frac{(-1)^k \delta^{(k-1)}(|y|^2 - \omega^2)}{(k-1)!}, \quad y \in \mathbb{R}^n. \tag{1.8}$$

1.2. An Application to Bessel Functions. Following Bateman and Erdélyi [1953], Whittaker and Watson [1927], we define

$$J_\nu(z) = \sum_0^\infty (-1)^r \frac{z^{2r+\nu}}{2^{2r+\nu} r! \Gamma(\nu + r + 1)} \tag{1.9}$$

for $-\pi < \arg z < \pi$ and $\nu \in \mathbb{C}$. Then we define $J_\nu(x_+) \in \mathscr{D}'(\mathbb{R}^n)$ for $\nu \in \mathbb{C}$ as an analytic continuation of

$$J_\nu(x_+) = \sum (-1)^r \frac{x_+^{2r+\nu}}{2^{2r+\nu} r! \Gamma(\nu + r + 1)}, \quad x \in \mathbb{R}, \tag{1.10}$$

from the domain $\operatorname{Re} \nu > 0$. By Theorem 1.1, each term $\dfrac{x_+^{2r+\nu}}{\Gamma(\nu + r + 1)}$ $(r = 0, 1, \ldots)$ is a holomorphic function of $\nu \in \mathbb{C}$ with values in $\mathscr{D}'(\mathbb{R})$. For integers $\nu = -1, -2, \ldots$, we have

$$\frac{x_+^{2r+\nu}}{\Gamma(\nu + r + 1)} = \begin{cases} \dfrac{(-1)^r \delta^{(\varkappa-1)}(x)}{(k-1)!} & \text{if } 2r + \nu = -k = -1, -2, \ldots; \\ 0 \quad \text{if } 2r + \nu \geqslant 0, \quad \nu + r + 1 = 0, -1, -2, \ldots, \\ \text{that is, if } -2r \leqslant \nu \leqslant -r - 1. \end{cases} \tag{1.11}$$

Hence, taking $\nu = -l$ $(l = 1, 2, \ldots)$ in (1.10), we obtain

$$J_{-l}(x_+) = 2^l \sum_{0 \leqslant 2r < l} \frac{(-1)^{l-r-1} \delta^{(l-2r-1)}(x)}{2^{2r} r! (l - 2r - 1)!} + (-1)^l J_l(x_+). \tag{1.12}$$

§2. The Function P_+^λ for the Case when $P(x)$ is a Quadratic Form of the Type $(m, n - m)$ with Real Coefficients

Let $P(x) = B(x)$ be a non-degenerate quadratic form in $\mathbb{R}^n$, with real coefficients. By a linear coordinate transformation, we can write

$$B(x) = x_1^2 + \cdots + x_m^2 - x_{m+1}^2 - \cdots - x_n^2, \tag{2.1}$$

where $0 \leqslant m \leqslant n$. We show that $P_+^\lambda(x)$ is analytic for $\operatorname{Re} \lambda > -1$ with values in $\mathscr{D}'(\mathbb{R}^n)$ and that it can be continued to be meromorphic for $\lambda \in \mathbb{C}$. If $m = 0$, then $B_+^\lambda(x) \equiv 0$. If $m \neq 0$, we consider separately the two cases $m = n$ and $1 \leqslant m \leqslant n - 1$. This is because when $m = n$, the set $B(x) = 0$ consists of only one point $x = 0$, while for $1 \leqslant m \leqslant n - 1$ this set is a conical surface $Q \subset \mathbb{R}^n$ which is smooth except at the vertex $x = 0$.

Remark 2.1. If $m = n$, the function $B_+^\lambda(x) = |x|$ is a holomorphic function of $\lambda \in \mathbb{C}$ for $x \neq 0$, with vaues in $\mathscr{D}'(\mathbb{R}^n \backslash \{0\})$. When $1 \leqslant m \leqslant n$, evidently grad $B(x) \neq 0$ for $x \neq 0$, and accordingly B_+^λ can be expressed as the composition of $(\cdot)_+^\lambda$ and $B(x)$:

$$B_+^\lambda(x) = (B(x))_+^\lambda, \quad x \neq 0. \tag{2.2}$$

Consequently, by Theorem 1.1, $B_+^\lambda(x)|_{x\neq 0}$ is a meromorphic function of $\lambda \in \mathbb{C}$ with values in $\mathscr{D}'(\mathbb{R}^n \backslash \{0\})$ and residues (see (1.2))

$$\operatorname*{res}_{\lambda=-k} B_+^\lambda(x) = \frac{(-1)^{k-1}\delta^{(k-1)}(B(x))}{(k-1)!}, \quad x \neq 0, \quad k = 1, 2, \ldots \tag{2.3}$$

2.1. The Case $m = n$. In this case $B_+(x) = |x|^2$ and $B_+^\lambda(x) = |x|^{2\lambda}$. We investigate only $|x|^\lambda$, all the corresponding results for $|x|^{2\lambda}$ being obtained by a simple scaling. We use spherical coordinates $x = r\omega$, where $|\omega| = 1$ and $r = |x| > 0$. With $\varphi(x) \in \mathscr{D}(\mathbb{R}^n)$, the integral

$$\langle |x|^\lambda, \varphi(x)\rangle = \int_0^\infty r^{\lambda+n-1}\overline{\varphi}(r)\, dr = \langle r_+^{\lambda+n-1}, \overline{\varphi}(r)\rangle \tag{2.4}$$

converges for Re $\lambda > -n$; here

$$\overline{\varphi}(r) = \int_{|\omega|=1} \varphi(r\omega)\, d\omega.$$

We expand $\varphi(x)$ in a Taylor series at the point $x = 0$ and see that, as $r \to 0$, $\overline{\varphi}(r)$ has a scrics cxpansion in which only even powers of r occur, that is, only the terms involving $r^0, r^2, \ldots$ are present. Accordingly, by Theorem 1.1, the integral (2.4) has poles only for $\lambda + n - 1 = -1, -3, \ldots, -1 - 2k, \ldots$ $(k = 0, 1, 2, \ldots)$ with residues (see Gel'fand and Shilov [1958a])

$$\operatorname*{res}_{\lambda=-n-2k} \langle |x|^\lambda, \varphi\rangle = \frac{\overline{\varphi}^{(2k)}(0)}{(2k)!} = \Omega_n \frac{\Delta^k\varphi(0)}{2^k k! n \ldots (n + 2k - 2)}, \tag{2.5}$$

where Ω_n is the surface area of the sphere $|x| = 1$ in $\mathbb{R}^n$. We note that the last relation is trivial for $k = 0$.

It follows from (2.5) that $|x|^\lambda$ is meromorphic in $\mathbb{C}$ with values in $\mathscr{D}'(\mathbb{R}^n)$ and simple poles at $\lambda = -n - 2k$, $k = 0, 1, 2, \ldots$ with residues

$$\operatorname*{res}_{\lambda=-n-2k} |x|^\lambda = \Omega_n \frac{\Delta^k\delta(x)}{2^k k! n \ldots (n + 2k - 2)}. \tag{2.6}$$

In particular, for $k = 0$

$$\operatorname*{res}_{\lambda=-n} |x|^\lambda = \Omega_n \delta(x). \tag{2.7}$$

Thus all the residues of $|x|^\lambda$ are concentrated at the point $x = 0$. Note that in the case $n = 1$, important for our later work, we have

$$\operatorname*{res}_{\lambda=-1-2k} |x|^\lambda = 2\frac{\delta^{(2k)}(x)}{2^k k!(2k-1)!}, \quad k = 0, 1, 2, \dots \tag{2.8}$$

From what has just been proved, we obtain

Theorem 2.1. *For $m = n$, the generalized function $B^\lambda = |x|^{2\lambda}$ is a meromorphic function of $\lambda \in \mathbb{C}$ with values in $\mathcal{D}'(\mathbb{R}^n)$ and simple poles at the points $\lambda = -\frac{n}{2} - k$. The residue at $\lambda = -\frac{n}{2}$ is given by*

$$\operatorname*{res}_{\lambda=-(n/2)} B^\lambda = \frac{1}{2}\Omega_n \delta(x). \tag{2.7'}$$

2.2. Application to Decomposition of δ-Function into Plane Waves. We take $a(x) = |\omega \cdot x|$, where $\omega \in \mathbb{R}^n \backslash 0$ and $\omega \cdot x = \omega_1 x_1 + \cdots + \omega_n x_n$ is a "plane wave". Then, for $\operatorname{Re} \lambda > -1$, we have

$$a_+^\lambda(x) = |\omega \cdot x|^\lambda \tag{2.9}$$

as a result of substituting $y = \omega \cdot x$ into $|y|^\lambda$, $y \in \mathbb{R}$. Since $\operatorname{grad} y \equiv \omega \neq 0$, it follows from (2.8) that $|\omega \cdot x|^\lambda$ can be continued from the domain $\operatorname{Re} \lambda > -1$ as a meromorphic function of $\lambda \in \mathbb{C}$ with values in $\mathcal{D}'(\mathbb{R}^n)$ and having simple poles at the points $\lambda = -1 - 2k$, $k = 0, 1, 2, \dots$ with the residues

$$\operatorname*{res}_{\lambda=-1-2k} |\omega \cdot x|^\lambda = 2\frac{\delta^{(2k)}(\omega \cdot x)}{2^k k!(2k-1)!}, \tag{2.10}$$

concentrated on the hyperplane $\omega \cdot x = 0$.

We take the mean value of (2.9) over the sphere $|\omega| = 1$. Then

$$\int_{|\omega|=1} |\omega \cdot x|^\lambda \, d\omega = C(\lambda, n) r^\lambda$$

for $\operatorname{Re} \lambda > -1$, where

$$C(\lambda, n) = 2\pi^{(n-1)/2} \frac{\Gamma\left(\frac{\lambda+1}{2}\right)}{\Gamma\left(\frac{\lambda+n}{2}\right)} \tag{2.11}$$

(see Gel'fand and Shilov [1958a]), whence

$$\frac{1}{\pi^{(n-1)/2}} \int_{|\omega|=1} \frac{|\omega \cdot x|^\lambda}{\Gamma\left(\frac{\lambda+1}{2}\right)} d\omega = \frac{2|x|^\lambda}{\Gamma\left(\frac{\lambda+n}{2}\right)}. \tag{2.12}$$

By (2.6) and (2.10), both sides of (2.12) can be continued to be analytic functions of $\lambda \in \mathbb{C}$. Hence the fact that they coincide for $\operatorname{Re} \lambda > -1$ implies that they coincide for all $\lambda \in \mathbb{C}$. The left-hand side has to be viewed as the integral of a function of a parameter ω with values in $\mathcal{D}'(\mathbb{R}^n)$.

When $\lambda = -n$, (2.12) and (2.6) yield

$$\frac{1}{\pi^{(n-1)/2}} \int_{|\omega|=1} \left(\frac{|\omega \cdot x|^{\lambda}}{\Gamma\left(\frac{\lambda+1}{2}\right)} \right) \Bigg|_{\lambda=-n} d\omega = \Omega_n \delta(x). \tag{2.13}$$

This is the decomposition of the δ-function into plane waves which has important applications (see Gel'fand and Shilov [1958a]).

2.3. The Case $1 \leqslant m \leqslant n-1$. In bipolar coordinates $(x_1, \ldots, x_m) = r\omega_m$, $(x_{m+1}, \ldots, x_n) = \rho\theta_{n-m}$, where $r, \rho > 0$ and $|\omega_m| = |\theta_{n-m}| = 1$, we have $B(x) = r^2 - \rho^2$. Therefore, with $\operatorname{Re} \lambda > -1$ and $\varphi \in \mathscr{D}(\mathbb{R}^n)$,

$$\langle B_+^{\lambda}(x), \varphi(x) \rangle = \int_0^{\infty} \left(\int_0^r (r^2 - \rho^2)^{\lambda} \overline{\varphi}(r, \rho) \rho^{n-m-1} \, d\rho \right) r^{m-1} \, dr, \tag{2.14}$$

where

$$\overline{\varphi}(r, \rho) = \int_{|\omega_m| = |\theta_{n-m}| = 1} \varphi(r\omega_m, \rho\theta_{n-m}) \, d\omega_m \, d\theta_{n-m}. \tag{2.15}$$

Just as before, $\overline{\varphi}(r, \rho)$ is expanded, as $r \to 0$ or $\rho \to 0$, in a series of only even powers of r and ρ.

The change of variables $\rho = \tau r$ and $1 - \tau = s$ yields

$$\begin{aligned}
\langle B_+^{\lambda}, \varphi \rangle &= \int_0^{\infty} \left(\int_0^1 (1 - \tau^2)^{\lambda} r^{2\lambda} \overline{\varphi}(r, \tau r) r^{n-1} \tau^{n-m-1} \, d\tau \right) dr \\
&= \int_0^1 s^{\lambda} \left[(2 - s)^{\lambda} \int_0^{\infty} r^{2\lambda+n-1} \overline{\varphi}(r, (1-s)r)(1-s)^{n-m-1} \, dr \right] ds \\
&= \int_0^1 s^{\lambda} \left[\int_0^{\infty} r^{2\lambda+n-1} I(r, s, \lambda) \, dr \right] ds \\
&= \langle s_+^{\lambda}, \langle r_+^{2\lambda+n-1}, I(r, s, \lambda) \rangle \rangle,
\end{aligned} \tag{2.16}$$

where

$$I(r, s, \lambda) = (2 - s)^{\lambda} \overline{\varphi}(r, (1-s)r)(1-s)_+^{n-m-1}. \tag{2.16$'$}$$

It follows from Theorem 1.1 that the internal integral (with respect to r) is a meromorphic function of $\lambda \in \mathbb{C}$ with simple poles at the points λ given by $2\lambda + n - 1 = -1, -3, \ldots, -1 - 2j, \ldots$ $(j = 0, 1, \ldots)$, that is, at the points $\lambda = \lambda_j^0 = -\frac{n}{2} - j$, because, as $r \to 0$, $I(r, s, \lambda)$ has a Taylor expansion in even powers of r only. Therefore, denoting $\mathscr{S}^0 = \left\{ -\frac{n}{2} - j : j = 0, 1, \ldots \right\}$, we have

$$\langle B_+^{\lambda}, \varphi \rangle = \left\langle s_+^{\lambda}, \sum_{\substack{\lambda_j^0 \in \mathscr{S}^0 \\ \lambda_j^0 > -N}} \frac{\partial_r^{2j} I(0+, s, \lambda)}{(2j)! 2(\lambda - \lambda_j^0)} + R_N(s, \lambda) \right\rangle, \tag{2.17}$$

where $\partial_r^{2j} I(0+, s, \lambda)$ and $R_N(s, \lambda)$ are smooth functions of $s \in [0, 1]$ which are holomorphic in λ for $\operatorname{Re} \lambda > -N$. Another applicaton of Theorem 1.1 (more precisely, the method of proof of this theorem) to (2.17) yields

$$\langle B_+^\lambda, \varphi \rangle = \sum_{\substack{\lambda_i^Q \in \mathscr{S}^Q \\ \lambda_i^Q > -N}} \left(\sum_{\substack{\lambda_j^0 \in \mathscr{S}^0 \\ \lambda_j^0 > -N}} \frac{\partial_s^{i-1} \partial_r^{2j} I(0+, 0, \lambda)}{2(i-1)!(2j)!(\lambda - \lambda_i^Q)(\lambda - \lambda_j^0)} + \frac{\partial_s^{i-1} R_N(0+, \lambda)}{(i-1)!(\lambda - \lambda_i^Q)} \right) + R_{NN}(\lambda), \tag{2.18}$$

where $\lambda_i^Q = -i$, $\mathscr{S}^Q = \{-1, -2, \dots\}$ and $R_{NN}(\lambda)$ is holomorphic for $\operatorname{Re} \lambda > -N$.

The above result shows that B_+^λ is meromorphic in $\mathbb{C}$ with poles at the points $\lambda \in \mathscr{S}^Q \cup \mathscr{S}^0$ which are simple if $\lambda \notin \mathscr{S}^Q \cap \mathscr{S}^0$. The residues of B_+^λ at the poles $\lambda \in \mathscr{S}^Q \setminus \mathscr{S}^0$ are concentrated at $s = 0$, that is, at $\rho = r$, or, in other words, on the "cone" $Q = \{x \in \mathbb{R}^n\colon B(x) = 0\}$. The residues at $\lambda \in \mathscr{S}^0 \setminus \mathscr{S}^Q$ are concentrated at $r = 0$, that is, at $\rho = 0$ and $\tau = 1 - s \in [0, 1]$. Thus, these residues are concentrated at the vertex $x = 0$ "on all angular directions $\tau \in [0, 1]$" (where $B(x) > 0$). At the points $\lambda_\Pi \in \mathscr{S}^Q \cap \mathscr{S}^0$ the poles are, in general, of the second order, and the first residue is concentrated on the cone Q (including the vertex $x = 0$), while the second residue $\left(\text{that is, the coefficient of } \dfrac{1}{(\lambda - \lambda_\Pi)^2}\right)$ is concentrated at the vertex $x = 0$ of Q. These observations lead to (Gel'fand and Shilov [1958a])

Theorem 2.2. 1) *If* $1 \leqslant m \leqslant n-1$ *and* $n \geqslant 2$, *the function* $\dfrac{B_+^\lambda(x)}{\Gamma(\lambda+1)}$ *is a holomorphic function of* $\lambda \in \mathbb{C}$ *with values in* $\mathscr{D}'(\mathbb{R}^n \setminus 0)$ *and a meromorphic function of* $\lambda \in \mathbb{C}$ *with values in* $\mathscr{D}'(\mathbb{R}^n)$.

2) *This function has only simple poles situated at the points*

$$\lambda \in \mathscr{S}^0 = \left\{ -\frac{n}{2} - j \colon j = 0, 1, 2, \dots \right\}. \tag{2.19}$$

The residues are concentrated at the vertex $x = 0$ *and the residue at the first pole* $\lambda = -\dfrac{n}{2}$ *is proportional to* $\delta(x)$:

$$\operatorname*{res}_{\lambda = -(n/2)} \frac{B_+^\lambda}{\Gamma(\lambda+1)} = \Omega_{m, n-m} \delta(x), \quad 1 \leqslant m \leqslant n-1, \quad n \geqslant 2; \tag{2.20}$$

$$\Omega_{m,n-m} = \frac{1}{2} \Omega_m \cdot \Omega_{n-m} \left\langle \frac{s_+^\lambda}{\Gamma(\lambda+1)}, (2-s)^\lambda (1-s)_+^{\frac{n-m-1}{2}} \right\rangle \Bigg|_{\lambda = -(n/2)}. \tag{2.21}$$

3) *For* $\lambda \in \mathscr{S}^Q = \{-1, -2, \dots\}$, *the values of* $\dfrac{B_+^\lambda}{\Gamma(\lambda+1)}$ *are generalized functions concentrated on the cone* Q:

$$\frac{B_+^\lambda}{\Gamma(\lambda+1)} \bigg|_{\lambda = -k} = \frac{\delta^{(k-1)}(B(x))}{(k-1)!}, \quad x \neq 0, \quad k = 1, 2, \dots. \tag{2.22}$$

In fact, (2.22) follows from (2.3), while (2.20) follows from (2.17) and (2.16):

$$\left\langle \operatorname*{res}_{\lambda=-(n/2)} \frac{B_+^\lambda}{\Gamma(\lambda+1)}, \varphi \right\rangle$$

$$= \left\langle \frac{s_+^\lambda}{\Gamma(\lambda+1)}, \frac{I(0+, s, \lambda)}{2} \right\rangle \Bigg|_{\lambda=-(n/2)}$$

$$= \frac{1}{2}\left\langle \frac{s_+^\lambda}{\Gamma(\lambda+1)}, (2-s)^\lambda \overline{\varphi}(0,0)(1-s)_+^{n-m-1} \right\rangle \Bigg|_{\lambda=-(n/2)}$$

$$= \frac{1}{2}\Omega_m \Omega_{n-m}\varphi(0) \left\langle \frac{s_+^\lambda}{\Gamma(\lambda+1)}, (2-s)^\lambda (1-s)_+^{n-m-1} \right\rangle \Bigg|_{\lambda=-(n/2)}. \tag{2.23}$$

Example 2.1. When $n = 4$, $m = 1$ and $B(x) = x_1^2 - x_2^2 - x_3^2 - x_4^2$ (a Lorentz interval), we obtain

$$\Omega_{1,3} = 4\pi(-[(2-s)^{-2}(1-s)_+^2]')|_{s=0} = \pi \tag{2.24}$$

from (2.21) and (1.2). Likewise, for $n = 2$, $m = 1$ and $B(x) = x_1^2 - x_2^2$,

$$\Omega_{1,1} = \tfrac{1}{2}2 \cdot 2 \cdot \tfrac{1}{2} = 1. \tag{2.24'}$$

2.4. Application to Bessel Functions. Let $B(x)$ be the quadratic form (2.1) and define the function $H_\nu(z) \equiv z^\nu J_\nu(z)$, with $\arg z \in]-\pi, \pi[$. Then, as for (1.10), we define the composition with $\sqrt{B_+(x)}$ for $\operatorname{Re} \nu > 0$ as the function

$$H_\nu(\sqrt{B_+(x)}) = B_+^{\nu/2} J_\nu(\sqrt{B_+}) \equiv \sum_0^\infty \frac{(-1)^r B_+^{r+\nu}(x)}{2^{2r+\nu} r! \Gamma(\nu+r+1)}, \tag{2.25}$$

and we note that it is continuous for $x \in \mathbb{R}^n$. When $\operatorname{Re} \nu \leqslant 0$, we define this function by its analytic continuation with respect to ν as a function of ν with values in $\mathscr{D}'(\mathbb{R}^n)$.

By Theorem 2.2, each term $\dfrac{B_+^{r+\nu}(x)}{\Gamma(\nu+r+1)}$ in (2.25) is a holomorphic function of $\nu \in \mathbb{C}$ with values in $\mathscr{D}'(\mathbb{R}^n \backslash \{0\})$, and a meromorphic function of $\nu \in \mathbb{C}$ with values in $\mathscr{D}'(\mathbb{R}^n)$ and simple poles at $r + \nu \in \mathscr{S}^0 = \left\{-\dfrac{n}{2} - j : j = 0, 1, 2, \ldots\right\}$.

The residue at $\nu = -r - \dfrac{n}{2}$ is given by (2.20):

$$\operatorname*{res}_{\nu=-r-(n/2)} \frac{B_+^{r+\nu}}{\Gamma(\nu+r+1)} = \Omega_{m,n-m}\delta(x), \quad 1 \leqslant m \leqslant n-1, \quad n \geqslant 2. \tag{2.26}$$

This leads to

Theorem 2.3. *$H_\nu(\sqrt{B_+(x)})$ is a meromorphic function of $\nu \in \mathbb{C}$ with values in $\mathscr{D}'(\mathbb{R}^n)$ and with simple poles at $\nu \in \mathscr{S}^0$.*

We shall use (2.26) in § 4 and § 5 to construct fundamental solutions of second-order differential equations (0.1); in particular, for the wave equation and the Klein-Gordon equation in $\mathbb{R}^k$ (in this case, $n = k + 1$ and $m = 1$).

We note that, as for (1.12), we have

$$H_{-l}(\sqrt{B_+(x)}) = 2^l \sum_{0 \leqslant r \leqslant l-1} \frac{(-1)^{lr}\delta^{(l-r-1)}(B(x))}{2^{2r}r!(l-r-1)!} + (-1)^l B_+^{-l/2} J_l(\sqrt{B_+}), \quad x \neq 0, \tag{2.27}$$

for $\nu = -l$ ($l = 1, 2, \ldots$).

Remark 2.1. The formulae (2.22) and (2.27) are valid only when $x \neq 0$. For (2.22) to be valid for "all" $x \in \mathbb{R}$, the function $\delta^{(k-1)}(B(x))$ must be suitably regularized at the point $x = 0$. Such a regularization is possible for (2.22) only in those cases where $\lambda = -k \notin \mathscr{S}^0 = \left\{-\frac{n}{2} - j : j = 0, 1, 2, \ldots\right\}$, that is, when the left-hand side of (2.22) is meaningful. In that case the right-hand side of (2.22) must be assumed, by definition, to be equal to the left-hand side. Then (2.27) continues to be valid for such a regularization of $\delta^{(l-r-1)}(B(x))$, as is clear from the derivation of (2.27), if

$$-l + r \notin \mathscr{S}^0, \quad r = 0, 1, 2, \ldots, l - 1. \tag{2.28}$$

This condition is satisfied when, for example, n is an odd integer or when

$$1 \leqslant l < n/2. \tag{2.29}$$

Thus for $k \neq \frac{n}{2} + j$, $j = 0, 1, \ldots$, we define a regularization of $\delta^{(k-1)}(B(x))$ using (2.22) which, by (2.16), we can write as

$$\begin{aligned}
&\left\langle \frac{(-1)^{k-1}\delta^{(k-1)}(B(x))}{(k-1)!}, \varphi(x) \right\rangle \\
&\equiv \left\langle \frac{B_+^\lambda}{\Gamma(\lambda+1)}\bigg|_{\lambda=-k}, \varphi \right\rangle \\
&= \left\langle \operatorname*{res}_{\lambda=-k} B_+^\lambda, \varphi \right\rangle \\
&= \frac{1}{(k-1)!}\left(\left(\frac{d}{ds}\right)^{k-1} \langle r_+^{-2k+n-1}, I(r, s, \lambda)\rangle\right)\bigg|_{s=0}
\end{aligned} \tag{2.30}$$

for $\varphi \in \mathscr{D}(\mathbb{R}^n)$. We note that for such k

$$\langle r_+^{-2k+n-1}, I(r, s, \lambda)\rangle = (2-s)^{-k} \int_0^\infty r^{-2k+n-1}\overline{\varphi}(r, (1-s)r)(1-s)^{n-k-1}\, dr \tag{2.31}$$

is a convergent integral when $-2k + n > 0$.

In particular, for $k = 1$ and $n \geqslant 3$, (2.22) holds if we put

$$\langle \delta(B(x), \varphi(x) \rangle = \frac{1}{2} \int_0^\infty r^{n-3} \overline{\varphi}(r, r)\, dr, \tag{2.32}$$

by (2.30) and (2.31). Note that for $n \geqslant 3$ and $k = 1$ the condition (1.24) in Chap. 1 holds and the expression (2.32) corresponds to (1.22) in Chap. 1, that is, it agrees with the "usual" practice of computation with the δ-function as with a unit measure. Thus, using (1.25) in Chap. 1 we obtain

$$\int_{\mathbb{R}^n} \delta(B(x))\varphi(x)\, dx = \int_{\mathbb{R}^n} r^{n-2} \delta(r^2 - \rho^2) \varphi(x) \frac{d^n x}{r^{m-1} \rho^{n-m-1}}$$

$$= \int_0^\infty \int_0^\infty r^{n-2} \frac{1}{2r} \delta(r - \rho) \overline{\varphi}(r, \rho)\, dr\, d\rho$$

$$= \frac{1}{2} \int_0^\infty r^{n-3} \overline{\varphi}(r, r)\, dr, \tag{2.33}$$

just like (1.26) of Chap. 1.

We note that when $k = 1$ this coincidence with the "usual" methods occurs only for $n \geqslant 3$ (as in Example 1.1, Chap. 1). When $n = 2$, the integral (2.32) diverges and (2.22) becomes false because its left-hand side does not exist. We further note that, under the condition (2.29), (2.22) remains valid when we compute with derivatives of the δ-function by the "usual" method because all the integrals occuring in these computations converge. For example, when $l = 2$, the condition (2.29) is satisfied for $n \geqslant 4$, etc. Later, we shall use (2.22) and (2.27) for the case $m = 1$, $n = 4$ and $k = 1$, $l = 1$ when (2.29) holds, these formulae then being

$$\left. \frac{B_+^\lambda}{\Gamma(\lambda + 1)} \right|_{\lambda = -1} = \delta(B(x)), \quad x \in \mathbb{R}^4, \tag{2.34}$$

$$H_{-1}(\sqrt{B_+(x)}) = -\delta(B(x)) - B_+^{-1/2}(x) J_1(\sqrt{B_+(x)}), \quad x \in \mathbb{R}^4. \tag{2.35}$$

Note that in this case $B(x) = x_1^2 - x_2^2 - x_3^2 - x_4^2$ is the Lorentz interval, and that the formulae (2.20), (2.24) and (2.25) lead to

$$\operatorname*{res}_{\lambda = -2} \frac{B_+^\lambda}{\Gamma(\lambda + 1)} = \pi \delta(x) \Rightarrow \operatorname*{res}_{\nu = -2} H_\nu(\sqrt{B_+(x)}) = \pi \delta(x). \tag{2.36}$$

Likewise, for $n = 2$ and $B = x_1^2 - x_2^2$, (2.20), (2.24′) and (2.25) yield

$$\operatorname*{res}_{\lambda = -1} \frac{B_+^\lambda}{\Gamma(\lambda + 1)} = \delta(x) \Rightarrow \operatorname*{res}_{\nu = -1} H_\nu(\sqrt{B_+(x)}) = \delta(x). \tag{2.37}$$

§ 3. Invariant Fundamental Solutions of Second-order Equations with Real Coefficients

In this section we construct fundamental solutions when the equation (0.1), with real coefficients, is of the *second-order*, that is, when $m = 2$. Such equations can be written in the form

$$Pu(x) = \sum_{i,j=1}^{n} a_{ij} \frac{\partial^2 u}{\partial x_i \partial x_j} + \sum_{i=1}^{n} a_i \frac{\partial u}{\partial x_i} + a_0 u = f(x), \quad x \in \mathbb{R}^n. \tag{3.1}$$

We assume that this equation has a non-degenerate quadratic form, that is,

$$\det A \neq 0, \quad A = (a_{ij})_{i,j=1,\ldots,n}, \quad a_{ij} = a_{ji}. \tag{3.2}$$

The substitution $u(x) = v(x)e^{b \cdot x}$, $b \in \mathbb{R}^n$, transforms (3.1) into

$$\begin{aligned} P_1 v := \sum_{i,j=1}^{n} a_{ij} \frac{\partial^2 v}{\partial x_i \partial x_j} + \sum_{i=1}^{n} (2(Ab)_i + a_i) \frac{\partial v}{\partial x_i} + (a_0 + (Ab, b) + (a, b))v \\ = fe^{-bx}. \end{aligned} \tag{3.3}$$

We determine the vector b by the condition that $2Ab + a = 0$, that is, $b = -\frac{1}{2}A^{-1}a$. Then (3.3) becomes

$$\begin{aligned} P_1 v = \sum_{i,j=1}^{n} a_{ij} \frac{\partial^2 v}{\partial x_i \partial x_j} + qv = f(x)e^{-bx} := g(x); \\ q = a_0 - \frac{1}{4}(a, A^{-1}a). \end{aligned} \tag{3.4}$$

Thus (3.1) and (3.4) are equivalent. Further, they have the same fundamental solutions because $\delta(x)e^{-b \cdot x} = \delta(x)$.

We reduce (3.4) to the canonical form by the linear change of variables $x = Cy$. Then A becomes the matrix

$$B = C^{-1}A(C^{-1})^t, \tag{3.5}$$

just as it does in the transfomation of a quadratic form when $(C^{-1})^t$ is applied. Therefore we can choose C so that B becomes diagonal:

$$B = \begin{pmatrix} b_{11} & & 0 \\ & \ddots & \\ 0 & & b_{nn} \end{pmatrix}; \quad b_{ii} = \pm 1, \quad i = 1, \ldots, n. \tag{3.6}$$

Then (3.4) takes the form

$$\begin{aligned} P_1 v = \frac{\partial^2 v}{\partial y_1^2} + \cdots + \frac{\partial^2 v}{\partial y_m^2} - \frac{\partial^2 v}{\partial y_{m+1}^2} - \cdots - \frac{\partial^2 v}{\partial y_n^2} + qv \\ = g(Cy), \quad y \in \mathbb{R}^n. \end{aligned} \tag{3.7}$$

For (3.7) we construct a fundamental solution $\mathscr{E}(y)$ satisfying

$$P_1\mathscr{E}(y) = \frac{\partial^2\mathscr{E}(y)}{\partial y_1^2} + \cdots + \frac{\partial^2\mathscr{E}(y)}{\partial y_m^2} - \frac{\partial^2\mathscr{E}(y)}{\partial y_{m+1}^2} - \cdots - \frac{\partial^2\mathscr{E}(y)}{\partial y_n^2} + q\mathscr{E}(y)$$
$$= \delta(y), \quad y \in \mathbb{R}^n. \tag{3.8}$$

Remark 3.1. If $g(x) = \delta(x)$, then $g(Cy) = \frac{1}{|C|}\delta(y)$ when $C \in GL(n)$. Therefore if $\mathscr{E}(y)$ is a fundamental solution of (3.8), then the function

$$\frac{\mathscr{E}(C^{-1}x)}{|C|} \in \mathscr{D}'(\mathbb{R}^n) \tag{3.8'}$$

is a fundamental solution of (3.4) (and (3.1)).

3.1. Analysis of Invariance Properties of the Equation. Consider the quadratic form

$$B(y) = \sum_{i=1}^{n} b_{ii}y_i^2 = y_1^2 + \cdots + y_m^2 - y_{m+1}^2 - \cdots - y_n^2, \quad y \in \mathbb{R}^n, \tag{3.9}$$

of equation (3.8). This form is invariant under "orthogonal" transformations $C \in O(n, m)$ (for which $C^tBC = B$, that is, $B(Cy) = B(y)$, $y \in \mathbb{R}^n$). The equation (3.8) is itself invariant under the transformation $y \mapsto C^t y$ because $C^tB(C^t)^t = B$ (compare with (3.5)). Thus, if $\mathscr{E}(y)$ is a solution of (3.8), $\mathscr{E}(C^t y)$ is also a solution of the same equation. This can also be seen directly since

$$\sum_k b_{kk}\frac{\partial^2(\mathscr{E}(C^t y))}{\partial y_k^2} = \sum_{k,i,j} b_{kk}\left(\frac{\partial^2\mathscr{E}}{\partial y_i \partial y_j}\right)(C^t y)C^t_{ik}C^t_{jk}$$
$$= \sum_{i,j}(C^tBC)_{ij}\left(\frac{\partial^2\mathscr{E}}{\partial y_i \partial y_j}\right)(C^t y)$$
$$= \sum b_{ii}\frac{\partial^2\mathscr{E}}{\partial y_i^2} = \delta(y), \quad y \in \mathbb{R}^n. \tag{3.10}$$

Moreover, since $B^{-1} = B$, it also follows that $C^{-1}B(C^t)^{-1} = B$, that is, C^t also preserves the form (3.9) and $\mathscr{E}(Cy)$ is a solution of (3.8).

These considerations of invariance suggest the idea of the existence of an *invariant fundamental solution*, that is, a solution for which $\mathscr{E}(Cx) = \mathscr{E}(x)$, $x \in \mathbb{R}^n$, for all $C \in O(n, m)$. In each connected component of the set $\{x \in \mathbb{R}^n\colon B(x) \neq 0\}$ such distributions have the form

$$\mathscr{E}(y) = E(\rho), \tag{3.11}$$

where $\rho := \sqrt{B(y)}$ and $E(\rho)$ is a generalized function. The branch $\sqrt{B(y)}$ is, for the time being, arbitrary.

We can choose $\rho = \sqrt{B(y)} \in C^\infty(\mathbb{R}^n \backslash Q)$, where Q is the "cone" $B(y) = 0$.

The range of values of the function $\rho(y) = \sqrt{B(y)}$, $y \in \mathbb{R}^n \backslash Q$, will in the sequel be denoted by $\mathscr{R}$ for brevity. Clearly $0 \notin \mathscr{R}$.

3.2. Determination of the Regular Part of an Invariant Fundamental Solution. We now find an ordinary differential equation satisfied by $E(\rho)$ assuming that (3.11) defines a fundamental solution, that is, (3.8) holds.

Substituting (3.11) into (3.8), we have

$$\frac{\partial \mathscr{E}(y)}{\partial y_k} = E'(\rho)\frac{\partial \rho}{\partial y_k}, \quad \frac{\partial^2 \mathscr{E}(y)}{\partial y_k^2} = E''(\rho)\left(\frac{\partial \rho}{\partial y_k}\right)^2 + E'(\rho)\frac{\partial^2 \rho}{\partial y_k^2} \tag{3.12}$$

for $y \in \mathbb{R}^n \setminus Q$. But

$$\frac{\partial \rho}{\partial y_k} = \pm\frac{y_k}{\rho}, \quad \frac{\partial^2 \rho}{\partial y_k^2} = \pm\frac{1}{\rho} - \frac{y_k^2}{\rho^3},$$

where the positive sign is taken for $1 \leqslant k \leqslant m$ and the negative sign for $m + 1 \leqslant k \leqslant n$. Hence

$$\frac{\partial^2 \mathscr{E}}{\partial y_k^2} = E''(\rho)\frac{y_k^2}{\rho^2} + E'(\rho)\left(\pm\frac{1}{\rho} - \frac{y_k^2}{\rho^3}\right), \quad \rho \in \mathscr{R}. \tag{3.13}$$

Adding these equations for $k = 1, \ldots, n$, we obtain

$$P_1\mathscr{E}(y) = E''(\rho) + \frac{n-1}{\rho}E'(\rho) + qE(\rho), \quad y \in \mathbb{R}^n \setminus Q. \tag{3.14}$$

Hence, by (3.8),

$$P_1\mathscr{E}(y) = E''(\rho) + \frac{n-1}{\rho}E'(\rho) + qE(\rho) = 0 \tag{3.14'}$$

for $\rho \in \mathscr{R}$, because $\delta(y) = 0$ when $y \neq 0$ and hence when $\rho \in \mathscr{R}$.

For further discussion, it is necessary to distinguish between the two cases $q = 0$ and $q \neq 0$. When $q = 0$, (3.14′) can be easily solved, and we have

$$E(\rho) = \begin{cases} c_1\dfrac{1}{\rho^{n-2}} + c_2, & \rho \in \mathscr{R}, \quad n \neq 2, \\ c_1 + c_2 \ln \rho, & \rho \in \mathscr{R}, \quad n = 2 \end{cases} \tag{3.15}$$

(compare with Riesz [1949]). When $q \neq 0$, the substitution $z = \sqrt{q}\rho$ reduces (3.14′) to the case $q = 1$:

$$P_1\mathscr{E}(y) = q\left(\frac{d^2E}{dz^2} + \frac{n-1}{z}\frac{dE}{dz} + E(z)\right) = 0, \quad z \in Z \equiv \sqrt{q}\mathscr{R}, \tag{3.16}$$

which in turn is transformed by the substitution $E(z) = z^{p_0}w(z)$ to a Bessel equation. Indeed,

$$\begin{aligned} \frac{dE(z)}{dz} &= z^{p_0}\frac{dw(z)}{dz} + p_0 z^{p_0-1}w, \quad z \in Z, \\ \frac{d^2E(z)}{dz^2} &= z^{p_0}\frac{d^2w(z)}{dz^2} + 2p_0 z^{p_0-1}\frac{dw}{dz} + p_0(p_0-1)z^{p_0-2}w(z), \end{aligned} \tag{3.17}$$

and substitution of (3.17) into (3.16) gives

$$P_1\mathscr{E}(y) = q\left[z^{p_0}\frac{d^2w(z)}{dz^2} + \frac{1}{z}z^{p_0}\frac{dw}{dz}(n-1+2p_0)\right.$$
$$\left. + w[z^{p_0} + z^{p_0-2}((n-1)p_0 + p_0(p_0-1))]\right]$$
$$= 0, \quad z \in Z. \tag{3.18}$$

Hence

$$P_1\mathscr{E}(y) = qz^{p_0}\left[\frac{d^2w}{dz^2} + \frac{1}{z}(n-1+2p_0)\frac{dw}{dz}\right.$$
$$\left. + \left(1 + \frac{(n-1)p_0 + p_0(p_0-1)}{z^2}\right)w(z)\right]$$
$$= 0, \quad z \in Z. \tag{3.19}$$

This equation coincides with the standard Bessel equation of order ν_0 if $n-1+2p_0 = 1$ and $\nu_0^2 = -((n-1)p_0 + p_0(p_0-1)) = -p_0(n-2+p_0)$ (Bateman and Erdélyi [1953], Whittaker and Watson [1927]). These equations yield the values of p_0 and ν_0:

$$p_0 = -\frac{n-2}{2}, \quad \nu_0^2 = \left(\frac{n-2}{2}\right)^2. \tag{3.20}$$

For these p_0 and ν_0, (3.19) takes the form

$$\frac{d^2w}{dz^2} + \frac{1}{z}\frac{dw}{dz} + \left(1 - \frac{\nu_0^2}{z^2}\right)w(z) = 0, \quad z \in Z. \tag{3.21}$$

(Bateman and Erdélyi [1953], Whittaker and Watson [1927].) We can take $\nu_0 = p_0 = -\dfrac{n-2}{2}$. The general solution of (3.21) is

$$w(z) = c_1 J_{\nu_0}(z) + c_2 Y_{\nu_0}(z), \quad \nu_0 = -\frac{n-2}{2}. \tag{3.22}$$

(Bateman and Erdélyi [1953], Whittaker and Watson [1927].) Therefore we find that in a domain where $B(y) \neq 0$

$$\mathscr{E}(y) = \mathscr{E}_{\nu_0}(y) = E_{\nu_0}(\rho) = z^{\nu^0}w(z)$$
$$= (\sqrt{q\rho})^{\nu_0}[C_1 J_{\nu_0}(\sqrt{q\rho}) + C_2 Y_{\nu_0}(\sqrt{q\rho})], \quad y \in \mathbb{R}^n \backslash Q. \tag{3.23}$$

(Compare with Hadamard [1932], Leray [1953].) Here $\sqrt{q\rho} = \sqrt{qB(x)}$ is a real-valued function in a domain where $qB(x) > 0$ and is pure imaginary where $qB(x) < 0$. Also, C_1 and C_2 are arbitrary constants in each connected component of the domain $\mathbb{R}^n \backslash Q$ (where $B(y) \neq 0$).

We have thus obtained the required invariant fundamental solution in a domain where $B(y) \neq 0$ in the explicit form (3.23) and (3.15). The functions (3.23) and (3.15) are smooth in $\mathbb{R}^n \backslash Q$ but in the case $1 \leqslant m \leqslant n-1$ and $n \geqslant 4$ they fail to be locally integrable, as $y \to Q$, in a neighbourhood of the surface Q where $B(y) = 0$. Therefore the required fundamental solutions are certain regularizations of these functions, for a suitable choice of C_1 and C_2. For brevity, the functions (3.23) and (3.15) will be referred to as *formal fundamental solutions.*

§4. Regularization of the Formal Fundamental Solution for the Case $q = 0$

When $q = 0$, the formal fundamental solutions are given by (3.15). We now show how to regularize them in order to obtain actual fundamental solutions. In doing so we distinguish between the two cases where, first, $m = 0$ or $m = n$ and, second, $1 \leqslant m \leqslant n-1$.

4.1. The Case $m = 0$ or $m = n$. For $m = 0$ and $m = n$, the fundamental solutions of (3.8) with $q = 0$ differ only in sign. Therefore it suffices to consider the case $m = n$. Then $P_1(\partial_y) = \Delta$ is the Laplace operator in $\mathbb{R}^n$ and $B(y) > 0$ for $y \neq 0$. We now show that for $n \neq 2$ the desired regularization of (3.15) can be constructed as an analytic continuation of the function

$$\hat{\mathscr{E}}_\lambda(y) = C_1 B^\lambda(y) + C_2 = C_1 |y|^{2\lambda} + C_2 \tag{4.1}$$

from the domain $\operatorname{Re} \lambda > 0$ to the point $\lambda = \lambda_0 \equiv -\dfrac{n-2}{2}$. We put $C_2 = 0$. If $\operatorname{Re} \lambda > 1$, then $\hat{\mathscr{E}}_\lambda(y) \in C^2(\mathbb{R}^n)$ and (3.14′) holds, with $\hat{\mathscr{E}}_\lambda$ instead of $\mathscr{E}$, for all $y \in \mathbb{R}^n$ and not only for $y \in \mathbb{R}^n \backslash Q$. Accordingly, for $\operatorname{Re} \lambda > 1$,

$$P_1 \hat{\mathscr{E}}_\lambda(y) = C_1 2\lambda(2\lambda + n - 2) B^{\lambda-1}(y), \quad y \in \mathbb{R}^n. \tag{4.2}$$

Here, by Theorem 2.1, the right-hand side is a meromorphic function of $\lambda \in \mathbb{C}$ with values in $\mathscr{D}'(\mathbb{R}^n)$ and so also is the left-hand side. Exploiting the uniqueness of an analytic continuation, we find that (4.2) is true for all $\lambda \in \mathbb{C}$ except for the discrete set of poles of the functions $\hat{\mathscr{E}}_\lambda$ and $B^{\lambda-1}$.

Remark 4.1. In a similar fashion we can show the validity of various remarkable results of classical analysis for generalized functions defined as analytic continuations with respect to a parameter (or parameters). This is the principal advantage of regularization using analytic continuation with respect to a parameter.

At the point $\lambda = \lambda_0 = -\dfrac{n}{2} + 1$ the function $B^\lambda(y)$ is, by Theorem 2.1, holomorphic while the function $B^{\lambda-1}(y) = |y|^{2(\lambda-1)}$ has a simple pole with residue equal to $\frac{1}{2}\Omega_n \delta(y)$ (see (2.7′)). Moreover, $2\lambda + n - 2 = 0$ when $\lambda = \lambda_0$. Therefore it follows from (4.2) that, as $\lambda \to \lambda_0$,

$$P_1\hat{\mathscr{E}}_{\lambda_0}(y) = C_1 4\lambda_0 \cdot \operatorname*{res}_{\lambda=\lambda_0} B^{\lambda-1} = C_1 2\lambda_0 \Omega_n \delta(y) = -C_1(n-2)\Omega_n\delta(y). \quad (4.3)$$

This equation yields C_1 when $n \neq 2$, and then the formula (4.1) leads to the desired fundamental solutions (2.10′) of Chap. 1.

The procedure just described does not go through for $n = 2$. This is due to the fact that for $\lambda = \lambda_0$ the multiplier of the right-hand side of (4.2) has a second-order zero and $B^{\lambda-1}$ has a first-order pole. Therefore in order to obtain a fundamental solution for $n = 2$ we differentiate (4.2) with respect to λ for $\lambda = \lambda_0 = 0$. This results in

$$\begin{aligned} P_1 \left.\frac{d\hat{\mathscr{E}}_\lambda}{d\lambda}\right|_{\lambda=0} &= C_1\left[(2(2\lambda + n - 2) + 2\lambda\cdot 2)B^{\lambda-1} + 2\lambda(2\lambda + n - 2)\frac{dB^{\lambda-1}}{d\lambda}\right]\Bigg|_{\lambda=0} \\ &= C_1\left[8 \operatorname*{res}_{\lambda=0} B^{\lambda-1} + 4 \operatorname*{res_2}_{\lambda=0} \frac{dB^{\lambda-1}}{d\lambda}\right] \\ &= C_1\left[8\frac{\Omega_2}{2}\delta(y) - 4\frac{\Omega_2}{2}\delta(y)\right] \\ &= C_1 4\pi\delta(y), \end{aligned} \quad (4.4)$$

where $\operatorname*{res_2}_{\lambda=0} \frac{dB^{\lambda-1}}{d\lambda} = \operatorname*{res}_{\lambda=0} B^{\lambda-1}$ is the "second" residue of $\frac{dB^{\lambda-1}}{d\lambda}$ at the point $\lambda = 0$, that is, is the coefficient of λ^{-2}. This leads to the formula (2.10′) of Chap. 1 when $n = 2$.

4.2. The Case $1 \leqslant m \leqslant n - 1$. In this case the functions (3.15) are smooth off the surface Q where $B(y) = 0$. We shall show that we can take $C_2 = 0$, as above, as well as $C_1 = 0$ in a domain where $B(y) < 0$. Namely, we construct a regularization of (3.15) as an analytic continuation of the function

$$\mathscr{E}_\lambda(y) = C_1\frac{B_+^\lambda}{\Gamma(\lambda+1)} = C_1\frac{\rho_+^{2\lambda}}{\Gamma(\lambda+1)}, \quad \rho_+(y) \equiv \sqrt{B_+(y)}, \quad (4.5)$$

to the point $\lambda = \lambda_0 \equiv -\dfrac{n-2}{2}$.

Remark 4.2. It is necessary to introduce the factor $\Gamma(\lambda + 1)$ in the denominator of (4.5) because of the presence of the poles of B_+^λ for $\lambda = -1, -2, \ldots$. These poles are connected with the singularities of B_+^λ on the surface Q, and it may happen that the function B_+^λ has a pole at the point λ_0 (for even $n \geqslant 4$). For the case $m = 0$ or $m = n$ examined above, B_+^λ had a pole only for $\lambda = -\frac{n}{2}, -\frac{n}{2} - 1, \ldots$ and the point $\lambda = \lambda_0$ was non-singular because $Q = \{0\}$ contained just one point.

Clearly, $\mathscr{E}_\lambda(y) \in C^2(\mathbb{R}^n)$ for Re $\lambda > 2$. Therefore (3.14) yields, as for (4.2),

$$P_1 \mathscr{E}_\lambda(y) = C_1 \left[2\lambda(2\lambda - 1) \frac{\rho_+^{2\lambda-2}}{\Gamma(\lambda+1)} + \frac{(n-1)2\lambda\rho_+^{2\lambda-2}}{\Gamma(\lambda+1)} \right]$$

$$= C_1 \frac{2(2\lambda + n - 2)}{\Gamma(\lambda)} B_+^{\lambda-1}(y), \quad y \in \mathbb{R}^n. \tag{4.6}$$

This relation is true for all λ as before, except for the discrete set of poles. Therefore, as $\lambda \to \lambda_0 = -\frac{n}{2} + 1$, we obtain

$$P_1 \mathscr{E}_{\lambda_0}(y) = C_1 4 \operatorname*{res}_{\lambda=\lambda_0} \frac{B_+^{\lambda-1}(y)}{\Gamma(\lambda)} = C_1 4\Omega_{m,n-m}\delta(y) \tag{4.7}$$

as for (4.3), where we have used (2.20). We thus find that $C_1 = \dfrac{1}{4\Omega_{m,n-m}}$ and the desired fundamental solution is

$$\mathscr{E}_{\lambda_0}(y) = \frac{1}{4\Omega_{m,n-m}} \frac{B_+^\lambda(y)}{\Gamma(\lambda+1)}\bigg|_{\lambda=-((n-2)/2)}. \tag{4.8}$$

Example 4.1. For $a = 1$, a fundamental solution of the wave equation (2.6) in Chap. 1 is given by

$$\mathscr{E}_k(t, x) = \frac{1}{4\Omega_{1,k}} \frac{(t^2 - |x|^2)_+^\lambda}{\Gamma(\lambda+1)}\bigg|_{\lambda=-((k-1)/2)} \tag{4.9}$$

in $\mathbb{R}^k$ because $n = k + 1$ and $-\dfrac{n-2}{2} = -\dfrac{k-1}{2}$.

Remark 4.3. In view of Remark 3.1, a fundamental solution of equation (2.6) in Chap. 1, with $a > 0$, is the function (compare with Riesz [1949])

$$\mathscr{E}_k(t, x) = \frac{1}{4\Omega_{1,k} \cdot a^k} \frac{\left(t^2 - \left|\frac{x}{a}\right|^2\right)_+^\lambda}{\Gamma(\lambda+1)}\Bigg|_{\lambda=-((k-1)/2)}$$

$$= \frac{1}{4\Omega_{1,k} a} \frac{(a^2t^2 - |x|^2)_+^\lambda}{\Gamma(\lambda+1)}\bigg|_{\lambda=-((k-1)/2)} \tag{4.9'}$$

since the transformation $(t, x) \mapsto C^{-1}(t, x) = (t, x/a)$ reduces equation (2.6), Chap. 1, to the case $a = 1$, and det $C = a^k$.

Remark 4.4. Let $a = 1$ in the wave equation (2.6) in Chap. 1. Then an invariant fundamental solution is also the function

$$\mathscr{E}_k^\pm(t, x) = \frac{\theta(\pm t)}{2\Omega_{1,k}} \frac{(t^2 - |x|^2)_+^\lambda}{\Gamma(\lambda+1)}\bigg|_{\lambda=-((k-1)/2)} \tag{4.10}$$

which is obtained, by definition, by analytic continuation of the function $\frac{\theta(\pm t)}{2\Omega_{1,k}} \frac{B_+^\lambda}{\Gamma(\lambda+1)}$ to the point $\lambda = -\frac{k-1}{2}$. Indeed, for this function (4.6) holds also for Re $\lambda > 2$. Moreover, this function can be continued as a meromorphic function of $\lambda \in \mathbb{C}$, a fact which can be established by repeating the proof of Theorem 2.2 word for word. However, the poles are now situated at the points $\lambda = -\frac{n}{2}, -\frac{n}{2} - \frac{1}{2}, \ldots$. The reason for this fact is that in defining $\overline{\varphi}$ in (2.15) the integration over $|\omega_m| = 1$ has now to be replaced simply by the substitution $\omega_m \equiv t = \pm 1$, and therefore, as $r \to 0$, $\overline{\varphi}(r, \rho)$ contains all non-negative integral powers of t. Also we now have $\overline{\varphi}(0, 0) = \frac{\Omega_1}{2} \Omega_k \varphi(0)$ and thus the residue (2.20) for $\theta(\pm t) \frac{B_+^\lambda}{\Gamma(\lambda+1)}$ is small by half. Hence in (4.10) the denominator is half that of (4.9).

Corollary 4.1. *By* (4.9), (4.10) *and* (2.22) *we find that for odd* $k = 2l + 1 \geqslant 3$

$$\mathscr{E}_k(t, x) = \frac{1}{4\Omega_{1,k}} \frac{\delta^{(l-1)}(t^2 - |x|^2)}{(l-1)!}, \quad (t, x) \neq 0;$$
$$\mathscr{E}_k^\pm(t, x) = \frac{\theta(\pm t)}{2\Omega_{1,k}} \frac{\delta^{(l-1)}(t^2 - |x|^2)}{(l-1)!}, \quad (t, x) \neq 0. \tag{4.11}$$

This means that for odd $k \geqslant 3$ the fundamental solution $\mathscr{E}_k$ of the wave equation in $\mathbb{R}^k$ is concentrated on the "***light cone***" $t^2 = |x|^2$ while $\mathscr{E}_k^\pm$ is concentrated on the "***future light cone***" (where $t \geqslant 0$) or on the "***past light cone***" (where $t \leqslant 0$), respectively. This implies that for odd $k \geqslant 3$ the solutions of the wave equation in $\mathbb{R}^k$ have sharp front and rear edges (see § 2, Chap. 1). For even $k \geqslant 2$ and for $k = 1$ the fundamental solutions $\mathscr{E}_k$ and $\mathscr{E}_k^\pm$ are concentrated in the interior of the light cone, that is, in the domain $t^2 - |x|^2 \geqslant 0$ (and $t \gtrless 0$), and consequently the solutions of the wave equation have sharp front edge but do not have sharp rear edge, that is, "***wave diffusion***" takes place.

Remark 4.5. With $k = 3$, (4.11) yields (2.6′) in Chap. 1 for $\mathscr{E}_3^\pm$ because $\Omega_{1,3} = \pi$, by (2.24). Furthermore, with $k = 1$, (4.10) leads to (2.6′) in Chap. 1 for $\mathscr{E}_1^\pm$ because $\Omega_{1,1} = 1$, by (2.24′). Finally, the formula (2.6′) in Chap. 1, for $\mathscr{E}_2^\pm$, can also be obtained from (4.10) if we compute $\Omega_{1,2}$. However, it is easier to obtain $\mathscr{E}_2^\pm$ by the *method of descent* (Petrovskij [1961], Vladimirov [1981]): we have, formally,

$$\mathscr{E}_2^\pm(t, x_1, x_2) = \int_{-\infty}^{+\infty} \mathscr{E}_3^\pm(t, x_1, x_2, x_3)\, dx_3. \tag{4.12}$$

A similar formula connects $\mathscr{E}_k^\pm$ and $\mathscr{E}_{k+1}^\pm$ for any $k \geqslant 1$. Note that for odd k we can easily compute $\Omega_{1,k}$ by means of differentiation just as we did in (2.24) and (2.24′). Therefore $\mathscr{E}_k^\pm$ can be obtained from (4.11) for odd k, and from the method of descent for even k, as in (4.12). As a corollary we also have $\Omega_{1,2} = \sqrt{\pi}$.

§ 5. Regularization of the Fundamental Solution for the Case $q \neq 0$

If $q \neq 0$, $\mathscr{E}(y)$ is expressed by (3.23) in a region where $B(y) \neq 0$. Let us show that the fundamental solution $\mathscr{E}(y)$ can be constructed as a regularization of the function (3.23). In doing so we can take $C_2 = 0$, at least when $n \neq 2$. Thus we seek the fundamental solution as a regularization of the function (see (2.25))

$$\mathscr{E}_{\nu_0}(y) = C_1(\sqrt{qB})^{\nu_0} J_{\nu_0}(\sqrt{qB}) = C_1 H_{\nu_0}(\sqrt{qB(y)}), \quad y \in \mathbb{R}^n \backslash Q, \tag{5.1}$$

where C_1 remains constant in connected components of the domain $\mathbb{R}^n \backslash Q$. As we mentioned earlier, when $q \neq 0$ we must distinguish the two cases in which, first, $1 \leqslant m \leqslant n-1$ and, second, $m = 0$ or $m = n$. By multiplying the equation (3.8) by -1, we can assume that $q > 0$.

5.1. The Case $1 \leqslant m \leqslant n - 1$. We assume that $C_1 = 0$ in a domain where $B(y) < 0$ and that C_1 is a constant in the domain where $B(y) > 0$, although this domain may have more than one connected component (as is the case, for example, with the Lorentz interval $B(y) = y_1^2 - y_2^2 - \cdots - y_n^2$). Then (5.1) (with $q > 0$) becomes

$$\mathscr{E}_{\nu_0}(y) = C_1 H_{\nu_0}(\sqrt{qB_+(y)}), \quad y \in \mathbb{R}^n \backslash Q. \tag{5.2}$$

We shall show that the desired regularization of the function (5.2) can be constructed as an analytic continuation of the function

$$\mathscr{E}_\nu(y) := C_1 H_\nu(\sqrt{qB_+(y)}), \quad y \in \mathbb{R}^n, \tag{5.3}$$

with respect to the parameter ν from the domain $\operatorname{Re} \nu > 0$ to the point $\nu = \nu_0 \equiv -\dfrac{n-2}{2}$. It is clear from (2.25) that $|H_\nu(\sqrt{qB_+(y)})| \sim |B_+(y)|^{\operatorname{Re} \nu}$ as $y \to Q$. Therefore $\mathscr{E}_\nu(y) \in C^2(\mathbb{R}^n)$ for $\operatorname{Re} \nu > 2$, and all the formulae (3.16)–(3.19) for $\mathscr{E}$ continue to be valid for $\mathscr{E}_\nu$, with ν and J_ν replacing p_0 and w. This holds for all $y \in \mathbb{R}^n$ and not only for $y \in \mathbb{R}^n \backslash Q$.

Therefore (3.19) (with $p_0 = \nu$) implies that, for $\operatorname{Re} \nu > 2$,

$$\begin{aligned} P_1 \mathscr{E}_\nu(y) &= C_1 q z^\nu \left[\frac{d^2 J_\nu}{dz^2} + \frac{1}{z}(n - 1 + 2\nu)\frac{dJ_\nu}{dz} + \left(1 + \frac{\nu(n-2+\nu)}{z^2}\right) J_\nu(z) \right] \\ &= C_1 q(n - 2 + 2\nu) \left[z^{\nu-1} \frac{dJ_\nu}{dz} + \nu z^{\nu-2} J_\nu \right], \quad y \in \mathbb{R}^n; \\ z &= z(y) \equiv \sqrt{qB_+(y)}. \end{aligned} \tag{5.4}$$

As before, this equality is also valid for all $\nu \in \mathbb{C}$ except for the discrete set of poles if, on the right-hand side, the functions $z^{\nu-1} \dfrac{dJ_\nu}{dz}$ and $z^{\nu-2} J_\nu$ are defined in the domain $\operatorname{Re} \nu < 2$ as analytic continuations with respect to ν with values in

$\mathscr{D}'(\mathbb{R}^n)$. The existence of such continuations follows obviously from Theorem 2.2 and formula (1.9) just as Theorem 2.3 did.

Theorem 2.3 implies, in particular, that the left-hand side of (5.4) is analytic in ν at the point $\nu = \nu_0 \equiv -\frac{n-2}{2}$. Let us find the limit of the right-hand side of (5.4) as $\nu \to \nu_0$.

We note that the terms $z^{\nu-1}\frac{dJ_\nu}{dz}$ and $z^{\nu-2}J_\nu$ on the right-hand side of (5.4) have simple poles at the point $\nu = \nu_0$ and the multiplier $n - 2 + 2\nu$ vanishes at this point. Therefore the right-hand side of (5.4) has a limit as $\nu \to \nu_0$ and is expressed in terms of the residues of $z^{\nu-1}\frac{dJ_\nu}{dz}$ and $z^{\nu-2}J_\nu$ at $\nu = \nu_0$. These residues can be found easily by using (2.20) and (1.19). Thus, since $\nu_0 = -\frac{n}{2} + 1$, it follows that

$$\begin{aligned}\operatorname*{res}_{\nu=\nu_0} z^{\nu-1}\frac{dJ_\nu}{dz} &= \operatorname*{res}_{\nu=\nu_0} (\sqrt{qB_+})^{\nu-1}2^{-\nu}\frac{\nu(\sqrt{qB_+})^{\nu-1}}{\Gamma(\nu+1)} \\ &= 2^{-\nu_0}q^{\nu_0-1}\operatorname*{res}_{\nu=\nu_0}\frac{B_+^{\nu-1}}{\Gamma(\nu)} = \frac{1}{2}\left(\frac{2}{q}\right)^{n/2}\operatorname*{res}_{\mu=-(n/2)}\frac{B_+^{\mu}}{\Gamma(\mu+1)} \\ &= \frac{1}{2}\left(\frac{2}{q}\right)^{n/2}\Omega_{m,n-m}\delta(y).\end{aligned} \tag{5.5}$$

Likewise,

$$\begin{aligned}\operatorname*{res}_{\nu=\nu_0} \nu z^{\nu-2}J_\nu &= \operatorname*{res}_{\nu=\nu_0} (\sqrt{qB_+})^{\nu-2}2^{-\nu}\frac{\nu(\sqrt{qB_+})^{\nu}}{\Gamma(\nu+1)} \\ &= \frac{1}{2}\left(\frac{2}{q}\right)^{n/2}\Omega_{m,n-m}\delta(y).\end{aligned} \tag{5.6}$$

Letting $\nu \to \nu_0$ in (5.4) and using (5.5) and (5.6), we obtain

$$\begin{aligned}P_1\mathscr{E}_{\nu_0}(y) &= C_1 q 2 \operatorname*{res}_{\nu=\nu_0}\left(z^{\nu-1}\frac{dJ_\nu}{dz} + \nu z^{\nu-2}J_\nu\right) \\ &= C_1 2q\left(\frac{2}{q}\right)^{n/2}\Omega_{m,n-m}\delta(y) = C_1 q^{\nu_0}2^{(n/2)+1}\Omega_{m,n-m}\delta(y).\end{aligned} \tag{5.7}$$

We thus find C_1 and the desired fundamental solution (see (2.25)):

$$\begin{aligned}\mathscr{E}_{\nu_0}(y) &= C_1 H_{\nu_0}(\sqrt{qB_+}) \\ &= \frac{q^{-\nu_0}}{2^{(n/2)+1}\Omega_{m,n-m}}[(qB_+(y))^{\nu/2}J_\nu(\sqrt{qB_+(y)})]|_{\nu=-((n-2)/2)} \\ &= \frac{q^{(n-2)/4}}{2^{(n/2)+1}\Omega_{m,n-m}}[B_+^{\nu/2}(y)J_\nu(\sqrt{qB_+(y)})]|_{\nu=-((n-2)/2)}.\end{aligned} \tag{5.8}$$

It can be seen easily that when $q_0 \to 0$ this formula reduces to (4.8). The above formula has been obtained under the assumption that $q > 0$. When $q < 0$, we reduce the problem to the preceding case by multiplying (3.8) by -1. Moreover, (5.8) provides a fundamental solution by replacing C_1 by $-C_1$, q by $-q$, B by $-B$ and interchanging m and $n-m$:

$$\mathscr{E}(x) = -\frac{|q|^{(n-2)/4}}{2^{(n/2)+1}\Omega_{n-m,m}}[(-B)_+^{\nu/2}J_\nu(\sqrt{(qB)_+})]|_{\nu=-((n-2)/2)}, \quad q < 0. \tag{5.9}$$

Corollary 5.1. *With $a = 1$, the Klein-Gordon equation (the equation* (2.7), Chap. 1) *in $\mathbb{R}^k$ has a fundamental solution*

$$\mathscr{E}_k(t, x) = \frac{m_0^{(k-1)/2}}{2^{(k+3)/2}\Omega_{1,k}}[(t^2 - |x|^2)_+^{\nu/2}J_\nu(m_0\sqrt{(t^2 - |x|^2)_+})]|_{\nu=-((k-1)/2)}. \tag{5.10}$$

The case $a > 0$ is reduced to the case $a = 1$ as mentioned in Remark 4.3.

Therefore for $a \neq 1$ a fundamental solution (compare with (4.9′)) of the Klein-Gordon equation is the function (compare with Hadamard [1932], Leray [1953]):

$$\mathscr{E}_k(t, x) = \frac{m_0^{(k-1)/2}}{2^{(k+3)/2}\Omega_{1,k}a^k} \times \left[\left(t^2 - \left|\frac{x}{a}\right|^2\right)_+^{\nu/2} J_\nu\left(m_0\sqrt{\left(t^2 - \left|\frac{x}{a}\right|^2\right)_+}\right)\right]\Bigg|_{\nu=-((k-1)/2)}. \tag{5.11}$$

Further, in analogy with (4.10), fundamental solutions are also given by

$$\mathscr{E}_k^{\pm}(t, x) = \frac{\theta(\pm t)m_0^{(k-1)/2}}{2^{(k+1)/2}\Omega_{1,k}a^k} \times \left[\left(t^2 - \left|\frac{x}{a}\right|^2\right)_+^{\nu/2} J_\nu\left(m_0\sqrt{\left(t^2 - \left|\frac{x}{a}\right|^2\right)_+}\right)\right]\Bigg|_{\nu=-((k-1)/2)} \tag{5.12}$$

which are defined by analytic continuation with respect to ν from the domain $\operatorname{Re}\nu > 0$ to the point $\nu = \nu_0 \equiv -\dfrac{k-1}{2}$; $\mathscr{E}_k^{\pm}(t, x)$ are the so-called "retarded" and "advanced" fundamental solutions of the Klein-Gordon equation (Bogolyubov and Shirkov [1973]).

Example 5.1. With $k = 1$, (5.12) leads to (2.7′) in Chap. 1 for $\mathscr{E}_1^{\pm}$, because $\Omega_{1,1} = 1$, by (2.24′), and $\nu_0 = 0$. When $k = 2$, (5.12) again yields (2.7′) in Chap. 1 for $\mathscr{E}_2^{\pm}$, because $\Omega_{1,2} = \sqrt{\pi}$ (see Remark 4.5) and $\nu_0 = -\frac{1}{2}$. Finally, for $k = 3$, (5.12), together with (2.35), yields

$$\mathscr{E}_3^{\pm}(t, x) = \frac{\theta(\pm t)m_0^2}{4\Omega_{1,3}a^3}\left(\delta\left(m_0^2\left(t^2 - \left|\frac{x}{a}\right|^2\right)\right) - \left(m_0^2\left(t^2 - \left|\frac{x}{a}\right|^2\right)\right)^{-(1/2)} J_1\left(m_0\sqrt{\left(t^2 - \left|\frac{x}{a}\right|^2\right)_+}\right)\right), \tag{5.13}$$

which coincides with (2.7′) in Chap. 1 for $\mathscr{E}_3^{\pm}$, because $\Omega_{1,3} = \pi$, by (2.24).

Note that, in view of Remark 2.1, the expression $\delta(a^2t^2 - |x|^2)$ in formula (2.7′), Chap. 1, is to be understood in the sense of Definition 1.4, Chap. 1, because in this case the condition (2.29) is satisfied ($l = 1$ and $n = 4$).

5.2. The Case $m = 0$ or $m = n$. As before, it suffices to examine the case $q > 0$.

We first take up the case $m = n$ and $q > 0$. Then $P_1 = \Delta + q$ is a Helmholtz operator. We shall show that for $n \neq 2$ a fundamental solution of P_1 can be constructed as a regularization of a function of the form (5.3). However, when $m = n$, (5.5)–(5.7) are not valid, since their proof uses (2.20) which is true only for the case $1 \leqslant m \leqslant n - 1$. Indeed, when $m = n$ the function B_+^{ν} has no poles at $\nu = -1, -2, \ldots$ if $\nu \notin \mathscr{S}^0$. Therefore the factor $\Gamma(\nu + 1)$ occuring in the "denominator" of the Bessel function J_ν (see (2.25)) must be dropped. The situation in § 4 was exactly similar for the case $q = 0$. This can be seen by comparing (4.5) and (4.1) (see also Remark 4.2). Thus we shall construct a fundamental solution as an analytic continuation with respect to ν of the function

$$\begin{aligned}\hat{\mathscr{E}}_\nu(y) &:= C_1 \Gamma(\nu + 1) H_\nu(\sqrt{qB}) \\ &= C_1 \sum_{r=0}^{\infty} (-1)^r \frac{(qB(y))^{r+\nu}}{2^{2r+\nu} r!(\nu + r)\ldots(\nu + 1)}\end{aligned} \tag{5.14}$$

to the point $\nu = \nu_0 := -\dfrac{n-2}{2}$. Here we have used (2.25) and the identity $\Gamma(\nu + r + 1) = (\nu + r)\ldots(\nu + 1)\Gamma(\nu + 1)$. Clearly, $B(y) \equiv B_+(y)$ now.

It is clear from (5.14) that $\hat{\mathscr{E}}_\nu(y) \in C^2(\mathbb{R}^n)$ for $\operatorname{Re} \nu > 2$, and Theorem 2.1 implies that $\hat{\mathscr{E}}_\nu(y)$ can be continued from the domain $\operatorname{Re} \nu > 0$ as a meromorphic function of $\nu \in \mathbb{C}$ with simple poles at the points $\nu \in \mathscr{S}^0$. For $\operatorname{Re} \nu > 2$, the function (5.14) differs from (5.3) only in the non-zero multiplier $\Gamma(\nu + 1)$. Therefore a formula of the form (5.4) remains valid for the function (5.14) provided that J_ν is replaced by the function

$$\mathscr{J}_\nu(z) \equiv \Gamma(\nu + 1) J_\nu(z) = \sum (-1)^r \frac{z^{2r+\nu}}{2^{2r+\nu} r!(\nu + r)\ldots(\nu + 1)}. \tag{5.15}$$

Thus, for $\operatorname{Re} \nu > 2$,

$$\begin{gathered} P_1 \hat{\mathscr{E}}_\nu(y) = C_1 q(n - 2 + 2\nu)\left[z^{\nu-1} \frac{d\mathscr{J}_\nu}{dz} + \nu z^{\nu-2} \mathscr{J}_\nu(z)\right], \\ y \in \mathbb{R}^n; \quad z \equiv \sqrt{qB(y)}. \end{gathered} \tag{5.16}$$

As before, this formula is valid for all $\nu \in \mathbb{C}$ except for the discrete set of poles of each side. The left-hand side is holomorphic for $\nu = \nu_0$. Let us find the limit of the right-hand side as $\nu \to \nu_0$. This can be obtained in exactly the same way as (5.7). Thus, as for (5.5) and (5.6), we obtain from (2.7′)

$$\operatorname*{res}_{\nu=\nu_0} z^{\nu-1}\frac{d\mathscr{J}_\nu}{dz} = 2^{-\nu_0}q^{\nu_0-1}\operatorname*{res}_{\nu=\nu_0} \nu B_+^{\nu-1} = 2^{-\nu_0}q^{\nu_0-1}\nu_0 \operatorname*{res}_{\mu=-(n/2)} B^\mu$$
$$= \frac{1}{4}\left(\frac{2}{q}\right)^{n/2} \nu_0 \Omega_n \delta(y) \tag{5.17}$$

and

$$\operatorname*{res}_{\nu=\nu_0} \nu z^{\nu-2}\mathscr{J}_\nu = 2^{-\nu_0}q^{\nu_0-1}\operatorname*{res}_{\nu=\nu_0} \nu B^{\nu-1} = \frac{1}{4}\left(\frac{2}{q}\right)^{n/2} \nu_0 \Omega_n \delta(y). \tag{5.18}$$

Letting $\nu \to \nu_0$ in (5.16) and using (5.17) and (5.18), we obtain

$$P_1 \hat{\mathscr{E}}_{\nu_0}(y) = C_1 q\left(\frac{2}{q}\right)^{n/2} \nu_0 \Omega_n \delta(y) = -C_1(n-2)\left(\frac{2}{q}\right)^{n/2-1} \Omega_n \delta(y). \tag{5.19}$$

This formula, with $n \neq 2$, yields C_1 and the desired fundamental solution of the operator $\Delta + q$ $(q > 0)$ in the form

$$\hat{\mathscr{E}}_{\nu_0}(y) = C_1[\Gamma(\nu+1)H_\nu(\sqrt{qB(y)})]|_{\nu=\nu_0}$$
$$= -\frac{q^{n/2-1}}{(n-2)\Omega_n 2^{n/2-1}}[\Gamma(\nu+1)(\sqrt{qB})^\nu J_\nu(\sqrt{qB})]|_{\nu=\nu_0}. \tag{5.20}$$

In order to obtain a fundamental solution for $n = 2$, it is enough to differentiate (5.16) with respect to ν for $\nu = \nu_0$. Then, as for (4.4), we obtain

$$P_1 \frac{d\hat{\mathscr{E}}_\nu}{d\nu} = C_1 q\left\{2\left[z^{\nu-1}\frac{d\mathscr{J}_\nu}{dz} + \nu z^{\nu-2}\mathscr{J}_\nu\right]\right.$$
$$\left. + (n-2+2\nu)\frac{d}{d\nu}\left[z^{\nu-1}\frac{d\mathscr{J}_\nu}{dz} + \nu z^{\nu-2}\mathscr{J}_\nu\right]\right\} \tag{5.21}$$

for $\operatorname{Re}\nu > 2$. Since an analytic continuation is unique, the last identity is also valid for all $\nu \in \mathbb{C}$ except for the discrete set of poles. At the point $\nu = \nu_0 \equiv -\dfrac{n-2}{2} = 0$ the left-hand side of (5.21) is holomorphic. Therefore letting $\nu \to \nu_0 = 0$ in (5.21), as we did to obtain (5.19), we find that, as for (5.17) and (5.18),

$$P_1 \left.\frac{d\hat{\mathscr{E}}_\nu}{d\nu}\right|_{\nu=0} = C_1 q\left\{2\left[\frac{1}{2}\left(\frac{2}{q}\right)^{n/2}\Omega_2\delta(y)\right] + 2\operatorname*{res}_{\nu=0}\left[z^{\nu-1}\frac{d\mathscr{J}_\nu}{dz} + \nu z^{\nu-2}\mathscr{J}_\nu\right]\right\}$$
$$= C_1 q 2\left(\frac{2}{q}\right)^{n/2}\Omega_2\delta(y) = C_1 8\pi\delta(y), \tag{5.22}$$

because $n = 2$ and $\Omega_2 = 2\pi$. Hence we obtain C_1 and the fundamental solution of $\Delta + q$ when $n = 2$ and $q > 0$ in the form

$$\hat{\mathscr{E}}(y) = \frac{1}{8\pi}\left.\frac{d\hat{\mathscr{E}}_\nu(y)}{d\nu}\right|_{\nu=0}. \tag{5.23}$$

This fundamental solution corresponds, as it should, to (3.23) but now $C_1 = 0$ and $C_2 \neq 0$ (in contrast to (5.1) and (5.14)). Thus, using relations between Bessel functions (Bateman and Erdélyi [1953]), we can write (5.23) as

$$\mathscr{E}(y) = \frac{1}{4} Y_0(\sqrt{qB(y)})$$

$$= -\frac{i}{8}(H_0^{(1)}(\sqrt{qB(y)}) - H_0^{(2)}(\sqrt{qB(y)})), \tag{5.24}$$

where $H_0^{(1)}(z) = J_0(z) + iY_0(z)$ is the Hankel function of the first kind and order zero, and $H_0^{(2)}(z) = J_0(z) - iY_0(z)$ is the Hankel function of the second kind. Since, by (1.9), $J_0(\sqrt{qB}) = \frac{1}{2}(H_0^{(1)}(\sqrt{qB}) + H_0^{(2)}(\sqrt{qB}))$ is an entire function of B, it follows from (3.14′) and (3.23) that $(\Delta + q)J_0(\sqrt{qB(y)}) = 0$ for $y \in \mathbb{R}^n$. Therefore fundamental solutions of $\Delta + q$ are, in addition to (5.24), also given by (Vladimirov [1981])

$$\mathscr{E}(y) - \frac{i}{4}J_0(\sqrt{qB}) = -\frac{i}{4}H_0^{(1)}(\sqrt{qB(y)}) \tag{5.25}$$

and

$$\mathscr{E}(y) + \frac{i}{4}J_0(\sqrt{qB}) = \frac{i}{4}H_0^{(2)}(\sqrt{qB(y)}),$$

these being complex conjugates. With $q = \omega^2$, (5.25) yield the formulae (2.11′) in Chap. 1 for $\mathscr{E}_2^{\pm}$.

Example 5.2. 1) For the one-dimensional Helmholtz operator $\frac{d^2}{dy^2} + \omega^2$ the general formula (5.20) yields the fundamental solution

$$\mathscr{E}(y) = \frac{\sqrt{2}}{2\omega}\Gamma\left(\frac{3}{2}\right)\sqrt{\omega|y|}J_{1/2}(\omega|y|) = \frac{\sin \omega|y|}{2\omega}, \quad y \in \mathbb{R} \tag{5.26}$$

(see formula (1.11), Chap. 1) because $\nu_0 = \frac{1}{2}$, $\Gamma(3/2) = \frac{\sqrt{\pi}}{2}$ and $J_{1/2}(z) = \sqrt{\frac{2}{\pi z}} \sin z$ (Whittaker and Watson [1927]). On the other hand, $\frac{\cos \omega|y|}{2\omega}$ is an entire function of y and hence satisfies the homogeneous Helmholtz equation (for $n = 1$). Therefore fundamental solutions of $\frac{d^2}{dy^2} + \omega^2$ are, in addition to (5.26), also given by

$$\mathscr{E}_1^{\pm}(y) = \frac{\sin \omega|y|}{2\omega} \mp i\frac{\cos \omega|y|}{2\omega} = \pm\frac{e^{\pm i\omega|y|}}{2i\omega}. \tag{5.27}$$

This formula coincides with formula (2.11′), Chap. 1, for $k = 1$.

2) Similarly, for $n = 3$ (5.20) (with $\nu_0 = -\frac{1}{2}$) yields

$$\mathscr{E}(y) = -\frac{\omega}{4\pi\sqrt{2}}\Gamma\left(\frac{1}{2}\right)(\omega|y|)^{-1/2}J_{-1/2}(\omega|y|)$$

$$= -\frac{1}{4\pi|y|}\cos\omega|y|. \tag{5.28}$$

Since $\frac{\sin\omega|y|}{|y|}$ is an entire function of y, it satisfies the homogeneous Helmholtz equation (for $k = 3$) and we also obtain fundamental solutions for $\Delta + \omega^2$ in $\mathbb{R}^3$ of the form $\mathscr{E}_3^{\pm} = -\frac{e^{\pm i\omega|y|}}{4\pi|y|}$, which coincides with (2.11′) in Chap. 1.

We now take up the case where $m = 0$, $q > 0$ and $Q = -\Delta + q$. Then $qB(y) < 0$ for $y \in \mathbb{R}^n\backslash 0$ and $\sqrt{qB(y)}$ is an imaginary quantity for $y \neq 0$. Therefore the fundamental solution (3.23) contains Bessel functions of an imaginary argument. We shall show that for $n \neq 2$ the fundamental solution can be constructed, as before, in the form (5.14). We have only to decide which branch of $(qB)^{r+\nu}$ to take in (5.14) when $qB < 0$ and $\nu \in \mathbb{C}$. To be definite, we write

$$(qB)^{r+\nu} = (-q)^{r+\nu}|B|^{r+\nu},$$

where

$$(-q)^{r+\nu} := e^{\pi i(r+\nu)}|q|^{r+\nu}, \tag{5.29}$$

that is, a cut along the negative side of the imaginary axis from $-i\infty$ to 0 is chosen for the power function $z^{r+\nu}$. Then all the computations (5.16)–(5.25) remain valid to within sign when q is replaced by $-q$ and B by $|B|$.

Consequently, noting that $J_\nu(iz) = e^{\pi i\nu/2}I_\nu(z)$, we obtain from (5.20) the fundamental solution for the operator $-\Delta + q$, with $n \neq 2$, in the form

$$\hat{\mathscr{E}}_{\nu_0}(y) = \frac{q^{n/2-1}}{(n-2)\Omega_n 2^{n/2-1}} \times [\Gamma(\nu+1)(\sqrt{q|B|})^\nu I_\nu(\sqrt{q|B(y)|})]|_{\nu=\nu_0}, \tag{5.30}$$

where the right-hand side is defined at the point $\nu = \nu_0 := -\frac{n-2}{2}$ as an analytic continuation from the domain $\operatorname{Re}\nu > 0$.

With $n = 2$, we replace q by $-q$ and B by $|B|$ in (5.24) and (5.25) and obtain the following fundamental solutions of the operator $\Delta - q$ on the plane:

$$\mathscr{E}(y) = \frac{1}{4}Y_0(i\sqrt{q|B(y)|}), \quad -\frac{i}{4}H_0^{(1)}(i\sqrt{q|B|}) = -\frac{1}{2\pi}K_0(\sqrt{q|B|}),$$
$$\frac{i}{4}H_0^{(2)}(i\sqrt{q|B|}) = -\frac{1}{2\pi}K_0(-\sqrt{q|B|}) \tag{5.31}$$

(see Bateman and Erdélyi [1953]).

Example 5.3. Similarly, for $n = 1$, by taking $\omega = im_0$ in (5.26) and (5.27), we obtain the fundamental solutions of $\frac{d^2}{dy^2} - m_0^2$, $m_0 > 0$:

$$\mathscr{E}(y) = \frac{\sin im_0|y|}{2im_0} = \frac{\operatorname{sh} m_0|y|}{2m_0}, \quad \mathscr{E}_1^{\pm}(y) = \mp\frac{e^{\mp m_0|y|}}{2m_0}. \tag{5.32}$$

Finally, for $n = 3$, from (5.28) and (2.11′) in Chap. 1 (for $k = 3$), we obtain the fundamental solutions for the operator $\Delta - m_0^2$ in $\mathbb{R}^3$:

$$\mathscr{E}(y) = -\frac{1}{4\pi|y|}\operatorname{ch} m_0|y|, \quad \mathscr{E}_3^{\pm}(y) = -\frac{1}{4\pi|y|}e^{\mp m_0|y|}. \tag{5.33}$$

§6. On Singularities of Fundamental Solutions of Second-order Equations with Real Coefficients and with Non-degenerate Quadratic Form

Let us analyse the singularities of the fundamental solutions of (3.1), with a non-degenerate quadratic form, which we obtained in §4 and §5. Under the condition (3.2), the operator P in (3.1) satisfies the hypotheses of Theorem 4.2′ in Chap. 3. Therefore Corollary 4.2 in Chap. 3 applies to the fundamental solutions of P. We now verify that this is indeed the case for the fundamental solutions $\mathscr{E}$ obtained in §4 and §5. To do this, we find the singular supports of the functions $\mathscr{E}(x)$.

The results in §4 and §5 show that in y-coordinates the set sing supp $\mathscr{E}$ is contained in the set

$$Q = \{y \in \mathbb{R}^n : B(y) := (By, y) = 0\}. \tag{6.1}$$

If $m = 0$ or $m = n$, the equation (3.8) (and (3.1)) is elliptic and the fundamental solution $\mathscr{E}(y)$ ($y \neq 0$) is smooth; this statement agrees with Theorem 5.1 in Chap. 3. Now take the case $1 \leqslant m \leqslant n - 1$, for which Q is a conical surface in $\mathbb{R}^n$. In the original coordinates $x = Cy$, the equation of Q is

$$Q = \{x \in \mathbb{R}^n : B(C^{-1}x) := ((C^{-1})^t BC^{-1}x, x) = 0\}. \tag{6.2}$$

But $B = B^{-1}$, by (3.6). Therefore (3.5) implies that $(C^{-1})^t BC^{-1} = A^{-1}$, and thus (6.2) can be written as

$$Q = \{x \in \mathbb{R}^n : (A^{-1}x, x) = 0\}. \tag{6.2′}$$

Proposition 6.1. *The set Q is the projection onto $\mathbb{R}^n$ of the set of all bicharacteristics of equation* (3.1) *that emanate from the points of the fibre $T_0^*\mathbb{R}^n$ above the point $x = 0$.*

Proof. The equation (4.5), Chap. 3, for the characteristic cone K of the operator P is

$$P_2(\xi) := (A\xi, \xi) = 0. \tag{6.3}$$

In view of Remark 4.3, Chap. 3, we must prove that the vectors $x \in Q$ are perpendicular to the tangent planes to the "cone" (6.3). But when $\xi \in K\backslash 0$, the perpendicular to $T_\xi K$ is proportional to $\text{grad}(A\xi, \xi) = 2A\xi$. It remains to show that the vector $x = A\xi$ satisfies (6.2′) if ξ satisfies (6.3).But this is obvious because

$$(A^{-1}x, x) = (\xi, A\xi) = 0. \quad \blacksquare \tag{6.4}$$

Thus sing supp $\mathscr{E} \subset Q$ and Q is the projection of the family of bicharacteristics onto $\mathbb{R}^n$. This agrees with Corollary 4.2, Chap. 3. We also note that the fundamental solutions (4.10) and (5.12) have singularities only on the one of two halves of the projections of bicharacteristics. This also agrees with Corollary 4.2, Chap. 3. Finally, the fundamental solutions constructed in § 4 and § 5 are the most regular in the sense that sing supp $\mathscr{E}$ does not contain projections of bicharacteristics which do not pass through the point $y = 0$. It turns out that this is a general phenomenon.

Theorem 6.1 (Grushin [1963b], Hörmander [1983, 1985]). *For every homogeneous operator P satisfying the hypotheses of Theorem 4.2′, Chap. 3, there is a fundamental solution $\mathscr{E}$ such that $WF(\mathscr{E})$ consists of the union of semi-bicharacteristics emanating from $0 \times \mathbb{R}^{n*}$ and no bicharacteristic lies wholly in $WF(\mathscr{E})$.*

Accordingly, sing supp $\mathscr{E}$ consists of the union of projections of these semi-bicharacteristics.

Chapter 5
Boundary-value Problems in Half-space

Boundary-value problems in the half-space $\mathbb{R}^n_+ := \{x \in \mathbb{R}^n \colon x_n > 0\}$ serve as models for general boundary-value problems in domains with smooth boundary. For equations with constant coefficients the solutions can be found in an explicit form. By analysing such solutions we can understand the basic features of the theory of boundary-value problems. In particular, we can determine the number of boundary conditions to be prescribed on the boundary $x_n = 0$ so that the boundary-value problem has a solution, and indeed a unique solution, for any boundary data from the class of functions in question. This number, roughly speaking, is equal to the number of well-defined roots of the corresponding equation in the sense of Petrovskij. For example, the Cauchy problem, where the number of boundary conditions is the order of the equation in $\dfrac{\partial}{\partial x_n}$, is well-posed only if all its roots are well defined in the sense of Petrovskij. Such equations are known as *Petrovskij well-defined equations.* To this type belong all parabolic and hyperbolic equations as well as the Schrödinger equation. On the other hand, elliptic equations of order m in $\mathbb{R}^n$, with $n \geqslant 3$, have exactly $m/2$ Petrovskij well-defined roots (when $n \geqslant 3$, m is even, by Proposition 5.1, Chap. 3), and hence $m/2$ boundary conditions must be prescribed for such equations.

The basic tool for investigating boundary-value problems in a half-space is the *contact Fourier transformation*, that is, a partial Fourier transformation with respect to the variables $(x_1, \ldots, x_{n-1}) := x'$ in the boundary plane $x_n = 0$ of $\mathbb{R}^n_+$.

§ 1. Equations with Constant Coefficients in a Half-space

1.1. General Solution of Equation (0.1) **in a Half-space.** We examine the equation (0.1), with $f(x) \equiv 0$, in the half-space $\mathbb{R}^n_+$. Thus we consider

$$P(\partial_x)u(x) = 0, \quad x_n > 0. \tag{1.1}$$

The solution u is sought in the following classes of functions (see (3.1), Chap. 2):

$$\begin{aligned} U^{(r)} &= \bigcup_{s \in \mathbb{R}} C^{(r)}(0, \infty; H_s(\mathbb{R}^{n-1})) \subset \mathscr{D}'(\mathbb{R}^n_+), \\ U^{(r)}_\alpha &= \bigcup_{s \in \mathbb{R}} C^{(r)}_\alpha(0, \infty; H_s(\mathbb{R}^{n-1})) \subset \mathscr{D}'(\mathbb{R}^n_+), \end{aligned} \tag{1.2}$$

where $r = 0, 1, 2, \ldots, \alpha \in \mathbb{R}$ and

$$C^{(r)}_\alpha = \left\{ u \in C^{(r)}(0, \infty; H_s(\mathbb{R}^{n-1})) \colon \sup_{x_n \geqslant 0} e^{-\alpha x_n} \|u(\cdot, x_n)\|_s < \infty \right\}. \tag{1.3}$$

Roughly speaking, the functions $u \in U^{(r)}_0$ are bounded with respect to x_n and the functions $u \subset U^{(r)}_\alpha$ grow (or decrease) like $e^{\alpha x_n}$ as $x_n \to \infty$.

We find the general solution of (1.1) in the classes $U^{(r)}$. To do this, we apply to (1.1) the generalized Fourier transformation $F_{x' \to \xi'}$ with respect to the "contact" variables $x' = (x_1, \ldots, x_{n-1})$ for $x_n > 0$. For example, if $u \in C_0^\infty(\mathbb{R}^n)$,

$$\begin{gathered} \tilde{u}(\xi', x_n) = F_{x' \to \xi'} u(x', x_n) = \int_{\mathbb{R}^{n-1}} e^{i\xi' \cdot x'} u(x', x_n)\, dx', \quad \xi' \in \mathbb{R}^{n-1} \\ x_n > 0. \end{gathered} \tag{1.4}$$

Equation (1.1) then becomes

$$P(-i\xi', \partial_{x_n})\tilde{u}(\xi', x_n) = 0 \tag{1.5}$$

for almost all $\xi' \in \mathbb{R}^{n-1}$, in the sense of generalized functions of $x_n > 0$. Thus (1.5) is an ordinary differential equation on the semi-axis $x_n > 0$ which depends on the parameter $\xi' \in \mathbb{R}^{n-1}$. This is the principal technique for solving equations in a half-space.

We write the characteristic equation of (1.5) in the form

$$P(-i\xi', \lambda) = \sum_{j=0}^{\nu} p_{(j)}(\xi')\lambda^j = 0, \quad \text{where } \nu \leqslant m \text{ and } p_{(\nu)}(\xi') \not\equiv 0. \tag{1.6}$$

This means that ν is the order of the operator $P(\partial_x)$ with respect to ∂_{x_n}.

The equality $\nu = m$ is equivalent to the fact that $P_m(0, \ldots, 0, 1) \neq 0$, that is, the boundary $\partial\mathbb{R}^n_+$ of $\mathbb{R}^n_+$ is not characteristic for $P(\partial_x)$ (see Definition 4.1, Chap. 3).

For almost all $\xi' \in \mathbb{R}^{n-1}$ (for which $p_{(\nu)}(\xi') \neq 0$), the equation (1.6) has roots

$$\lambda = \lambda_1(\xi'), \ldots, \lambda_\nu(\xi'), \tag{1.7}$$

taking multiplicities into account. For simplicity throughout the discussion, we assume that the roots $\lambda_k(\xi')$ are simple for almost all $\xi' \in \mathbb{R}^{n-1}$, that is,

$$\lambda_k(\xi') \neq \lambda_j(\xi') \quad (k \neq j) \tag{1.8}$$

for almost all $\xi' \in \mathbb{R}^{n-1}$. Sometimes we also require that

$$\lambda_k(\xi') \neq \lambda_j(\xi') \quad (k \neq j) \quad \text{for all } \xi' \in \mathbb{R}^{n-1}. \tag{1.8'}$$

Then the general solution of (1.5) has the form

$$\tilde{u}(\xi', x_n) = \sum_{k=1}^{\lambda} C_k(\xi') e^{\lambda_k(\xi') x_n} \tag{1.9}$$

for almost all $\xi' \in \mathbb{R}^{n-1}$ and $x_n > 0$. The general case of multiple roots can be investigated along the same lines.

We arrange the $\lambda_k(\xi')$ with the real parts in increasing order, so that

$$\operatorname{Re} \lambda_1(\xi') \leqslant \cdots \leqslant \operatorname{Re} \lambda_\nu(\xi') \tag{1.10}$$

for almost all $\xi' \in \mathbb{C}^{n-1}$. It is known (Hörmander [1973]) that the $\lambda_k(\xi')$ can be chosen to be continuous functions of $\xi' \in \mathbb{C}^{n-1}$ in a domain where $p_\nu(\xi') \neq 0$.

Note that the modulus of the k-th term of (1.9) is bounded for $x_n > 0$ if $\operatorname{Re} \lambda_k(\xi') \leqslant 0$ and grows exponentially as $x_n \to +\infty$ if $\operatorname{Re} \lambda_k(\xi') > 0$.

With $\alpha \in \mathbb{R}$, let $\nu_\alpha = \nu_\alpha(\xi')$ denote the number of roots $\lambda_k(\xi')$ satisfying the condition that $\operatorname{Re} \lambda_k(\xi') \leqslant \alpha$:

$$\operatorname{Re} \lambda_1(\xi') \leqslant \cdots \leqslant \operatorname{Re} \lambda_{\nu_\alpha}(\xi') \leqslant \alpha < \operatorname{Re} \lambda_{\nu_\alpha+1}(\xi') \leqslant \cdots \leqslant \operatorname{Re} \lambda_\nu(\xi') \tag{1.11}$$

almost everywhere in $\mathbb{R}^{n-1}$. Then in order that u lies in $U_\alpha^{(r)}$, where u is given by (1.9), it is necessary that

$$C_k(\xi') = 0 \quad (k > \nu_\alpha(\xi')) \tag{1.12}$$

almost everywhere in $\mathbb{R}^{n-1}$. Consequently, for the general solution of (1.11) in the class $U_\alpha^{(r)}$, we have

$$\tilde{u}(\xi', x_n) = \sum_{k=1}^{\nu_\alpha(\xi')} C_k(\xi') e^{\lambda_k(\xi') x_n} \tag{1.12'}$$

almost everywhere in $\mathbb{R}^{n-1}$, for all $x_n > 0$ and certain $C_k(\xi')$.

Definition 1.1. 1) A *root* $\lambda_k(\xi')$ of (1.6) is said to be *Petrovskij well-defined* if

$$\bar{\alpha}_k \equiv \sup_{\xi' \in \mathbb{R}^{n-1}} \operatorname{Re} \lambda_k(\xi') < \infty, \tag{1.13}$$

and *α-well-defined* if $\bar{\alpha}_k \leqslant \alpha$. The root is said to be *stable* if $\bar{\alpha}_k \leqslant 0$ and *unstable* if $\bar{\alpha}_k > 0$.

2) The *equation* (1.1) and the *operator* $P(\partial_x)$ are said to be *Petrovskij well-defined* if all the roots $\lambda_k(\xi')$ $(k = 1, \ldots, \nu)$ are Petrovskij well-defined, and to be *α-regular* if $\nu_\alpha(\xi')$ is independent of ξ' for almost all $\xi' \in \mathbb{R}^{n-1}$, that is, if

$$\nu_\alpha(\xi') \equiv \nu_\alpha \tag{1.14}$$

almost everywhere in $\mathbb{R}^{n-1}$.

If an equation is Petrovskij well-defined, then it is α-regular for all $\alpha \geqslant \bar{\alpha}_\nu$. Note that $\bar{\alpha}_1 \leqslant \cdots \leqslant \bar{\alpha}_\nu$.

1.2. Classification of Equations in Half-space

Definition 1.2. The *operator* $P(\partial_x)$ is said to be (*Gårding*) *hyperbolic* in $\mathbb{R}^n_+$ if it is Petrovskij well-defined and the boundary $x_n = 0$ is non-characteristic for P (that is, $\nu = m$).

Proposition 1.1 (Hörmander [1983, 1985]). *If P is a hyperbolic operator in $\mathbb{R}^n_+$ then its principal part P_m is also a hyperbolic operator in $\mathbb{R}^n_+$.*

Indeed, $P_m(-i\xi, \lambda) = \lim_{t\to\infty} t^{-m}P(-it\xi, t\lambda)$. Since all the roots $t\lambda$ of the polynomial $t^{-m}P(-it\xi, t\lambda)$ lie in the half-plane $\operatorname{Re}\lambda \leqslant \bar{\alpha}_m$, it follows that all the roots λ of the polynomial $P_m(-i\xi, \lambda)$ lie in the half-plane $\operatorname{Re}\lambda \leqslant 0$.

Proposition 1.2 (Hörmander [1983, 1985]). *If $P = P_m$ is a homogeneous hyperbolic operator, then its roots $\lambda_k^0(\xi')$ are pure imaginary for $\xi' \in \mathbb{R}^{n-1}\backslash 0$.*

In fact, for the homogeneous polynomial we have

$$P_m(-it\xi', t\lambda) = t^m P_m(-i\xi', \lambda) \tag{1.15}$$

for all $t \in \mathbb{C}$ and $(\xi', \lambda) \in \mathbb{C}^n$. Hence if λ_k^0 is a root of the polynomial $P_m(-i\xi', \cdot)$, it follows that $t\lambda_k^0$ is a root of $P_m(-it\xi', \cdot)$ for all $t \in \mathbb{R}$. But $\operatorname{Re} t\lambda_k^0$ cannot be bounded for $t \in \mathbb{R}$ if $\operatorname{Re}\lambda_k^0 \neq 0$.

Corollary 1.1. *For all $\alpha \geqslant 0$ all the roots of the homogeneous hyperbolic operator P are α-well-defined and P is regular for all $\alpha \in \mathbb{R}$.*

Note that (1.15) and (1.10) imply that

$$\lambda_k^0(t\xi') = t\lambda_k^0(\xi') \; \forall t > 0, \quad \xi' \in \mathbb{C}^{n-1}\backslash 0. \tag{1.15'}$$

The Definition 4.2 of strict hyperbolicity given in Chap. 3 is obviously equivalent to the following definition.

Definition 1.3 (Hörmander [1983, 1985], Petrovskij [1961]). The operator $P(\partial_x)$ is said to be *strictly Petrovskij hyperbolic* if the roots $\lambda_k^0(\xi')$ $(k = 1, \ldots, m)$ of the corresponding principal part $P_m(\partial_x)$ are distinct and pure imaginary for all $\xi' \in \mathbb{R}^{n-1}\backslash 0$.

Proposition 1.3 (Hörmander [1983, 1985]). *For an arbitrary operator $P(\partial_x)$ with a given principal part $P_m(\partial_x)$ to be hyperbolic, it is necessary and sufficient that P_m is strictly hyperbolic.*

Examples of strictly hyperbolic equations.

1) With $n = k + 1$ and $x_n \equiv t$, the wave equation

$$\Box u = \frac{\partial^2}{\partial t^2} u - a^2 \Delta u(x, t) = 0, \quad t > 0, \quad x \in \mathbb{R}^k, \tag{1.16}$$

has the form (1.1). Therefore the corresponding characteristic equation (1.6) becomes $\lambda^2 + a^2|\xi|^2 = 0$, where $\xi \in \mathbb{R}^k$ and ξ has been written for ξ'. Then $\lambda_{1,2}(\xi) = \pm ia|\xi|$. For the general solution $u \in U_0^{(r)}$, we have, by (1.12′),

$$\tilde{u}(\xi, t) = C_1(\xi)e^{ia|\xi|t} + C_2(\xi)e^{-ia|\xi|t} = A(\xi)\cos a|\xi|t + B(\xi)\sin a|\xi|t. \tag{1.16′}$$

2) Similarly, for the Klein-Gordon equation

$$(\Box + m_0^2)u(x, t) = 0, \quad t > 0, \quad x \in \mathbb{R}^k, \tag{1.17}$$

we have the roots $\lambda_{1,2} = \pm i\omega(\xi)$, where $\omega(\xi) = \sqrt{a^2|\xi|^2 + m_0^2}$ and, for the general solution $u \in U_0^{(r)}$, we have

$$\tilde{u}(\xi, t) = A(\xi)\cos\omega(\xi)t + B(\xi)\sin\omega(\xi)t. \tag{1.17′}$$

3) For the *wave equation with friction*

$$\left(\Box + K\frac{\partial}{\partial t}\right)u(x, t) = 0, \quad t > 0, \quad x \in \mathbb{R}^k; \quad K > 0, \tag{1.18}$$

the characteristic equation $\lambda^2 + a^2|\xi|^2 + K\lambda = 0$ has the roots $\lambda_{1,2} = -\frac{K}{2} + \pm\sqrt{\frac{K^2}{4} - a^2|\xi|^2}$. Therefore for the general solution $u \in U_0^{(r)}$ we obtain

$$\tilde{u}(\xi, t) = \begin{cases} C_1(\xi)e^{((-(K/2))-\mu(\xi))t} + C_2(\xi)e^{((-(K/2))+\mu(\xi))t} & \text{if } a|\xi| < \frac{K}{2}; \\ e^{-(K/2)t}(A(\xi)\cos\omega(\xi)t + B(\xi)\sin\omega(\xi)t) & \text{if } a|\xi| > \frac{K}{2}, \end{cases} \tag{1.18′}$$

where $\mu(\xi) = \sqrt{\frac{K^2}{4} - a^2|\xi|^2}$, $\omega(\xi) = \sqrt{a^2|\xi|^2 - \frac{K^2}{4}}$.

4) In an alternative way of introducing friction, the equation is

$$\left(\frac{\partial}{\partial t} + K\right)^2 u(x, t) = a^2 \Delta u, \quad t > 0, \quad x \in \mathbb{R}^k, \tag{1.19}$$

and the roots are $\lambda_{1,2} = -K \pm ia|\xi|$. For the general solution $u \in U_0^{(r)}$ we have

$$\tilde{u}(\xi, t) = e^{-Kt}(C_1(\xi)e^{ia|\xi|t} + C_2(\xi)e^{-ia|\xi|t}). \tag{1.19′}$$

Notice that all the equations (1.16)–(1.19) are α-regular for $\alpha \geqslant 0$, with $\nu_\alpha = 2$ for $\alpha \geqslant 0$. But (1.18) fails to be α-regular if $\alpha \in]-K, 0[$.

Definition 5.1, Chap. 3, of an elliptic operator is obviously equivalent to the following definition.

Definition 1.4. The operator $P(\partial_x)$ is said to be *elliptic* if the boundary $x_n = 0$ is non-characteristic, that is, if $v = m$ and

$$\operatorname{Re} \lambda_k^0(\xi') \neq 0 \quad (k = 1, \ldots, m) \quad \text{for } \xi' \in \mathbb{R}^{n-1}\backslash 0, \tag{1.20}$$

where $\lambda_k^0(\xi')$ are roots of the corresponding principal part $P_m(\partial_x)$ of the operator P (see (4.1) of Chap. 3).

Proposition 1.4 (Hörmander [1983, 1985], Vishik and Ehskin [1965]). *If the operator $P(\partial_x)$ is elliptic and $n \geqslant 3$, then the munber m is even, $m = 2l$, and the number $v_-(\xi')$ of Petrovskij well-defined roots λ_k equals l.*

That m is even when $n \geqslant 3$ follows from Proposition 5.1 in Chap. 3.

By (1.15′), all the roots λ_k^0 of the homogeneous operator $P_m(\partial_x)$ with $\operatorname{Re} \lambda_k^0 < 0$ are Petrovskij well-defined, while those with $\operatorname{Re} \lambda_k^0(\xi') > 0$ are Petrovskij ill-defined. This result leads easily to the conclusion of Proposition 1.4 when $P = P_m$. The general case where $P \neq P_m$ reduces to the previous case on applying the following lemma.

Lemma 1.1. *If the boundary $x_n = 0$ is non-characteristic for the operator $P(\partial_x)$, then the roots $\lambda_k(\xi')$ corresponding to the operator $P(\partial_x)$ and the roots $\lambda_k^0(\xi')$ corresponding to $P_m(\partial_x)$ have the same asymptotics as $|\xi'| \to \infty$ in the sense that*

$$\lambda_k(\xi') = \lambda_k^0(\xi') + o(|\xi'|), \quad |\xi'| \to \infty \quad \xi' \in \mathbb{C}^{n-1}. \tag{1.21}$$

Proof. We take some $\xi' \in \mathbb{C}^{n-1}$. Then the polynomial equations for $\lambda_k^0(\xi')$ and for the ratios $\lambda_k(t\xi')/t$ have asymptotically the same form (see (1.6)) since, first, these equations are

$$\begin{cases} 0 = \dfrac{P(-it\xi', \lambda_k(t\xi'))}{t^m} = \displaystyle\sum_{j=0}^{m} \frac{p_{(j)}(t\xi')}{t^{m-j}} \left(\frac{\lambda_k(t\xi')}{t}\right)^j, \\ 0 = P_m(-i\xi', \lambda_k^0(\xi')) = \displaystyle\sum_{j=0}^{m} p_{(j)}^0(\xi')(\lambda_k^0(\xi'))^i \end{cases} \tag{1.22}$$

and, second, as $t \to \infty$

$$\frac{p_{(j)}(t\xi')}{t^{m-j}} \to p_{(j)}^0(\xi') \tag{1.23}$$

and $p_{(m)}(\xi') = p_{(m)}^0(\xi') = P_m(0, \ldots, 0, 1) \neq 0$. Hence

$$\lambda_k(t\xi')/t \to \lambda_k^0(\xi') \tag{1.24}$$

as $t \to \infty$. ■

Corollary 1.2. *It follows from* (1.21) *that for the elliptic operator* $P(\partial x)$, *with* $n \geqslant 3$, *we have*

$$\operatorname{Re} \lambda_k(\xi') \to \begin{cases} -\infty, & \text{if } k \leqslant l \\ +\infty, & \text{if } k > l \end{cases} \tag{1.25}$$

with the ordering (1.10).

The result follows because it is true for the roots $\lambda_k^0(\xi')$. ■
Thus Proposition 1.4 is proved. ■

Examples of elliptic equations
1) For the Laplace equation

$$\Delta u(x) = 0, \quad x_n > 0, \quad x' \equiv (x_1, \ldots, x_{n-1}) \in \mathbb{R}^{n-1}, \tag{1.26}$$

the characteristic equation $-|\xi'|^2 + \lambda^2 = 0$ has roots $\lambda_{1,2} = \mp|\xi'|$ and, for the general solution $u \in U_0^{(r)}$, we have

$$\tilde{u}(\xi', x_n) = C_1(\xi')e^{-|\xi'|x_n}. \tag{1.26'}$$

2) Similarly, for the equation

$$\Delta u - qu(x) = 0, \quad x_n > 0, \quad x' \in \mathbb{R}^{n-1}; \quad q > 0, \tag{1.27}$$

the roots are $\lambda_{1,2} = \mp\sqrt{|\xi'|^2 + q}$ and, for the general solution $u \in U_0^{(r)}$, we have

$$\tilde{u}(\xi', x_n) = C_1(\xi')e^{-\sqrt{|\xi'|^2+q}x_n}. \tag{1.27'}$$

3) For the Helmholtz equation

$$\Delta u + \omega_0^2 u(x) = 0, \quad x_n > 0, \quad x' \in \mathbb{R}^{n-1}, \quad \omega_0 \in \mathbb{R}\backslash 0, \tag{1.28}$$

the roots are $\lambda_{1,2} = \mp\sqrt{|\xi'|^2 - \omega_0^2}$. Hence, for the general solution $u \in U_0^{(r)}$, we have

$$\tilde{u}(\xi', x_n) = \begin{cases} C_1(\xi')e^{i\omega x_n} + C_2(\xi')e^{-i\omega x_n} & \text{if } |\xi'| < |\omega_0|, \\ C_1(\xi')e^{-\sqrt{|\xi'|^2-\omega_0^2}x_n} & \text{if } |\xi'| > |\omega_0|, \end{cases} \tag{1.28'}$$

where $\omega \equiv \sqrt{\omega_0^2 - |\xi'|^2}$
4) For the *Cauchy-Riemann equation*

$$\frac{\partial}{\partial \bar{z}} u(x_1, x_2) \equiv \frac{1}{2}\left(\frac{\partial u}{\partial x_1} + \frac{i\partial u}{\partial x_2}\right) = 0, \quad x_2 > 0, \quad x_1 \in R, \tag{1.29}$$

the root is $\lambda = \xi_1$ and, for the general solution $u \in U_0^{(r)}$, we have

$$\tilde{u}(\xi_1, x_2) = \begin{cases} C_1(\xi_1)e^{\xi_1 x_2} & \text{if } \xi_1 \leqslant 0, \\ 0 & \text{if } \xi_1 > 0. \end{cases} \tag{1.29'}$$

The equation (1.26) is α-regular only for $\alpha = 0$ while (1.27) is regular only for $\alpha \in [-\sqrt{q}, \sqrt{q}]$. The equations (1.28) and (1.29) are not α-regular for any $\alpha \in \mathbb{R}$.
Corollary 1.2 leads to

Proposition 1.5. *The elliptic operator $P(\partial_x)$, with $n \geqslant 3$, is α-regular only for $\alpha \in [\bar{\alpha}_l, \underline{\alpha}_{l+1}]$ provided that $\bar{\alpha}_l \leqslant \underline{\alpha}_{l+1} \equiv \inf_{\xi' \in \mathbb{R}^{n-1}} \operatorname{Re} \lambda_{l+1}(\xi')$. Furthermore, $\nu_\alpha = m/2$.*

Definition 1.5. The equation (1.1) is said to be *Shilov h-parabolic* ($h \in \mathbb{R}$) if

$$\operatorname{Re} \lambda_k(\xi') \leqslant -C_1|\xi'|^h + C_2 \quad \text{for } \xi' \in \mathbb{R}^{n-1} \text{ and all } k = 1, \ldots, \nu \tag{1.30}$$

for some $C_1 > 0$ and $C_2 \in \mathbb{R}$.

Every Shilov parabolic equation is clearly Petrovskij well-defined and is α-regular for all $\alpha \geqslant \bar{\alpha}_\nu$.

Examples of parabolic equations

1) The equation of heat transfer (or diffusion)

$$\frac{\partial u}{\partial t} = a^2 \Delta u(x, t), \quad t > 0, \quad x \in \mathbb{R}^k; \quad a > 0, \tag{1.31}$$

has the form (1.1) with $n = k + 1$ and $x_n \equiv t$. It is Shilov 2-parabolic and $\lambda_1 = -a^2|\xi|^2$. For $u \in U_0^{(r)}$ we have

$$\tilde{u}(\xi, t) = C_k(\xi) e^{-a^2|\xi|^2 t}. \tag{1.31'}$$

2) The equation of *heat transfer with absorption* (or *generation* if $q < 0$)

$$\frac{\partial u}{\partial t} = a^2 \Delta u(x, t) - qu, \quad t > 0, \quad x \in \mathbb{R}^k; \quad a > 0, \quad q \in \mathbb{R}, \tag{1.32}$$

is also Shilov 2-parabolic and $\lambda_1 = -a^2|\xi|^2 - q$. For $u \in U_0^{(r)}$ we have

$$\tilde{u}(\xi, t) = \begin{cases} C_1(\xi) e^{-(a^2|\xi|^2 + q)t}, & a^2|\xi|^2 \geqslant -q, \\ 0, & a^2|\xi|^2 < -q. \end{cases} \tag{1.32'}$$

The equation (1.32) is α-regular if $\alpha \geqslant -q$, and then $\nu_\alpha = 1$, but it fails to be α-regular if $\alpha < -q$.

Definition 1.6. The equation (1.1) is said to be *Petrovskij β-parabolic* ($\beta = 1, 2, \ldots$) if, first,

$$p_\alpha = 0 \quad \text{for } |\alpha'| + \beta\alpha_n > \beta\nu \tag{1.33}$$

and, second, all the roots of the equation

$$P_{m,\beta}(\xi', \lambda^0) := \sum_{|\alpha'| + \beta\alpha_n = \nu\beta} p_\alpha(-i\xi')^{\alpha'}(\lambda^0)^{\alpha_n} = 0 \tag{1.34}$$

are *strictly stable* in the sense that

$$\operatorname{Re} \lambda^0(\xi') < 0, \quad |\xi'| = 1, \quad \xi' \in \mathbb{R}^{n-1}. \tag{1.35}$$

For example, the equations (1.31) and (1.32) are Petrovskij 2-parabolic.

Notice that (1.35) can hold only for even β, $\beta = 2b$, where $b = 1, 2, \ldots$. Indeed, if λ^0 is a root of (1.34) with $\xi' \in \mathbb{R}^{n-1} \backslash 0$, then $t^\beta \lambda^0$ is a root of the same equation in $t\xi'$. This contradicts (1.35) if $t = -1$ and β is odd.

Proposition 1.6. *If the equation* (1.1) *is Petrovskij 2b-parabolic, then it is also Shilov 2b-parabolic.*

Proof. Denote the roots of (1.34) by $\lambda_k^0(\xi')$. As for (1.15′), we have

$$\lambda_k^0(t\xi') = t^{2b} \lambda_k^0(\xi'), \quad t > 0, \quad \xi' \in \mathbb{R}^{n-1} \backslash 0. \tag{1.36}$$

Hence, by (1.35), the roots $\lambda_k^0(\xi')$ satisfy the required condition (1.30). It remains to note that

$$p_{(\nu)}(\xi') \equiv \sum_{\alpha_n=\nu} p_\alpha(-i\xi')^{\alpha'} = p_{(0,\dots,0,\nu)} \neq 0 \tag{1.37}$$

by (1.6) and (1.33). Hence, as $|\xi'| \to \infty$, the roots $\lambda_k(\xi')$ and $\lambda_k^0(\xi')$ have the same asymptotics in the sense that

$$\lambda_k(\xi') = \lambda_k^0(\xi') + o(|\xi'|^{2b}). \tag{1.37'}$$

This result follows by applying Lemma 1.1 to the operator $P(\partial_{x'}, \partial_s^{2b})$, for which the plane $s = 0$ is non-characteristic in view of (1.37). ■

Remark 1.1. In a half-space, the classification of equations (1.1) into hyperbolic, elliptic and parabolic equations is incomplete. For example, for the Schrödinger equation

$$\frac{\partial\psi}{\partial t} = ia^2\Delta\psi(x,t), \quad t \geqslant 0, \quad x \in \mathbb{R}^k; \quad a > 0, \tag{1.38}$$

the characteristic equation has the root $\lambda_1 = -ia^2|\xi|^2$. Therefore for $\psi \in U_0^{(r)}$ we have

$$\tilde{\psi}(\xi, t) = C_1(\xi)e^{-ia^2|\xi|^2 t}. \tag{1.38'}$$

The Schrödinger equation is neither hyperbolic nor elliptic nor Shilov parabolic. It is, however, α-regular for all $\alpha \in \mathbb{R}$.

Similarly, for the wave equation with strong friction

$$\frac{\partial^2 u}{\partial t^2} = a^2\Delta u(x,t) + K\frac{\partial}{\partial t}\Delta u, \quad t > 0, \quad x \in \mathbb{R}^k; \quad a > 0, \quad K > 0, \tag{1.39}$$

the roots of the characteristic equation $\lambda^2 = -a^2|\xi|^2 - K\lambda|\xi|^2$ are given by $\lambda_{1,2} = -\dfrac{K|\xi|^2}{2} \pm \sqrt{\dfrac{K^2|\xi|^4}{4} - a^2|\xi|^2}$. Hence for $u \in U_0^{(r)}$ we have

$$\tilde{u}(\xi, t) = C_1(\xi)e^{\lambda_1 t} + C_2(\xi)e^{\lambda_2 t}. \tag{1.39'}$$

The equation (1.39) is neither hyperbolic nor elliptic nor Shilov parabolic, but it is α-regular for all $\alpha \geqslant 0$ (while it fails to be so if $\alpha < 0$).

§2. Regular Boundary-value Problems in a Half-space in Classes of Bounded Functions

As (1.9) shows, the equation (1.1) in a half-space has, in general, an infinite number of linearly independent solutions. Indeed, if we take, for example, $C_k(\xi') \in C_0^\infty(\mathbb{R}^n)$, we see that

$$u(x', x_n) = F^{-1}_{\xi' \to x'}\tilde{u}(\xi', x_n) \in U_\alpha^{(r)} \tag{2.1}$$

for a sufficiently large α $\left(\alpha \geqslant \max\limits_{\xi' \in \operatorname{supp} C_k} \operatorname{Re}\lambda_k(\xi') \text{ for all } k = 1, \dots, \nu\right)$ and any $r = 1, 2, \dots$, and u is a solution of (1.1).

A regular, that is, a "properly posed" problem for equation (1.1) must have a unique solution. For this purpose additional conditions must be imposed on u in order to determine a unique solution from among all the solutions. In the problems of mathematical physics this is done by prescribing certain conditions to be satisfied on the boundary of the domain. In the case of equation (1.1) the boundary is the hypersurface $x_n = 0$. On this surface we prescribe boundary conditions of the form

$$B_j(\partial_x)u(x)|_{x_n=0+} = f_j(x'), \quad x' \in \mathbb{R}^{n-1}; \quad j = 1, \ldots, l, \tag{2.2}$$

where

$$B_j(\partial_x) = \sum_{|\alpha| \leqslant m_j} b_{j\alpha}\partial_x^\alpha. \tag{2.2'}$$

Take $r \geqslant \overline{m} := \max_j m_j$. Then, by Theorem 3.1 in Chap. 2, the conditions (2.2) are satisfied for $u \in U^{(r)}$ or $u \in U_\alpha^{(r)}$ as equalities in the space $H_{-\infty}(\mathbb{R}^{n-1}) := \bigcup_{s \in \mathbb{R}} H_s(\mathbb{R}^{n-1})$.

Our task is to find conditions on the operators $P, B_1, \ldots, B_l$ under which the boundary-value problem (1.1), (2.2) is "properly posed"; in particular, our task is to find l. Roughly speaking, the answer to this question consists of rejecting all the terms in (1.9) corresponding to unstable roots (for which $\operatorname{Re} \lambda_k > 0$) and showing that l is the number of stable roots (for which $\operatorname{Re} \lambda_k \leqslant 0$). We have also to show that the values $(B_j(-i\xi', \lambda_k(\xi)))_{k=1}^l$ of the symbols on the stable roots must be linearly independent.

2.1. Regular Boundary-value Problems

Definition 2.1. The boundary-value problem (1.1), (2.2) is said to be *regular* in $U_0^{(r)}$, $r \geqslant \overline{m}$, if, for all $f_j \in H_{-\infty}(\mathbb{R}^{n-1})$, it has a solution $u \in U_0^{(r)}$ and this solution is unique.

Let us find necessary conditions for the problem (1.1), (2.2) to be regular in $U_0^{(r)}$.

Assume that the problem (1.1), (2.2) is regular in $U_0^{(r)}$ and let $u \in U_0^{(r)}$ be the solution. Then, by (1.12'),

$$\tilde{u}(\xi', x_n) = \sum_{k=1}^{\nu_0(\xi')} C_k(\xi')e^{\lambda_k(\xi')x_n} \tag{2.3}$$

almost everywhere in $\mathbb{R}^{n-1}\backslash 0$, for $x_n > 0$. The functions $C_k(\xi')$ are determined from the boundary conditions (2.2). To do this we apply the contact Fourier transformation $F_{x' \to \xi'}$ to (2.2) with $x_n = 0$ (see (1.4)). This yields

$$B_j(-i\xi', \partial_{x_n})\tilde{u}(\xi', 0+) = \tilde{f}_j(\xi') \quad (j = 1, \ldots, l) \tag{2.4}$$

almost everywhere in $\mathbb{R}^{n-1}$. Substituting (2.3) into this last equation, we have

$$\sum_1^{\nu_0(\xi')} B_j(-i\xi', \lambda_k(\xi'))C_k(\xi') = \tilde{f}_j(\xi') \quad (j = 1, \ldots, l) \tag{2.5}$$

almost everywhere in $\mathbb{R}^{n-1}$. Since the problem (1.1), (2.2) is regular in $U_0^{(r)}$, it follows that, for any $\tilde{f}_j(\xi') \in \tilde{H}_{-\infty} \equiv F_{x' \to \xi'} H_{-\infty}(\mathbb{R}^n)$, the system (2.5) has a solution for almost all $\xi' \in \mathbb{R}^{n-1}$ and the solution is unique. Thus we have

Proposition 2.1. *For the boundary-value problem* (1.1), (2.2) *to be regular in* $U_0^{(r)}$, $r \geqslant \bar{m}$, *it is necessary that*

$$\nu_0(\xi') \equiv \nu_0 = l \tag{2.6}$$

almost everywhere in $\mathbb{R}^{n-1}$. *Thus, in particular, the equation* (1.1) *must be* 0-*regular; moreover the matrix* $B_{jk}(\xi') \equiv B_j(-i\xi', \lambda_k(\xi'))$ *must be non-degenerate for almost all* $\xi' \in \mathbb{R}^{n-1}$:

$$\det(B_{jk}(\xi'))_{j,k=1}^{l} \neq 0 \quad \textit{almost everywhere in } \mathbb{R}^{n-1}. \tag{2.6$'$}$$

It is a Shapiro-Lopatinskij type condition (Lopatinskij [1953]) which is necessary for the problem (1.1), (2.2) to be regular in $U_0^{(r)}$. Let us write (2.5) in the vector form

$$B(\xi')C(\xi') = \tilde{F}(\xi') \tag{2.7}$$

almost everywhere in $\mathbb{R}^{n-1}$, where $B(\xi') = (B_{jk}(\xi'))$ is an $l \times l$ matrix and $C(\xi')$ and $\tilde{F}(\xi')$ are column vectors with components $C_k(\xi')$ and $\tilde{f}_j(\xi')$. Then, by (2.6′),

$$C(\xi') = B^{-1}(\xi')\tilde{F}(\xi') \tag{2.8}$$

almost everywhere in $\mathbb{R}^{n-1}$.

Example 2.1. For any operator $P(\partial_x)$ satisfying the assumption (1.8), the boundary-value problem

$$\partial_{x_n}^{j-1} u(x', 0+) = f_j(x') \quad (j = 1, \ldots, l) \tag{2.9}$$

satisfies (2.6′). Indeed, here we have $B_{jk}(\xi') = \lambda_k^{j-1}(\xi')$ and by (1.8),

$$\det B(\xi') = \prod_{k>i} (\lambda_k(\xi') - \lambda_i(\xi')) \neq 0 \tag{2.10}$$

almost everywhere in $\mathbb{R}^{n-1}$.

We now present sufficient conditions for the regularity of the problem (1.1), (2.2). We assume that the necessary condition (2.6) is satisfied.

Proposition 2.2. *For the problem* (1.1), (2.2) *to be regular in* $U_0^{(r)}$ *it is sufficient that* (2.6′) *is satisfied for all* $\xi' \in \mathbb{R}^{n-1}$ *(and not only almost everywhere). Thus*

$$\det B(\xi') \neq 0 \quad \textit{for all } \xi' \in \mathbb{R}^{n-1}. \tag{2.6$''$}$$

Proof. We have to verify that the solution u found from (2.1), (2.3) and (2.8) belongs to the class $U_0^{(r)}$ $(r \geqslant \bar{m})$. We note first that (2.6″) leads to the estimate

$$|\det B(\xi')| \geqslant C(1 + |\xi'|)^a, \quad \xi' \in \mathbb{R}^{n-1}, \tag{2.11}$$

for some $a \in \mathbb{R}$ and some $C > 0$. This estimate follows on applying Theorem A.2.5 of Hörmander [1983, 1985] to

$$f(x) := \inf_{|\xi'|=x} |\det B(\xi')|, \tag{2.12}$$

which is a semi-algebraic function of x. Since $f(x)$ is semi-algebraic, the set

$$F := \{(x, y) \in \mathbb{R}^2 \colon y \leqslant f(x)\} \tag{2.13}$$

is also semi-algebraic in $\mathbb{R}^2$. The fact that F is semi-algebraic is in turn a consequence of Corollary A.2.4 (Hörmander [1983, 1985]) because $f(x)$ has a representation of the form (A.2.2) (Hörmander [1983, 1985]):

$$\begin{gathered} f(x) = \inf(y > 0 \colon y^2 = |\det(B_j(-i\xi', \lambda_k))_{j,k=1}^k|^2, \quad |\xi'| = x, \\ P(-i\xi', \lambda_k) = 0, \quad \operatorname{Re} \lambda_1 \leqslant \operatorname{Re} \lambda_2 \leqslant \cdots \leqslant \operatorname{Re} \lambda_l \leqslant 0). \end{gathered} \tag{2.14}$$

We note that all these results of Hörmander [1983, 1985] are established by using the Seidenberg-Tarski principle. As for (2.11), we can establish the estimate

$$|\lambda_k(\xi')| \leqslant C(1 + |\xi'|)^{a_k}, \quad \xi' \in \mathbb{R}^{n-1}, \tag{2.15}$$

for some $a_k \in \mathbb{R}$. This result, together with (2.11), implies that

$$\|B^{-1}(\xi')\| \leqslant C(1 + |\xi'|)^b, \quad \xi' \in \mathbb{R}^{n-1}, \tag{2.16}$$

for some $b \in \mathbb{R}$. Since $\tilde{f}_j(\xi') \in \tilde{H}_{-\infty} \equiv \tilde{H}_\infty(\mathbb{R}^{n-1})$, the last result and (2.8) together imply that

$$C_k(\xi') \in \tilde{H}_{-\infty}(\mathbb{R}^{n-1}). \tag{2.17}$$

It remains to verify that, for some $s \in \mathbb{R}$ and all $j = 0, 1, 2, \ldots$ (and not only for $j \leqslant r$),

$$\|\partial_{x_n}^j u(\cdot, x_n)\|_s \leqslant C_{j,s} < \infty \tag{2.18}$$

for $x_n > 0$. To see this, we note first that (2.3) gives

$$\partial_{x_n}^j \tilde{u}(\xi', x_n) = \sum_{k=1}^{\nu_0} C_k(\xi') \lambda_k^j(\xi') e^{\lambda_k x_n}. \tag{2.19}$$

In view of (2.15) and the fact that $\operatorname{Re} \lambda_k(\xi') \leqslant 0$ $(k \leqslant \nu_0)$ (2.19) implies that

$$|\partial_{x_n}^j \tilde{u}(\xi', x_n)|^2 \leqslant C \sum_{k=1}^{\nu_0} |C_k(\xi')|^2 (1 + |\xi'|)^{2ja_k}. \tag{2.20}$$

Hence, by Definition 3.1 in Chap. 2,

$$\begin{aligned} \|\partial_{x_n}^j u(\cdot, x_n)\|_s^2 &= \int (1 + |\xi'|)^{2s} |\partial_{x_n}^j \tilde{u}(\xi', x_n)|^2 \, d\xi' \\ &\leqslant C \sum_{k=1}^{\nu_0} \int (1 + |\xi'|)^{2(s + ja_k)} |C_k(\xi')|^2 \, d\xi'. \end{aligned} \tag{2.21}$$

for all $s \in \mathbb{R}$, and it remains only to recall that $C_k \in \tilde{H}_{-\infty}$. ■

We note that in (2.18) s depends on j, as is clear from (2.20).

Remark 2.1. If the condition (1.8′) holds, the boundary-value problem (2.9) satisfies (2.6″) because of (2.10).

Proposition 2.3. *If* (1.8′) *holds for* $k, j \leqslant \nu_0$, *then the condition* (2.6″) (*together with* (2.6)) *is necessary and sufficient for the boundary-value problem* (1.1), (2.2) *to be regular in* $U_0^{(r)}$, *where* $r \geqslant \max(\overline{m}, \nu_0 - 1)$.

Proof. The sufficiency has been proved in Proposition 2.2. To deal with the necessity, we note that, by (1.2),

$$\partial_{x_n}^{j-1} u(x', 0+) := v_j(x') \in H_{-\infty}(\mathbb{R}^n) \quad (j = 1, \ldots, \nu_0) \tag{2.22}$$

for $u \in U_0^{(r)}$, where $r \geqslant \nu_0 - 1$. Substituting (2.3) into (2.22) and noting that $\nu_0 = l$, by (2.6), we obtain a system of equations for $C_k(\xi')$ which is analogous to (2.5):

$$\sum_{k=1}^{l} \lambda_k^{j-1}(\xi') C_k(\xi') = \tilde{v}_j(\xi') \in \tilde{H}_{-\infty}. \tag{2.23}$$

The determinant (2.10) of the system satisfies (2.6″), by (1.8′). If the problem (1.1), (2.2) is regular in $U_0^{(r)}$, then, for all $f_j \in H_{-\infty}(\mathbb{R}^{n-1})$, it has a solution $u \in U_0^{(r)}$ and then $C_k(\xi') \in \tilde{H}_{-\infty}$ for all k, as shown above. But, by (2.8),

$$C_k(\xi') = \sum_j B_{kj}^{-1}(\xi') \tilde{f}_j(\xi') \tag{2.24}$$

almost everywhere in $\mathbb{R}^{n-1}$. Thus, by (2.24), the functions $B_{kj}^{-1}(\xi')$ are locally bounded, implying that (2.6″) holds. ■

2.2. Examples of Regular Boundary-value Problems

1) The *Klein-Gordon hyperbolic equation* (1.17) has the form (1.1) with $n = k + 1$ and $x_n = t$. For this equation $\lambda_{1,2}(\xi) = \pm i\sqrt{|\xi|^2 + m_0^2}$ for $\xi \in \mathbb{R}^k$ and therefore $\nu_0 = 2$. Also (1.8′) is satisfied with ξ in place of ξ'. Therefore a problem for (1.17) which is regular in $U_0^{(r)}$ is given by (2.9) with $l = 2$; this is the so-called *Cauchy problem* with initial conditions

$$u(x, 0+) = u_0(x), \quad \partial_t u(x, 0) = u_1(x) \tag{2.25}$$

for $x \in \mathbb{R}^k$. Substituting the general solution (1.17′) into (2.25), we find, as for (2.5) and (2.8), that $A(\xi) = u_0(\xi)$ and $B(\xi) = u_1(\xi)/\omega(\xi)$, whence

$$\begin{aligned} \tilde{u}(\xi, t) &= \tilde{u}_0(\xi) \cos \omega(\xi) t + \frac{\tilde{u}_1(\xi)}{\omega(\xi)} \sin \omega(\xi) t; \\ \omega(\xi) &:= \sqrt{a^2 |\xi|^2 + m_0^2}. \end{aligned} \tag{2.26}$$

2) For the elliptic equation (1.27), we have $\nu_0 = 1$ and the following problems are regular in $U_0^{(r)}$.

a) The *Dirichlet problem*:

$$u|_{x_n = 0+} = u_0(x') \Rightarrow \tilde{u}(\xi', x_n) = \tilde{u}_0(\xi') e^{-\sqrt{|\xi|^2 + q}\, x_n}. \tag{2.27}$$

b) The *Neumann problem*:

$$\left.\frac{\partial u}{\partial x_n}\right|_{x_n=0+} = u_1(x') \Rightarrow \tilde{u}(\xi', x_n) = -\frac{\tilde{u}_1(\xi')}{\sqrt{|\xi'|^2+q}} e^{-\sqrt{|\xi'|^2+q}x_n}. \tag{2.28}$$

c) The *third boundary-value problem* (with $\sigma \geqslant 0$ or with $\operatorname{Im} \sigma \neq 0$):

$$\left(\frac{\partial u}{\partial x_n} - \sigma u\right)_{x_n=0+} = f(x') \Rightarrow \tilde{u}(\xi', x_n)$$

$$= -\frac{\tilde{f}(\xi')}{\sqrt{|\xi'|^2+q}+\sigma} e^{-\sqrt{|\xi'|^2+q}x_n}. \tag{2.29}$$

d) The *oblique derivative problem*

$$\left.\left(\sum_{k=1}^{n} b_k \frac{\partial u}{\partial x_k}\right)\right|_{x_n=0+} = f(x') \Rightarrow \tilde{u}(\xi', x_n)$$

$$= -\frac{\tilde{f}(\xi') e^{-\sqrt{|\xi'|^2+q}x_n}}{i\sum_{k=1}^{n-1} b_k \xi_k + b_n\sqrt{|\xi'|^2+q}} \tag{2.30}$$

(where $b_k \in \mathbb{R}$ and not all b_k are zero) is regular in $U_0^{(r)}$ if $b_n \neq 0$ or if $n = 2$.

3) For the Laplace equation (1.26), we also have $\nu_0 = 1$ and the Dirichlet problem is regular in $U_0^{(r)}$:

$$u|_{x_n=0+} = u_0(x') \Rightarrow \tilde{u}(\xi', x_n) = \tilde{u}_0(\xi') e^{-|\xi'|x_n}. \tag{2.31}$$

The third boundary-value problem (2.29) with $\sigma > 0$ or with $\operatorname{Im} \sigma \neq 0$ gives a similar case. Here

$$\left(\frac{\partial u}{\partial x_n} - \sigma u\right)_{x_n=0+} = f(x') \Rightarrow \tilde{u}(\xi', x_n) = -\frac{\tilde{f}(\xi')}{|\xi'|+\sigma} e^{-|\xi'|x_n}. \tag{2.32}$$

4) The parabolic equation (1.32) has the form (1.1) with $n = k + 1$ and $x_n = t$. For this equation $\nu_0 = 1$ if $q \geqslant 0$, and a problem which is regular in $U_0^{(r)}$ is the Cauchy problem with initial condition

$$u(x, 0+) = u_0(x) \Rightarrow \tilde{u}(\xi, t) = \tilde{u}_0(\xi) e^{-(a^2|\xi|^2+q)t}. \tag{2.33}$$

5) For the Schrödinger equation (1.38), we also have $\nu_0 = 1$ and a problem which is regular in $U_0^{(r)}$ is again the Cauchy problem with initial condition

$$\psi(x, 0+) = \psi_0(x) \Rightarrow \tilde{\psi}(\xi, t) = \tilde{\psi}_0(\xi) e^{-ia^2|\xi|^2 t}. \tag{2.34}$$

6) **Definition 2.2.** By the *Cauchy problem* for the general equation (1.1) is meant a problem with conditions of the form (2.9) in which $l = \nu$.

By Proposition 2.1, it is necessary for this problem to be regular in $U_0^{(r)}$ that $\nu_0(\xi') \equiv \nu$ (condition (2.6) with $l = \nu$) and, by Remark 2.1 and Proposition 2.2, this condition is also sufficient if, in addition, (1.8′) holds.

§ 3. Regular Boundary-value Problems in Classes of Exponentially Growing Functions

3.1. Definition and Examples

Definition 3.1. The boundary-value problem (1.1), (2.2) is said to be *regular* in the class $U_\alpha^{(r)}$, where $\alpha \in \mathbb{R}$ and $r \geqslant \overline{m}$, if it has a solution $u \in U_\alpha^{(r)}$ for any $f_j \in H_{-\infty}(\mathbb{R}^{n-1})$ and the solution is unique.

Proposition 3.1. *For the problem* (1.1), (2.2) *to be regular in* $U_\alpha^{(r)}$, $r \geqslant \overline{m}$, *it is*
1) *necessary that, as for* (2.6),

$$\nu_\alpha(\xi') \equiv \nu_\alpha = l \tag{3.1}$$

almost everywhere in $\mathbb{R}^{n-1}$; *in particular, it is necessary that the equation* (1.1) *is* α*-regular and that the condition* (2.6′) *holds;*
2) *sufficient that the conditions* (3.1) *and* (2.6″) *hold,*
3) *necessary and sufficient that the conditions* (2.6″) *and* (3.1) *hold together provided that the condition* (1.8′) *is satisfied for* $k, j \leqslant \nu_\alpha$ *and* $r \geqslant \max(\nu_\alpha - 1, \overline{m})$.

This proposition can be proved in exactly the same way as Propositions 2.1–2.3.

Examples of boundary-value problems which are regular in $U_\alpha^{(r)}$

1) The hyperbolic wave equation (1.16) has the form (1.1) with $n = k + 1$ and $x_n = t$. For this equation $\nu_0 = 2$, but (1.8′) is satisfied only for $\xi \in \mathbb{R}^k \backslash 0$. Therefore the Cauchy problem for (1.16) with the conditions (2.25) satsfies the requirements (2.6) and (2.6′) but fails to satisfy (2.6″). Substituting (1.16′) into (2.25), we find that $A(\xi) = \tilde{u}_0(\xi)$ and $B(\xi) = \tilde{u}_1(\xi)/|\xi|$, whence we have

$$\tilde{u}(\xi, t) = \tilde{u}_0(\xi) \cos a|\xi|t + \tilde{u}_1(\xi)\frac{\sin a|\xi|t}{|\xi|}. \tag{3.2}$$

This result implies that, as for (2.18)–(2.21),

$$\|\partial_{x_n}^j u(\cdot, t)\|_s^2 \leqslant C(\|u_0\|_{s+j}^2 + \|u_1\|_{s+j-1}^2 \cdot t) \tag{3.3}$$

for $t > 0$ and $s \in \mathbb{R}$, provided that $\|u_0\|_{s+j} < \infty$ and $\|u_1\|_{s+j-1} < \infty$. It follows from (3.2) that the Cauchy problem (1.16), (2.25) is not regular in $U_0^{(r)}$ but (3.3) shows that it is regular in $U_\alpha^{(r)}$ for all $\alpha > 0$. This example shows that the conditions of Proposition 2.2 (and, in particular, the condition (1.8′)), which are sufficient for the problem (1.1), (2.2) to be regular in $U_0^{(r)}$, are close to being necessary.

2) Similarly, for the equations (1.18) and (1.39), the Cauchy problem with the conditions (2.22) is regular in $U_\alpha^{(r)}$ if $\alpha > 0$, while it is regular for the equation (1.19) if $\alpha > -K$.

3) The equations (1.16) and (1.17) have only one solution $u = 0$ in the class $U_\alpha^{(r)}$ if $\alpha < 0$. Therefore a regular problem for these equations in $U_\alpha^{(r)}$, with $\alpha < 0$, is one without any boundary conditions. The same is true for the equations (1.18) and (1.19) if $\alpha < -K$. The equation (1.18) is not α-regular for $\alpha \in]-K, 0[$

and therefore for these values of α there are no regular boundary-value problems for (1.18) in $U_\alpha^{(r)}$; the situation with the equation (1.39) is similar for $\alpha < 0$.

4) For the elliptic equation (1.27), all the boundary-value problems (2.27)–(2.30) are regular in $U_\alpha^{(r)}$ for $\alpha \in [-\sqrt{q}, \sqrt{q}]$. When $\alpha \notin [-\sqrt{q}, \sqrt{q}]$, there are no regular boundary-value problems for (1.27) in $U_\alpha^{(r)}$ because, for these α, the equation is not α-regular. The same is true for the Laplace equation when $\alpha \neq 0$ and for both the Helmholtz equation (1.28) and the Cauchy-Riemann equation (1.29) for all $\alpha \in \mathbb{R}$.

5) For the parabolic equation (1.32), the Cauchy problem (2.33) is regular in $U_\alpha^{(r)}$ if $\alpha \geqslant -q$. If $\alpha < -q$, there are no regular boundary-value problems in $U_\alpha^{(r)}$ for this equation.

Remark 3.1. The examples examined above show that the number l of boundary conditions in boundary-value problems which are regular in $U_\alpha^{(r)}$ for a given equation depends weakly on α. For example, for the above-mentioned hyperbolic equations we have $l = 0$ or $l = 2$; for the elliptic, parabolic and Schrödinger equations we have $l = 1$ only; while for the equation (1.39) we have $l = 2$ only. This situation is related to condition (2.6) and the fact that, as a rule, the equation is α-regular only for one (or two) values of $\alpha \in \mathbb{R}$ provided that such α's do exist.

For example, no such values of α exist for the Helmholtz equation (1.28) and the Cauchy-Riemann equation (1.29), and accordingly there are no regular boundary-value problems of the form (2.2) in $U_\alpha^{(r)}$ for these equations.

On the other hand, for example, the equation (compare with (1.19))

$$\left(\left(\frac{\partial}{\partial t} + K_1\right)^2 - a^2\Delta\right)\left(\left(\frac{\partial}{\partial t} + K_2\right)^2 - a^2\Delta\right)u(x, t) = 0, \qquad x \in \mathbb{R}^k, \quad t > 0, \tag{3.4}$$

with $K_2 < K_1$, has the roots

$$\lambda_{1,2} = -K_1 \pm ia|\xi|, \quad \lambda_{3,4} = -K_2 \pm ia|\xi|; \quad \xi \in \mathbb{R}^k.$$

Hence, for (3.4), we have the solution

$$\tilde{u}(\xi, t) = e^{-K_1 t}(C_1(\xi)e^{ia|\xi|t} + C_2(\xi)e^{-ia|\xi|t}) + e^{-K_2 t}(C_3(\xi)e^{ia|\xi|t} + C_4(\xi)e^{-ia|\xi|t})$$

in $U^{(r)}$, and

$$v_\alpha(\xi) = \begin{cases} 0, & \alpha < -K_1, \\ 2, & -K_1 \leqslant \alpha < -K_2, \\ 4, & \alpha \geqslant -K_2. \end{cases}$$

Therefore a regular problem for (3.4) in $U_\alpha^{(r)}$ must contain four initial conditions if $\alpha \geqslant -K_2$ (the Cauchy problem, for example), two conditions if $\alpha \in [-K_1, -K_2]$ while no condition need be prescribed if $\alpha < -K_1$.

3.2. The Cauchy Problem. In order that the Cauchy problem for the general equation (1.1) (see Definition 2.2) be regular in $U_\alpha^{(r)}$, it is necessary, by (3.1), that the condition

$$(v_\alpha(\xi') = v, \xi' \in \mathbb{R}^{n-1}) \Leftrightarrow \bar{\alpha}_v \leqslant \alpha \tag{3.5}$$

holds. This implies, in particular, that (1.1) must be Petrovskij well-defined.

Proposition 3.2. *Assume that the equation* (1.1) *is Petrovskij well-defined and that* (*see* (1.6))

$$p_{(v)}(\xi') \equiv \text{const} \neq 0. \tag{3.5'}$$

Then the Cauchy problem for (1.1) *is regular in* $U_\alpha^{(r)}$ *when* $\alpha > \bar{\alpha}_v$ (*even if* (1.8) *is not satisfied*).

Proof. When the condition (3.5) is satisfied, the Cauchy problem for (1.1) can be written as the system

$$\begin{cases} \dfrac{\partial u_0}{\partial t} = u_1(t, x'), \quad t > 0, \quad x' \in \mathbb{R}^{n-1}, \\ \vdots \\ \dfrac{\partial u_{v-2}}{\partial t} = u_{v-1}(t, x'), \\ \dfrac{\partial u_{v-1}}{\partial t} = -\dfrac{1}{p^v} \displaystyle\sum_{j=0}^{v-1} p_{(j)}(i\partial_{x'}) u_j; \end{cases} \tag{3.6}$$

$$\begin{cases} u_0(x', 0+) = f_1(x'), \\ \vdots \\ u_{v-1}(x', 0+) = f_v(x'), \end{cases}$$

where $t = x_n$. This problem can be expressed in the vector form

$$\frac{\partial U}{\partial t} = Q(\partial_{x'}) U(x', t), \quad t > 0, \quad x' \in \mathbb{R}^{n-1};$$
$$U|_{t=0+} = F \equiv \begin{bmatrix} f_1 \\ \vdots \\ f_v \end{bmatrix}; \quad U \equiv \begin{bmatrix} u_0 \\ \vdots \\ u_v \end{bmatrix}, \tag{3.7}$$

or, on applying the contact Fourier transformation, in the form

$$\frac{\partial \tilde{U}}{\partial t} = Q(-i\xi')\tilde{U}(\xi', t), \quad t > 0; \quad \tilde{U}(\xi', 0+) = \tilde{F}(\xi'), \tag{3.8}$$

almost everywhere in $\mathbb{R}^{n-1}$. Now it follows that

$$\tilde{U}(\xi', t) = e^{Q(-i\xi')t} \tilde{F}(\xi') \tag{3.9}$$

for $t \geqslant 0$, almost everywhere in $\mathbb{R}^{n-1}$. We note that

$$\det(Q(-i\xi') - \lambda E) = P(-i\xi', \lambda), \tag{3.10}$$

and so the eigenvalues of the matrix $Q(-i\xi')$ are $\lambda_k(\xi')$ ($k = 1, \ldots, v$). Therefore, in view of the estimates contained in Gel'fand and Shilov [1958c] and Shilov [1965], we obtain

$$\|e^{Q(-i\xi')t}\| \leqslant Ce^{\bar{\alpha}_v t}(1 + t\|Q(-i\xi')\|)^{v-1} \quad \text{for } t > 0, \quad \xi' \in \mathbb{C}^{n-1}. \tag{3.11}$$

This result implies that

$$\|\partial_t^l e^{Q(-i\xi')t}\| \leqslant Ce^{\bar{\alpha}_v t}(1 + |\xi'|)^{M_l}(1 + t)^{v-1}, \quad t > 0, \quad \xi' \in \mathbb{C}^{n-1}, \tag{3.12}$$

for all $l = 0, 1, 2, \ldots$. Therefore, as for (3.3),

$$|||\partial_t^l U(\cdot, t)|||_s \leqslant C_\alpha e^{\alpha t} \sum_{j=1}^{v} \|f_j\|_{s+Ml}, \quad t > 0, \tag{3.13}$$

for $\alpha > \bar{\alpha}_v$ and all $s \in \mathbb{R}$; here $|||\cdot|||_s$ denotes the norm in $[H_s(\mathbb{R}^{n-1})]^v$. ■

As an example, the Cauchy problem (2.25) for the equation (1.39) is regular in $U_\alpha^{(r)}$ when $\alpha > 0$.

Corollary 3.1. *The Cauchy problem is regular in* $U_\alpha^{(r)}$, *where* $\alpha > \bar{\alpha}_v$, *for those equations* (1.1) *that are Gårding hyperbolic or Petrovskij parabolic.*

Indeed, for such equations both conditions (3.5) and (3.5′) hold.

3.3. The Dirichlet Problem for Elliptic Equations. Let $P(\partial_x)$ be an elliptic operator in $\mathbb{R}^n$ of order $m = 2l$, where l is an integer (this is always true if $n \geqslant 3$, by Proposition 5.1 in Chap. 3).

Definition 3.1. By the *Dirichlet problem* for the operator $P(\partial_x)$ is meant a boundary-value problem with conditions of the form (2.9) in which $l = m/2$.

Proposition 3.3. 1) *If* $\bar{\alpha}_l \leqslant \underline{\alpha}_{l+1}$ *and the condition* (1.8′) *holds, then the Dirichlet problem for the operator P is regular in* $U_\alpha^{(r)}$ *for* $\alpha \in [\bar{\alpha}_l, \underline{\alpha}_{l+1}]$.

2) *If* $\bar{\alpha}_l > \underline{\alpha}_{l+1}$, *there are no regular boundary-value problems for* $P(\partial_x)$) *in* $U_\alpha^{(r)}$.

Proof. Statement 1) follows from Proposition 1.5 and (2.10) in view of Statement 2) in Proposition 3.1. Statement 2) also follows from Proposition 1.5 on account of Statement 1) in Proposition 3.1. ■

§4. Regular Boundary-value Problems in the Class of Functions of Arbitrary Growth

4.1. Let us examine the boundary-value problem (1.1), (2.2) in the function space $U^{(r)}$, where $r \geqslant \bar{m}$.

Definition 4.1. The boundary-value problem (1.1), (2.2) is said to be *regular in the class* $U^{(r)}$ $(r \geqslant \overline{m})$ if it has a solution $u \in U^{(r)}$ for all $f_j \in H_{-\infty}(\mathbb{R}^{n-1})$ and the solution is unique.

Let us find necessary conditions on the operators $P, B_1, \ldots, B_l$ so that the problem (1.1), (2.2) is regular in $U^{(r)}$. The function u given by (1.9) lies in $U^{(r)}$ for any $C_k(\xi') \in C_0^\infty(\mathbb{R}^{n-1})$ and is a solution of the problem (1.1), (2.2). This gives

Proposition 4.1. *For the problem* (1.1), (2.2) *to be regular in* $U^{(r)}$ *it is*

1) *necessary that* $l = v$ *and the condition* (2.6′) *is satisfied,*

2) *sufficient that the operator P is Petrovskij well-defined,* $l = v$ *and* (2.6″) *holds,*

3) *necessary and sufficient that, subject to* (1.8′) *and* $r \geqslant v - 1$, *P is Petrovskij well-defined,* $l = v$ *and* (2.6″) *holds.*

Proof. The proofs of 1) and 2) repeat word for word those of Propositions 2.1 and 2.2 respectively. The necessity of the Petrovskij well-defined condition in Statement 3) follows from the method of proof of Proposition 2.3. Thus, if the problem (1.1), (2.2) is regular in $U^{(r)}$ $(r \geqslant v)$ then from (1.8′) it can be shown, as in the proof of Proposition 2.3, that the functions $C_k(\xi')$ in (1.9) are arbitrary elements of $\tilde{H}_{-\infty}$. Taking $C_j(\xi') \equiv 0$, with $j \neq k$, we obtain from (1.9)

$$\tilde{u}(\xi', x_n) = C_k(\xi') e^{\lambda_k(\xi') x_n} \in H_{-\infty} \tag{4.1}$$

for all k, all $C_k(\xi') \in \tilde{H}_{-\infty}$, and all $x_n > 0$. But this is possible only if $\operatorname{Re} \lambda_k(\xi')$ is a bounded function of $\xi' \in \mathbb{R}^{n-1}$. Indeed, according to Theorem A.2.5 of Hörmander [1983, 1985], mentioned in § 2,

$$\max_{|\xi'|=\rho} \operatorname{Re} \lambda_k(\xi') \sim C\rho^{a_k} \tag{4.2}$$

as $\rho \to \infty$, for some C and $a_k \in \mathbb{R}$. If $a_k > 0$, then (4.1) cannot hold. Accordingly, $a_k \leqslant 0$ and thus all the roots $\lambda_k(\xi')$ are Petrovskij well-defined. The necessity of (2.6″) can be shown in the same way as in the proof of Proposition 2.3. ■

Remark 4.1. If $a_k > 0$ in (4.2) and $C_k(\xi') \in \tilde{H}_{-\infty}$, then for $x_n > 0$ the function $\tilde{u}(\cdot, x_n)$ lies in $\mathscr{D}'(\mathbb{R}^n)$ but not, in general, in $S'(\mathbb{R}^{n-1})$. Accordingly, for $x_n > 0$, $u(\cdot, x_n)$ lies in $Z'(\mathbb{C}^{n-1})$ but not, in general, in $\mathscr{D}'(\mathbb{R}^{n-1})$.

Corollary 4.1. *For elliptic equations, with* $n \geqslant 3$, *there are no regular boundary-value problems in* $U^{(r)}$.

Indeed, these equations are not Petrovskij well-defined.

Hadamard's example. The Cauchy problem (2.25), with $t \equiv x_n$, for the Laplace equation (1.26) is not regular in $U^{(r)}$. Indeed, by (1.9), for the general solution $u \in U^{(r)}$ we have

$$\begin{aligned} \tilde{u}(\xi', x_n) &= C_1(\xi') e^{-|\xi'| x_n} + C_2(\xi') e^{|\xi'| x_n} \\ &= A(\xi') \operatorname{ch} |\xi'| x_n + B(\xi') \operatorname{sh} |\xi'| x_n. \end{aligned} \tag{4.3}$$

Substitution of (4.3) into (2.25) yields $A(\xi') = \tilde{u}_0(\xi')$ and $B(\xi') = \tilde{u}_1(\xi')/|\xi'|$, whence

$$\tilde{u}(\xi', x_n) = \tilde{u}_0(\xi') \operatorname{ch}|\xi'|x_n + \tilde{u}_1(\xi') \frac{\operatorname{sh}|\xi'|x_n}{|\xi'|}. \tag{4.4}$$

If we take $\tilde{u}_1(\xi') \equiv 0$, then $\tilde{u}(\cdot, x_n) \notin \tilde{H}_{-\infty}(\mathbb{R}^{n-1})$ for $x_n > 0$, as shown if, for example, we choose $\tilde{u}_0(\xi') \equiv 1$ or $\tilde{u}_0(\xi') \equiv (1 + |\xi'|^2)^{-N}$ for any $N > 0$. Therefore the Cauchy problem (2.25) for the Laplace equation, with $u_1(x') \equiv 0$, has no solution $u \in U^{(r)}$ for $u_0(x') = \delta(x')$ or for $u_0(x') = F^{-1}_{\xi' \to x'}(1 + |\xi'|^2)^{-N}$.

Note that the latter function $u_0(x')$ has $2N - n$ continuous derivatives for $2N - n > 0$ and decreases faster than any power of x' as $|x'| \to \infty$.

Remark 4.2. If the Cauchy problem for the operator P is regular in $U_\alpha^{(r)}$ for some $\alpha \in \mathbb{R}$, where $r \geqslant \nu - 1$, then (3.1) implies (3.5). This Cauchy problem is therefore regular in $U^{(r)}$ also. In particular, the Cauchy problem for P is regular in $U^{(r)}$ $(r \geqslant \nu - 1)$ under the hypotheses of Proposition 3.2 (for example, for Gårding hyperbolic or Petrovskij parabolic operators P (see Corollary 3.1)).

Thus the introduction of the solution space $U^{(r)}$, in place of $U_\alpha^{(r)}$, does not lead to any new regular boundary-value problems.

§ 5. Well-posed and Continuous Boundary-value Problems in a Half-space

5.1. Well-posed Boundary Value Problems. Let there be given normed spaces $H \subset U^{(r)}$ $(r \geqslant \overline{m})$ and $H^j \subset H_{-\infty}(\mathbb{R}^{n-1})$ for $j = 1, \ldots, l$. We write $\mathscr{H} = H^1 \oplus \cdots \oplus H^l$.

Definition 5.1. The boundary-value problem (1.1), (2.2) is said to be *well-posed* in the spaces $H, H^1, \ldots, H^l$ if

1) it has a unique solution $u \equiv \mathscr{R}(f_1, \ldots, f_l) \in H$ for all $f_j \in H^j$;
2) the map $\mathscr{R}\colon \mathscr{H} \to H$ is continuous, that is,

$$\|u\|_H \leqslant C \sum_{j=1}^{l} \|f_j\|_{H^j}. \tag{5.1}$$

We introduce the normed space

$$C_\alpha(0, \infty; H_s) := \left\{ u \in C(0, \infty; H_s) \colon \|u\|_{s,\alpha} := \sup_{x_n > 0} e^{-\alpha x_n} \|u(\cdot, x_n)\|_s < \infty \right\}; \quad H_s \equiv H_s(\mathbb{R}^{n-1}). \tag{5.2}$$

Proposition 5.1. *All the boundary-value problems which have been examined in § 2 and § 3 and which are regular in $U_\alpha^{(r)}$ are well-posed in the spaces $H := C_\alpha(0, \infty; H_s)$ and $H^j = H_{s_j} := H_{s_j}(\mathbb{R}^{n-1})$ for all $s_j \in \mathbb{R}$ with some $s \leqslant s_j$.*

Proof. For the boundary-value problems of § 2 and § 3 which are regular in $U_\alpha^{(r)}$, we have the estimate

$$\|u\|_{s,\alpha} \leqslant C \sum_j \|u_j\|_{s_j} \tag{5.1'}$$

for any $s_j \in \mathbb{R}$ with some $s \leqslant s_j$. For example, such an estimate for $\alpha = 0$ can be obtained from the arguments (2.15)–(2.21) (with $j = 0$). For $\alpha \neq 0$, estimates of the form (5.1′) are proved in exactly the same way. For instance, under the hypotheses of Proposition 3.2, an estimate of the form (5.1′) has been obtained for the Cauchy problem in (3.13).

The estimate (5.1′) implies that the operator $\mathscr{R}$ is bounded. Since $\mathscr{R}$ is linear, it follows that it is also continuous. ■

5.2. Continuous Well-posed Boundary-value Problems. Let $\mathfrak{A}: U^{(r)} \to [H_{-\infty}(\mathbb{R}^{n-1})]^l$ denote the operator corresponding to the boundary-value problem (1.1), (2.2):

$$\mathfrak{A}u = (\gamma B_1 u, \ldots, \gamma B_l u), \quad u \in U^{(r)}, \quad r \geqslant \overline{m}; \quad (\gamma u)(x') := u(x', 0+), \quad u \in U^{(0)}. \tag{5.3}$$

Of special interest are the boundary-value problems which are well-posed in certain spaces $H, H^1, \ldots, H^l$ and for which the operator $\mathfrak{A}: H \to \mathscr{H} \equiv H^1 \oplus \cdots \oplus H^l$ is continuous. In fact, the operator $\mathscr{R}: \mathscr{H} \to H$ is a right inverse of $\mathfrak{A}$ and is continuous, because the boundary-value problem is well-posed. Therefore the operator $\mathfrak{A} + \delta$ also has a continuous right inverse $\mathscr{B}_\delta: \mathscr{H} \to H$ provided that $\mathfrak{A}$ is continuous and the operator $\delta: H \to \mathscr{H}$ has a sufficiently small norm. Roughly speaking, the boundary-value problem remains well-posed in function spaces in question under small perturbations of the operators P and B_j. This enables us to develop the theory of boundary-value problems for differential operators with small changes in the coefficients and then with more general variable coefficients (C^∞ etc.). The method of freezing of coefficients (Agranovich and Vishik [1964], Vishik and Ehskin [1965]) is based on this idea.

We now give conditions for the continuous boundary-value problems (1.1), (2.2) to be well-posed in the spaces

$$\begin{cases} H \equiv H^{(r)}_{\alpha,s} \equiv \left\{ u \in U^{(r)}: \|u\|^{(r)}_{\alpha,s} := \sup\limits_{\substack{x_n > 0 \\ i \leqslant r}} e^{-\alpha x_n} \|\partial^i_{x_n} u(\cdot, x_n)\|_{s - i\mu} < \infty \right\}, \\ H^j := H_{s-\mu_j}(\mathbb{R}^{n-1}), \quad j = 1, \ldots, l. \end{cases} \tag{5.4}$$

In these spaces the continuity of the operator $\mathfrak{A}: H \to \mathscr{H}$ for all $s \in \mathbb{R}$ follows from Proposition 3.1 in Chap. 2, if $\mu = 1$ and $\mu_j = m_j$, and $r \geqslant \overline{m}$.

We denote by λ^0_k the roots of the characteristic equation (1.6) of the leading part $P_m(\partial_x)$ of the operator $P(\partial_x)$. Also, let

$$B^0_j(\partial_x) = \sum_{|\alpha| = m_j} b_{j\alpha} \partial^\alpha_x \tag{5.5}$$

denote the leading part of the operator $B_j(\partial_x)$.

Proposition 5.2. *Assume that the boundary $x_n = 0$ is non-characteristic for the operator P and that, for the boundary-value problem* (1.1), (2.2), *the conditions* (3.1), (2.6″) *and the* ***Shapiro-Lopatinskij condition***

$$\det B_j^0(-i\xi', \lambda_k^0(\xi'))_{j,k=1}^l \neq 0 \quad \text{for } |\xi'| = 1, \quad \xi' \in \mathbb{R}^{n-1}, \tag{5.6}$$

hold (Lopatinskij [1953]). *Then the problem* (1.1), (2.2) *is well-posed in the spaces* (5.4) *for* $\mu = 1$, $\mu_j = m_j$, $r \geqslant \overline{m}$ *and all* $s \in \mathbb{R}$.

Proof. Since the boundary $x_n = 0$ is non-characteristic,

$$B_j(-i\xi', \lambda_k(\xi')) = B_j^0(-i\xi', \lambda_k^0(\xi')) + o(|\xi'|^{m_j}) \tag{5.7}$$

as $|\xi'| \to \infty$, by Lemma 1.1. By (1.15′) and (5.6), this gives

$$\det B_{jk}(\xi')| \sim |\det B_j^0(-i\xi', \lambda_k^0(\xi'))| \sim |\xi'|^{m_1 + \cdots + m_l} \tag{5.8}$$

as $|\xi'| \to \infty$. Consequently, by (2.6″), the entries of the inverse matrix $B^{-1}(\xi')$ satisfy the asymptotics

$$|B_{kj}^{-1}(\xi')| \sim (1 + |\xi'|)^{-m_j}, \quad |\xi'| \to \infty. \tag{5.9}$$

Hence (2.8) implies that

$$|C_k(\xi')| \leqslant C \sum_{j=1}^{l} (1 + |\xi'|)^{-m_j} |\tilde{f}_j(\xi')|. \tag{5.10}$$

But (1.12′) yields the estimate

$$|\tilde{u}(\xi', x_n)| \leqslant C \sum_{k=1}^{l} |C_k(\xi')| e^{\alpha x_n}, \tag{5.11}$$

and so (5.10) leads to

$$\sup_{x_n > 0} e^{-\alpha x_n} \|u(\cdot, x_n)\|_s \leqslant C \sum_{j=1}^{l} \|f_j\|_{s-m_j}. \tag{5.11′}$$

Similarly, (1.12′) yields

$$|\partial_{x_n}^i \tilde{u}(\xi', x_n)| \leqslant C \sum_{k=1}^{l} |\lambda_k^i(\xi')| \cdot |C_k(\xi')| e^{\alpha x_n}. \tag{5.12}$$

Hence, by (1.21), (1.15′) and (5.10), we find that

$$\sup_{x_n > 0} e^{-\alpha x_n} \|\partial_{x_n}^i u(\cdot, x_n)\|_{s-i} \leqslant C \sum_{j=1}^{l} \|f_j\|_{s-m_j}. \quad \blacksquare \tag{5.13}$$

This proposition can also be extended to operators P that are Petrovskij parabolic. If P is Petrovskij β-parabolic, then (5.5) must be modified as follows. Assuming that $b_{j\alpha} = 0$ for $|\alpha'| + \beta\alpha_n > \beta\nu^j$, we write

$$B_j^0(\partial_x) := \sum_{|\alpha'| + \beta\alpha_n = \beta\nu^j} b_{j\alpha} \partial_x^\alpha. \tag{5.14}$$

Then Proposition 3.1, Chap. 2, shows that the operator $\mathfrak{A}$ is continuous in the spaces (5.4) where we take $\mu = \beta$, $\mu_j = \beta\nu^j$ and $r \geqslant \bar{\nu} = \max \nu^j$.

Proposition 5.2′. *Assume that the operator* P *is Petrovskij* β*-parabolic and that the conditions* (2.6″) *and* (5.6) *hold. Let* $\lambda_k^0(\xi')$ *be the roots of the equation* (1.34). *Then the boundary-value problem* (1.1), (2.2) *is well-posed in the spaces* (5.4) *for* $r \geqslant \bar{\nu}$, *all* $s \in \mathbb{R}$, $\mu = \beta$ *and* $\mu_j = \beta\nu^j$.

The proof of this proposition almost coincides with that of the preceding proposition; the only difference is that the conditions (1.36) and (1.37′) are used in place of (1.15′) and (1.21).

Corollary 5.1. *Let the condition* (1.8′) *be satisfied and, similarly, let*

$$\lambda_k^0(\xi') \neq \lambda_j^0(\xi') \quad \text{for } k \neq j, \quad |\xi'| = 1, \quad \xi' \in \mathbb{R}^{n-1}. \tag{5.15}$$

Then the Cauchy problem for the Petrovskij hyperbolic (*or* β*-parabolic*) *operator* $P(\partial_x)$ *is well posed in the spaces* (5.4) *for* $r \geqslant m - 1$, $\mu = 1$ *and* $\mu_j = j - 1$ (*or* $\mu = \beta$ *and* $\mu_j = \beta(j-1)$).

Indeed, the conditions (2.6″) and (5.6) are satisfied in view of (2.10).

Similary, under conditions (1.8′) and (5.15), the Dirichlet problem (see Definition 3.1) for the elliptic operator $P(\partial_x)$ is well-posed in the spaces (5.4) for $n \geqslant 3$, $r \geqslant \frac{m}{2} - 1$, $\mu = 1$ and $\mu_j = j - 1$ if $\bar{\alpha}_l < \underline{\alpha}_{l+1}$ and $\alpha \in [\bar{\alpha}_l, \underline{\alpha}_{l+1}]$. A voluminous literature is available on the question of boundary-value problems being well-posed in the most diverse spaces. In Gel'fand and Shilov [1958c] a detailed account can be found of the results on the regularity and well-posedness of the Cauchy problem for general differential equations with constant coefficients in classes of functions that grow like $e^{a|x|^b}$ for various values of a and b. The book by Gel'fand and Shilov [1958c] contains precise bounds for the growth exponents a and b of the solution $u(x)$ and of initial conditions under which the Cauchy problem is regular and well-posed.

Fundamental results on the well-posedness of the Cauchy problem in L_2-type spaces were obtained by Petrovskij [1937a, 1937b].

§ 6. The Poisson Kernel for the Boundary-value Problem in a Half-space

6.1. The Poisson Kernel and the Fundamental Solution of the Boundary-value Problem

Definition 6.1. By a *fundamental solution* of the boundary-value problem (1.1), (2.2) is meant a vector function $E(x) = (E_k(x))_{k=1,\ldots,l}$, where $E_k \in U^{(r)}$ are the solutions in $\mathbb{R}^n_+$ of the following boundary-value problems

$$\begin{cases} P(\partial_x)E_k(x) = 0, & x_n > 0, \quad x' = (x_1, \ldots, x_{n-1}) \in \mathbb{R}^{n-1}, \\ \gamma B_j E_k(x) = \delta_{jk}\delta(x'), & j = 1, \ldots, l, \end{cases} \tag{6.1}$$

in which δ_{jk} denotes the Kronecker symbol and $\delta(x') = \delta_{n-1}(x')$ is the δ-function in $\mathbb{R}^{n-1}$.

If the problem (1.1), (2.2) is regular in $U_\alpha^{(r)}$ ($r \geqslant \bar{m}$) for some α, then such a function $E_k \in U_\alpha^{(r)}$ exists for each $k = 1, \ldots, l$, and is unique, $\left(\text{because } \delta(x') \in H_s(\mathbb{R}^{n-1}), \text{ where } s < -\frac{n-1}{2}\right)$.

Definition 6.2. The *Poisson kernel of the boundary-value problem* (1.1), (2.2) is the vector function

$$\mathscr{P}(x; y') = E(x' - y'; x_n), \quad x_n > 0, \quad x', y' \in \mathbb{R}^{n-1}. \tag{6.2}$$

Note that $\mathscr{P}(x; y') \in U_\alpha^{(r)} \subset \mathscr{D}'(\mathbb{R}_+^n)$ for all $y' \in \mathbb{R}^{n-1}$. It follows from (6.1) that $\mathscr{P}_k(x; y') \equiv E_k(x' - y'; x_n)$ is a solution of the boundary-value problem

$$\begin{cases} P(\partial_x)\mathscr{P}_k(x, y') = 0, \quad x_n > 0, \\ \gamma B_j(\partial_x)\mathscr{P}_k(x, y') = \delta_{jk}\delta(x' - y'), \quad j = 1, \ldots, l, \end{cases} \tag{6.3}$$

for all $y' \in \mathbb{R}^{n-1}$ where y' plays the role of a parameter.

Proposition 6.1. *Let the boundary-value problem* (1.1), (2.2) *be regular in* $U_\alpha^{(r)}$, *where* $r \geqslant \max(m, \bar{m})$. *Then, for any boundary data* $f_j(x') \in C_0^\infty(\mathbb{R}^{n-1})$ *its solution* u *is expressed by the formula (in which* $F = (f_1, \ldots, f_l)$*)*

$$\begin{aligned} u(x', x_n) &= E(\cdot, x_n) * F = \sum_k E_k(\cdot, x_n) * f_k(\cdot) \\ &= \sum_k \langle E_k(x' - y'; x_n), f_k(y') \rangle. \end{aligned} \tag{6.4}$$

Proof. With $\alpha_n \leqslant m$, we have

$$\begin{aligned} \partial_{x_n}^{\alpha_n} u(x', x_n) &= \sum_k \langle \partial_{x_n}^{\alpha_n} E_k(x' - y', x_n), f_k(y') \rangle \\ &= \sum_k (\partial_{x_n}^{\alpha_n} E_k(\cdot, x_n) * f_k(\cdot))(x') \end{aligned} \tag{6.5}$$

for the function (6.4), because $E_k \in C^{(r)}(0, \infty; H_s)$ for some $s \in \mathbb{R}$ and $r \geqslant m$. Hence the property (1.47), Chap. 1, of the convolution and (6.1) together yield

$$P(\partial_x)u(x', x_n) = \sum_k ((PE_k)(\cdot, x_n) * f_k(\cdot))(x') = 0 \tag{6.6}$$

for $x_n > 0$. Similarly,

$$\gamma B_j u(x') = \sum_k \gamma B_j E_k * f_k = \delta_{n-1} * f_j = f_j. \quad \blacksquare \tag{6.7}$$

We note that the Poisson kernel (6.2) enables us to write (6.4) in the form

$$u(x', x_n) = \langle \mathscr{P}(x; y'), F(y') \rangle = \sum_k \langle \mathscr{P}_k(x, y'), f_k(y') \rangle. \tag{6.8}$$

6.2. The Connection Between the Fundamental Solution of the Cauchy Problem and the Retarded Fundamental Solution of the Operator $P(\partial_x)$. Suppose that, for an operator $P(\partial_x)$ of order v in ∂_{x_n}, the Cauchy problem is regular in $U^{(r)}$, where $r \geqslant v - 1$, and that $p_{(v)} \equiv \text{const} \neq 0$ while $p_{(j)} = 0$ for $1 \leqslant j \leqslant v - 1$ in (1.6). Then this Cauchy problem has a fundamental solution $E(x) \in [C^{(v-1)}(0, \infty; H_s(\mathbb{R}^{n-1}))]^v$ for some $s \in \mathbb{R}$. In this case, the system (6.1) takes the form

$$\begin{cases} p_{(v)} \dfrac{\partial^v E_k}{\partial x_n^v} + p_{(0)}(i\partial_{x'})E_k = 0, \quad x_n > 0, \quad x' \in \mathbb{R}^{n-1}, \\ \gamma \partial_{x_n}^{j-1} E_k = \delta_{jk}\delta(x'), \quad j = 1, \ldots, v. \end{cases} \tag{6.9}$$

We define a generalized function $\mathscr{E}(x) \in \mathscr{D}'(\mathbb{R}^n)$ by the formula

$$\mathscr{E}(x', x_n) = \begin{cases} \dfrac{1}{p_{(\nu)}} E_\nu(x', x_n), & x_n > 0, \\ 0, \quad x_n < 0, \end{cases}$$

that is,

$$\langle \mathscr{E}, \varphi \rangle := \int_0^\infty \frac{1}{p_{(\nu)}} \langle E_\nu(\cdot, x_n), \varphi(\cdot, x_n) \rangle \, dx_n, \quad \forall \varphi \in \mathscr{D}(\mathbb{R}^n). \tag{6.10}$$

Then $\mathscr{E}(x)$ is a retarded fundamental solution of P, that is, $\mathscr{E}(x)$ is zero for $x_n < 0$ and $P(\partial_x)\mathscr{E}(x) = \delta(x)$. As for Lemma 1.1 in Chap. 1, this result is a consequence of (6.9). Conversely, the fundamental solution $E(x)$ can be expressed in terms of $\mathscr{E}(x)$ in (6.10) as follows:

$$E_k(x) = \partial_{x_n}^{\nu-k} E_\nu = p_{(\nu)} \partial_{x_n}^{\nu-k} \mathscr{E}(x), \quad x_n > 0; \quad 1 \leqslant k \leqslant \nu - 1. \tag{6.11}$$

Indeed, the function

$$\hat{E}_k(x', x_n) = \partial_{x_n}^{\nu-k} E_1, \quad 2 \leqslant k \leqslant \nu, \tag{6.12}$$

lies in $U^{(r)}$ $(r \geqslant \nu - 1)$ and is a solution of the problem (6.9), just as E_k is. Therefore, since the Cauchy problem is regular, $\hat{E}_k = E_k$, leading to (6.11).

Thus the fundamental solution (6.9) is expressed in terms of the retarded fundamental solution (6.10) of the operator $P(\partial_x)$ by means of (6.11). A retarded fundamental solution of $P(\partial_x)$ may not be *a priori* unique. Therefore the question arises as to how to identify the function (6.10) in the set of all retarded fundamental solutions. For example, for the Gårding hyperbolic operators the function (6.10) can be identified as follows.

Proposition 6.2. *For a Gårding hyperbolic operator P the retarded fundamental solution* (6.10) *has support in a cone $|x'| \leqslant Ax_n$ and such a fundamental solution is unique.*

Proof. It follows from (6.10), (6.9) and (3.9) that

$$\tilde{\mathscr{E}}(\xi', x_n) = \frac{1}{p_{(\nu)}} \tilde{E}_\nu(\xi', x_n) = \frac{1}{p_{(\nu)}} e^{Q(-i\xi')x_n} \begin{pmatrix} 0 \\ \vdots \\ 0 \\ 1 \end{pmatrix}, \tag{6.13}$$

$$\xi' \in \mathbb{R}^{n-1}, \quad x_n \geqslant 0; \quad \nu = m.$$

Therefore for all $x_n > 0$, $\tilde{\mathscr{E}}(\xi', x_n)$ is an entire function of $\xi' \in \mathbb{C}^{n-1}$. Also, by (3.10), the eigenvalues of the matrix $Q(-i\xi')$ are $\lambda_k(\xi')$. Finally, since the operator is Gårding hyperbolic, Lemma 1.1 yields

$$|\mathrm{Re}\, \lambda_k^0(\xi')| \sim |\mathrm{Re}\, \lambda_k(\xi')| \leqslant A(|\mathrm{Im}\, \xi'| + 1)$$

for $\xi' \in \mathbb{C}^{n-1}$. Using (6.13) and the estimate obtained in Gel'fand and Shilov [1958c] and Shilov [1965], we find, as for (3.11), that

$$|\tilde{\mathscr{E}}(\xi', x_n)| \leqslant C\|e^{Q(-i\xi')x_n}\| \leqslant Ce^{A|\mathrm{Im}\,\xi'|x_n}(1 + x_n\|Q(-i\xi')\|)^{\nu-1}, \qquad (6.14)$$

for $\xi' \in \mathbb{C}^{n-1}$ and $x_n > 0$. By Theorem 5.1 in Chap. 2, it follows from the last result that $\operatorname{supp} \mathscr{E}(\cdot, x_n) \subset \{x' \in \mathbb{R}^{n-1}: |x'| \leqslant Ax_n\} \Rightarrow \operatorname{supp} \mathscr{E} \subset Q \equiv \{x \in \mathbb{R}^n: |x'| \leqslant Ax_n\}$. It remains to show that the fundamental solution $\mathscr{E}$ with support in a cone Q is unique. But this result is obvious since, if $P\mathscr{E}_1 = \delta(x)$, $x \in \mathbb{R}^n$, and $\operatorname{supp} \mathscr{E}_1 \subset Q$, we have

$$\mathscr{E}_1 = (P\mathscr{E}) * \mathscr{E}_1 = \mathscr{E} * (P\mathscr{E}_1) = \mathscr{E}, \qquad (6.15)$$

on applying (1.47), Chap. 1, since the convolution $\mathscr{E} * \mathscr{E}_1$ is defined. ∎

Corollary 6.1. *For the wave equation* (1.6), *Chap.* 1, *in* $\mathbb{R}^k$, *the fundamental solution* $E_{(k)}$ *of the Cauchy problem* (2.21) *has the form*

$$E_{(1)}(x, t) = \begin{bmatrix} \dfrac{1}{2}\delta(at - |x|) \\ \dfrac{1}{2a}\theta(at - |x|) \end{bmatrix}, \quad E_{(2)}(x, t) = \begin{bmatrix} \dfrac{\partial}{\partial t}\mathscr{E}_2^+(x, t) \\ \mathscr{E}_2^+(x, t) \end{bmatrix},$$

$$E_{(3)}(x, t) = \begin{bmatrix} \dfrac{\partial}{\partial t}\mathscr{E}_3^+(x, t) \\ \mathscr{E}_3^+(x, t) \end{bmatrix},$$

by (6.11), *where the functions* $\mathscr{E}_2^+$ *and* $\mathscr{E}_3^+$ *are defined in* (1.6′), *Chap.* 1. *Accordingly, the solution* (6.4) *of this Cauchy problem, with* $k = 1$, *is expressed by the* ***d'Alembert formula***

$$u(x, t) = \frac{1}{2}(u_0(x - at) + u_0(x + at)) + \frac{1}{2a}\int_{x-at}^{x+at} u_1(y)\,dy$$

$$\forall u_0, u_1 \in C_0^\infty(\mathbb{R}).$$

Similarly, for $k = 2$ *and* 3 *respectively,* $u(x, t)$ *is given by the* ***Poisson and Kirchhoff formulae***

$$u(x, t) = \frac{\partial}{\partial t}\langle \mathscr{E}_k^+(x - y, t), u_0(y)\rangle + \langle \mathscr{E}_k^+(x - y, t), u_1(y)\rangle$$

$$\forall u_0, u_1 \in C_0^\infty(\mathbb{R}^k).$$

For Petrovskij parabolic equations the function (6.10) is identified in the following manner.

Proposition 6.3 (Ehjdel'man [1964]). *For a Petrovskij* 2*b-parabolic operator* $P(\partial_x)$ *the retarded fundamental solution* (6.10) *is a smooth function for* $x \neq 0$ *and for all* α *satisfies the estimate*

$$|\partial_x^\alpha \mathscr{E}(x)| \leqslant C_\alpha x_n^{\nu-\alpha_0-1-((n-1+|\alpha|)/2b)} e^{-(|x'|^{2b}/x_n)^{1/(2b-1)}}, \quad x_n > 0. \tag{6.16}$$

Such a fundamental solution is unique.

The estimates (6.16) have been established in Ehjdel'man [1964] and the uniqueness can be established as in (6.15).

Corollary 6.2. *For the heat equation* (1.31) *the fundamental solution of the Cauchy problem coincides with* (2.8′) *of Chap.* 1, *and the solution of the Cauchy problem has the form*

$$u(x, t) = \frac{1}{(2a\sqrt{\pi t})^k} \int e^{-|x-u|^2/4a^2t} u_0(y)\, dy \quad \forall u_0 \in C_0^\infty(\mathbb{R}^k), \quad t > 0.$$

A similar formula is obtained from (2.34) *for the solution of the Cauchy problem for the Schrödinger equation* (1.38). *Namely,*

$$\psi(x, t) = \frac{1}{(2a\sqrt{\pi i t})^k} \int e^{i|x-y|^2/4a^2t} \psi_0(y)\, dy.$$

That is, a fundamental solution of the Cauchy problem for the Schrödinger equation is given by the function (2.9′) *of Chap.* 1.

§ 7. Boundary-value Problems in a Half-space for Non-homogeneous Equations

7.1. Non-homogeneous Equations in a Half-space. Consider the non-homogeneous equation

$$P(\partial_x)v(x) = f(x), \quad x_n > 0, \tag{7.1}$$

where $f \in U_\alpha^{(r-\nu)_+}$ for some α and r. We construct a particular solution v of this equation in the class $U_\alpha^{(r)}$.

Proposition 7.1. *Assume that* $p_{(\nu)} \equiv \text{const} \neq 0$ *in* (1.6) *and that* $\bar{\alpha}_\mu < \underline{\alpha}_{\mu+1}$ *for some* $\mu \leqslant \nu$. *Then for* $\alpha \in]\bar{\alpha}_\mu, \underline{\alpha}_{\mu+1}[$ *the equation* (7.1) *has a solution* $v \in U_\alpha^{(r)}$ *for* $f(x) \in U_\alpha^{(r-\nu)_+}$.

Proof. The function $f \in U_\alpha^{(r-\nu)_+}$ can be extended to the domain $x_n < 0$ as a function $Lf(x) \in C^{(r-\nu)_+}(-\infty, \infty; H_s(\mathbb{R}^{n-1}))$ for some $s \in \mathbb{R}$ in such a manner that

$$\begin{gathered} Lf(\cdot, x_n) = f(\cdot, x_n) \quad \text{if } x_n > 0, \quad Lf(\cdot, x_n) = 0 \quad \text{if } x_n < -1, \\ \|\partial_{x_n}^j Lf(\cdot, x_n)\|_s \leqslant Ce^{\alpha x_n}, \quad x_n \in \mathbb{R}, \quad j \leqslant (r-\nu)_+. \end{gathered} \tag{7.2}$$

To do this we use the *Stein extension operator* (Vishik and Ehskin [1965]).

We construct a solution $Lv \in C^{(r)}(0, \infty; H_s(\mathbb{R}^{n-1}))$ of the equation

$$P(\partial_x)Lv(x) = Lf(x), \quad x \in \mathbb{R}^n, \tag{7.3}$$

such that, for some $s_j \ll s$,

$$\|\partial_{x_n}^j Lv(\cdot, x_n)\|_{s_j} \leqslant Ce^{\alpha x_n}, \quad x_n > 0, \quad j \leqslant r. \tag{7.4}$$

Then $v := Lv|_{x_n > 0} \in U_\alpha^{(r)}$ is the solution of (7.1).

On applying the contact Fourier transformation, (7.3) becomes

$$P(-i\xi', \partial_{x_n})\widetilde{Lv}(\xi', x_n) = \widetilde{Lf}(\xi', x_n) \tag{7.5}$$

for all $x_n \in \mathbb{R}$ and almost all $\xi' \in \mathbb{R}^{n-1}$. Now, for all $\xi' \in \mathbb{R}^{n-1}$ the operator $P(-i\xi', \partial_{x_n})$ has the symbol

$$P(-i\xi', -i\xi_n) \neq 0 \quad \text{for } \operatorname{Im} \xi_n \in]\bar{\alpha}_\mu, \underline{\alpha}_{\mu+1}[. \tag{7.6}$$

Therefore, for $\alpha \in]\bar{\alpha}_\mu, \underline{\alpha}_{\mu+1}[$, the function

$$\tilde{\mathscr{E}}(\xi', x_n) \equiv F^{-1}_{\xi_n \to x_n}\left(\frac{1}{P(-i\xi', -i\xi_n)}\bigg|_{\operatorname{Im} \xi_n = \alpha}\right) \tag{7.7}$$

is a fundamental solution of the operator $P(-i\xi', \partial_{x_n})$. It follows from (7.6) that

$$|\tilde{\mathscr{E}}(\xi', x_n)| \leqslant \begin{cases} C_\varepsilon e^{(\underline{\alpha}_{\mu+1} - \varepsilon)x_n}, & x_n < 0, \\ C_\varepsilon e^{(\bar{\alpha}_\mu + \varepsilon)x_n}, & x_n > 0, \end{cases} \tag{7.8}$$

for all $\varepsilon > 0$, where C_ε is independent of ξ'. These estimates follow from the decomposition

$$P(-i\xi) = p_{(\nu)}(-i\xi_n - \lambda_1(\xi'))\dots(-i\xi_n - \lambda_\nu(\xi')), \tag{7.9}$$

because $p_{(\nu)} \neq 0$ and $\operatorname{Re} \lambda_k(\xi') < \bar{\alpha}_\mu + \varepsilon$ for $k \leqslant \mu$, while $\operatorname{Re} \lambda_k(\xi') > \underline{\alpha}_{\mu+1} - \varepsilon$ for $k \geqslant \mu + 1$. To make the discussion simple, we assume that $\nu \geqslant 1$. Define a function $\widetilde{Lv}(\xi', x_n)$ by the formula

$$\begin{aligned} \widetilde{Lv}(\xi', x_n) &= \int_{-\infty}^{+\infty} \tilde{\mathscr{E}}(\xi', x_n - y_n)\widetilde{Lf}(\xi', y_n)\, dy_n \\ &= F^{-1}_{\xi_n \to x_n}\left(\frac{F_{x_n \to \xi_n}(\widetilde{Lf})}{P(-i\xi', -i\xi_n)}\bigg|_{\operatorname{Im} \xi_n = \alpha}\right). \end{aligned} \tag{7.10}$$

Here the integral of a function of the parameter y_n and having values in $H_s(\mathbb{R}^{n-1})$ is taken in the sense of Bochner. The convergence of this Bochner integral follows from (7.8) and (7.2):

$$\begin{aligned} \|\widetilde{Lv}(\cdot, x_n)\|_{\tilde{H}_s} &\leqslant \int_{-\infty}^{x_n} C_\varepsilon e^{(\bar{\alpha}_\mu + \varepsilon)(x_n - y_n)} e^{\alpha y_n}\, dy_n + \int_{x_n}^{+\infty} C_\varepsilon e^{(\underline{\alpha}_{\mu+1} - \varepsilon)(x_n - y_n)} e^{\alpha y_n}\, dy_n \\ &\leqslant C_\varepsilon^1 e^{\alpha x_n}, \end{aligned} \tag{7.11}$$

provided that $\alpha \in]\bar{\alpha}_\mu + \varepsilon, \underline{\alpha}_{\mu+1} - \varepsilon[$. Clearly, the function

$$Lv(x', x_n) := F^{-1}_{\xi' \to x'} \widetilde{Lv}(\xi', x_n)$$

satisfies (7.3); the proof of the estimates (7.4) in this case is similar to that of (7.11). ■

7.2. Boundary-value Problems for Non-homogeneous Equations. We examine the boundary-value problem

$$\begin{cases} P(\partial_x)u(x) = f(x), & x_n > 0, \\ \gamma B_j(\partial_x)u(x') = f_j(x'), & x' \in \mathbb{R}^{n-1}; \quad j = 1, \dots, l. \end{cases} \tag{7.12}$$

Proposition 7.2. *Let α and the operator P satisfy the hypotheses of Proposition* 7.1, *and let the boundary-value problem* (1.1), (2.2) *be regular in $U_\alpha^{(r)}$, where $r \geqslant \overline{m}$. Then, for any $f \in U_\alpha^{(r-\nu)_+}$ and $f_j \in H_{-\infty}(\mathbb{R}^{n-1})$, the boundary-value problem* (7.12) *has a unique solution $u \in U_\alpha^{(r)}$.*

Proof. Suppose that $u \in U_\alpha^{(r)}$ is a solution of the problem (7.12) and that $v \in U_\alpha^{(r)}$ is a solution of the equation (7.1) constructed in Proposition 7.1. Then $w \equiv u - v \in U_\alpha^{(r)}$ is a solution of the problem

$$\begin{cases} Pw = 0, & x_n > 0, \\ \gamma B_j w = g_j := f_j - \gamma B_j w, & j = 1, \dots, l, \end{cases} \tag{7.13}$$

and w exists and is unique because the problem (1.1), (2.2) is regular in $U_\alpha^{(r)}$. ■

Remark 7.1. The hypotheses of Proposition 7.2 are satisfied by

a) the Cauchy problem for the Gårding hyperbolic or for Petrovskij parabolic operator P provided that $\alpha > \overline{\alpha}_\nu$ and $r \geqslant \nu - 1$ (this statement follows from Corollary 3.1);

b) the Dirichlet problem for the elliptic operator $P(\partial_x)$ of order $m = 2l$, with $n \geqslant 3$, if $\overline{\alpha}_l < \alpha < \underline{\alpha}_{l+1}$ and $r \geqslant l - 1$ (this statement is a consequence of Proposition 3.3).

Chapter 6
Sharp and Diffusion Fronts of Hyperbolic Equations[1]

Hyperbolic equations constitute a large class of partial differential equations. The most well-known representative of this class is the wave equation

$$\frac{\partial^2 u}{\partial t^2} - k^2 \sum \frac{\partial^2 u}{\partial z_i^2} = 0,$$

which describes the propagation of waves with speed k. In analogy with this example the solutions of arbitrary hyperbolic equations are also referred to as waves. An elementary wave arising from instantaneous perturbation at a point has a singularity on a cone in space-time on the so-called wave front, is analytic outside it and vanishes outside its convex hull. For example, the front of the

[1] This chapter has been written by V.A. Vasil'ev.

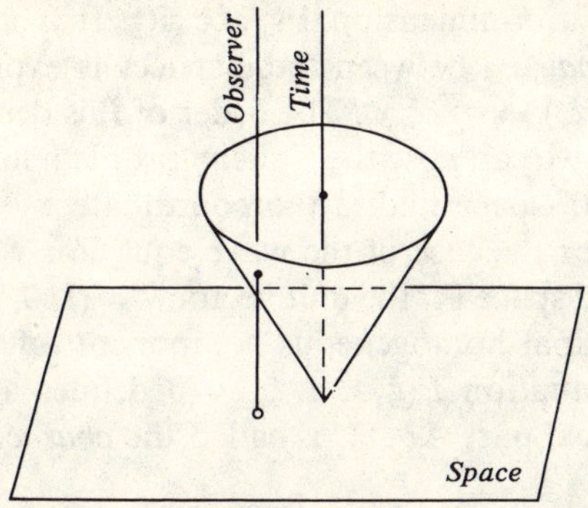

Fig. 1

wave equation is defined by the equation $k^2t^2 = \sum z_i^2$, $t \geqslant 0$. The subject matter of this chapter is the study of the qualitative behaviour of the wave as it approaches its front.

The wave equations themselves provide various examples of such qualitative behaviour. Thus in our four-dimensional space-time (and in any other $2l$-dimensional space, $l \geqslant 2$) the signal is observed only for an instant when it passes the observer. In contrast, in odd-dimensional space, the signal continues to make sound beyond the moment t_0 of reception (with intensity proportional to $1/\sqrt{t^2 - t_0^2}$). The first situation enables us to communicate by means of sound while the second explains the fact that the "acoustic layer" in an ocean, which is otherwise an excellent conductor of separate signals, is unsuitable for transmitting even slightly complicated information. Both examples of the behaviour of sound waves have analogues for arbitrary hyperbolic equations. In the language of the general theory we say that, in the first case, the internal component of the complement of the front is a lacuna while, in the second case, diffusion of the wave occurs on the side of such a component; the external component is a lacuna for all dimensions (and for all hyperbolic equations).

§ 1. Basic Notions

1.1. Hyperbolic Operators. Considerr the linear space $\mathbb{R}_x^n$ with coordinates x_i $(i = 1, \ldots, n)$ and a *differential operator P with constant coefficients* on $\mathbb{R}_x^n$, that is, a finite sum of the form

$$\sum P_\alpha(v - i\partial/\partial x_1)^{\alpha_1} \cdot \ldots \cdot (v - i\partial/\partial x_n)^{\alpha_n},$$

where the P_α are constants enumerated with respect to the multi-index $\alpha = (\alpha_1, \ldots, \alpha_n)$. To such an operator there corresponds the *characteristic polynomial*

$$P = \sum P_\alpha \xi^\alpha, \quad \xi^\alpha = \xi_1^{\alpha_1} \cdot \ldots \cdot \xi_n^{\alpha_n},$$

in the coordinates ξ_i of n-dimensional space $\mathbb{R}^n_\xi$. It is useful to regard $\mathbb{R}^n_\xi$ as dual to $\mathbb{R}^n_x$, and the *pairing* between these spaces is expressed by the formula $\langle(\xi_1, \ldots, \xi_n), (x_1, \ldots, x_n)\rangle = \sum \xi_i x_i$. The order of P is denoted by deg P.

For the operator P we can pose the Cauchy problem in the half-space $x_1 \geqslant 0$. The choice of this half-space makes the coordinate x_1 more important than the remaining ones (for the case of the wave equation x_1 denotes time); correspondingly, in the dual space $\mathbb{R}^n_\xi$ the unit vector $\vartheta = (1, 0, \ldots, 0)$ is distinguished.

Let $\bar{P}$ be the principal homogeneous component (of highest degree) of the polynomial P. The equation $\bar{P}(\xi_1, \ldots, \xi_n) = 0$ defines a conical hypersurface $A = A(P)$ in $\mathbb{C}^n_\xi$; the real part $A \cap \mathbb{R}^n_\xi$ is called the *characteristic cone* for P and is denoted by Re A.

Definition 1.1. The operator P is said to be *Petrovskij hyperbolic* (or *strictly hyperbolic*) if, first, the set Re A is non-singular outside the point 0 and, second, any line in $\mathbb{R}^n_\xi$, parallel to the unit vector ϑ and not passing through 0, intersects Re A in exactly deg P distinct points.

Example 1.1. 1) The wave operator.

2) Any operator in $\mathbb{R}^2$ such that $P(\vartheta) \neq 0$.

3) A non-degenerate cubic curve in $\mathbb{RP}^2$ can have two or one component (and is represented either by both the curves in Fig. 2a or by only the right-hand one). The corresponding third-degree polynomial in $\mathbb{R}^3_\xi$ is never hyperbolic in the second case while, in the first case, it is hyperbolic precisely when ϑ is directed towards the interior of the cone-like component of the surface Re A. Numerous other examples of hyperbolic polynomials can be found in Atiyah, Bott and Gårding [1970, 1973].

In the space of operators of a given order acting in $\mathbb{R}^n_x$ the Petrovskij operators constitute an open domain.

Theorem 1.1 (Nuij [1968]). *For any natural integers k and n, the set of polynomials of degree k in* $\mathbb{R}^n_\xi$ *which are Petrovskij hyperbolic with respect to a fixed coordinate system consists of two connected components each of which is contractible.*

1.2. Wave Fronts. The fundamental geometric object involved in the qualitative description of the solutions of a hyperbolic equation is the wave front

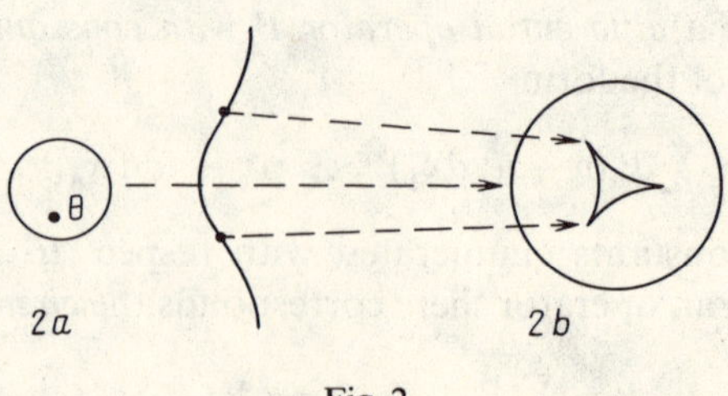

Fig. 2

which we now define. Let $\xi \neq 0$ be an arbitrary point of the cone Re A. The tangent plane to Re A at this point will be regarded as a subspace of $\mathbb{R}^n_\xi$. We consider in the original space $\mathbb{R}^n_x$ the set of all vectors x, with positive first coordinate x_1, which are orthogonal to this plane. The set of such x for a given ξ constitutes a ray in $\mathbb{R}^n_x$.

Definition 1.2. The union of such rays taken over all the points $\xi \in \text{Re } A\setminus\{0\}$ is called the *wave front* of the operator P and is denoted by $W(P)$.

The singularities of the wave front can even lie far away from the origin: they correspond to the flattening points of Re A, that is, to points where the curvature form of Re A has rank less than $n - 2$. For example, the projectivization of the wave front corresponding to the cubic surface in Fig. 2a is depicted in Fig. 2b. Furthermore, the points of inflexion of the original curve go over to cusps of the projectivization of the front (in Fig. 2a the third point of inflection lies at infinity).

1.3. Theorem 1.2 (see Atiyah, Bott and Gårding [1970, 1973], Petrovskij [1945]). *If P is a Petrovskij hyperbolic operator, then*

A) *P has a fundamental solution $E(P)$ whose support is a cone, with vertex* 0, *lying in the half-space $x_1 \geqslant 0$ and intersecting the plane $x_1 = 0$ only at the point* 0;

B) *such a fundamental solution $E(P)$ is unique*;

C) *the solution $E(P)$ is analytic outside the surface $W(P)$ and vanishes outside the convex hull of $W(P)$.*

Remark 1.1. Statement A of Theorem 1.2 can be formalized as the definition of a (non-strictly) hyperbolic operator: P is said to be *hyperbolic* if it has a fundamental solution whose support is contained in the cone described in Statement A. For general hyperbolic operator the wave front can also be defined (see Atiyah, Bott and Gårding [1970, 1973]) and all the assertions of Theorem 1.2 remain valid. Such operators also have an explicit algebraic characterization (see Atiyah, Bott and Gårding [1970, 1973]), but now hyperbolicity is defined not only by the principal part of the operator – the lower terms must also be known. In the space of all operators the hyperbolic, but not strictly hyperbolic, operators constitute a set of positive co-dimension; this set lies in the closure of the set of strictly hyperbolic operators.

1.4. Sharpness, Diffusion and Lacunae. Let P be a hyperbolic operator and W its wave front.

Definition 1.3. A. There is *holomorphic sharpness* at a point y of W on the side of a local (near y) component[2] l of the complement of W if $E(P)$ is continued from l as a holomorphic function in a neighbourhood of a point of l. Similarly, there is a C^∞*-sharpness* if $E(P)$ has a C^∞-extension from l to its closure $\bar{l}$. In these

[2] The term "component" everywhere signifies a "component of linear connectedness".

cases the component l is called a (*holomorphic or C^∞-*) *local lacuna* of the operator P near y.

B. If there is a no sharpness on the side l, we say that *diffusion of waves* takes place in l.

C. A component L of the complement to the front is called a *holomorphic lacuna* (*C^∞-lacuna*) if from its side there is holomorphic (C^∞-) sharpness at any point of its closure (or, what is the same, at the single point 0).

D. If $E(P) \equiv 0$ in L, then L is called a *strong lacuna* (Atiyah, Bott and Gårding [1970, 1973]) or simply a *lacuna* (Petrovskij [1945]).

Example 1.2. We again consider the wave operator

$$P_n = \frac{\partial^2}{\partial t^2} - k^2 \sum_{i=2}^{n} \frac{\partial^2}{\partial z_i^2}.$$

It is known that the corresponding fundamental solutions $E(P_n)$ for $n = 2, 3, 4$ are given by $E(P_2) = \theta(kt - |z|)/2k$, $E(P_3) = \theta(kt - |z|)/2\pi k\sqrt{k^2 t^2 - |z|^2}$ and $E(P_4) = \theta(t)\delta(k^2 t^2 - |z|^2)/2\pi h$, where θ is the Heaviside function. Evidently, for the case $n = 4$ the interior component of the complement to the front is a strong lacuna while, for the case $n = 2$, it is a holomorphic but not a strong lacuna, and diffusion takes place when $n = 3$. It turns out that, as n grows, the qualitative picture remains the same as for $n = 4$ when n is even and the same as for $n = 3$ when n is odd. The exceptional case $n = 2$ is explained by the fact that n does not exceed the order of the wave operator.

For the third-order equation in $\mathbb{R}_x^3$, the projectivization of the wave front of which is depicted in Fig. 2b, the extreme interior component is a holomorphic but not a strong lacuna while the intermediate component (where "edges" of the interior hang) is diffusive near any point of its boundary.

§ 2. The Petrovskij Criterion

2.1. In his work [1945], Petrovskij discovered a connection between the property of the component L being a lacuna and a certain topological property which is now referred to as the *Petrovskij criterion.* This property consists of the triviality of a class of homologies, the *Petrovskij class*, which we now describe. In this section P denotes a strictly hyperbolic operator in $\mathbb{R}_x^n$.

Let $x \in \mathbb{R}_x^n$ be any point of the component L, under investigation, of the complement of the front W, and let $X \subset \mathbb{C}_\xi^n$ be a hyperplane orthogonal to x. Let $\mathbb{CP}_\xi^{n-1}$ be the set of all one-dimensional complex subspaces of $\mathbb{C}_\xi^n$. We denote by X^* and A^* the hypersurfaces in $\mathbb{CP}_\xi^{n-1}$ which are obtained from the projectivization of the plane X and the cone A. A *Petrovskij cycle*, which we now construct, is an $(n-2)$-dimensional cycle in the set $X^* - A^*$. Let us describe the cycle.

Since x does not lie in the front, it follows that near the set $\mathbb{RP}_\xi^{n-1}$ the hypersurfaces X^* and A^* intersect transversally and their intersection is smooth. The set $X^* \cap \operatorname{Re} A^*$ of real points of this intersection is an $(n-3)$-dimensional submanifold in $X^* \cap A^*$. If n is even, this manifold is orientable. A special choice of this orientation that turns the manifold into a cycle has been specified by Atiyah, Bott and Gårding [1970, 1973]. A Petrovskij cycle $B(x) \subset X^* - A^*$ is defined as the image of this cycle under a tubular map. (Let us recall the construction of this map. Consider a tubular neighbourhood in X^* of the set of non-singular points of the manifold $X^* \cap A^*$ and fibre this neighbourhood in some way into two-dimensional discs which are transversal to the set $X^* \cap A^*$. Then, to any cycle V lying in the smooth part of the set $X^* \cap A^*$, we can assign the union of boundaries of those discs which have intersection with $X^* \cap A^*$ at points of V. The complex orientation of the sets X^* and $X^* \cap A^*$ and the original orientation of V enable us to define an orientation for the tube thus obtained.)

For n odd the cycle $B(x)$ coincides far from A^* with the manifold $\operatorname{Re} X^*$ having multiplicity 2, and near A^* is double to a pair of contours enclosing A^* in X^* from two sides. The non-real part of this cycle coincides geometrically with the above-mentioned tube in $X^* - A^*$ around the manifold $X^* \cap \operatorname{Re} A^*$ but the two halves of this tube separated by the set $\operatorname{Re} X^*$ have divergent orientations. For the case $n = 3$, the cycle $B(x)$ is depicted in Fig. 3, in which the set $A^* \cap X^*$ is indicated by crosses.

Definition 2.1. The class $\beta(x)$ defined by the cycle $B(x)$ in the group $H_{n-2}(X^* - A^*)$ of homologies of the space $X^* - A^*$ with complex coefficients is known as the *Petrovskij class*, and the condition $\beta(x) = 0$ as the *Petrovskij criterion.*

It can be seen easily that the Petrovskij criterion is always satisfied for $n = 2$, while for $n = 3$ it is equivalent to the fact that the set $X^* \cap A^*$ has no non-real points; see Fig. 3.

2.2. Theorem 2.1 (see Atiyah, Bott and Gårding [1970, 1973], Gabrichlov [1986], Petrovskij [1945]). *If $\beta(x) = 0$, then the component L, which contains x, of the complement of the front is a holomorphic lacuna for P and for all operators*

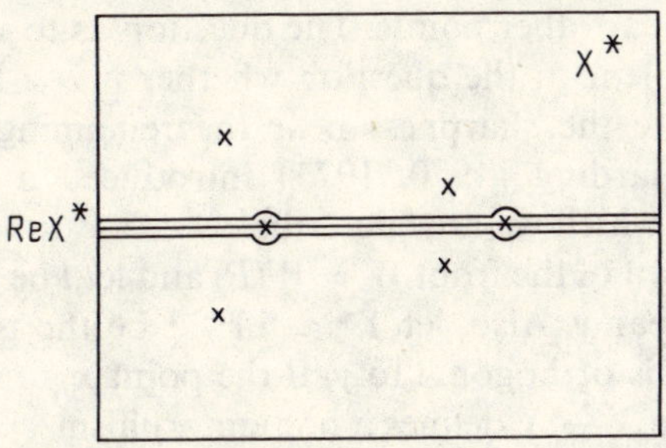

Fig. 3

having the same principal part $\bar{P}$ (if, in addition, deg $P < n$ *and* $P = \bar{P}$, *then it is also a strong lacuna). If the set* A^* *is smooth, then the converse is also true. Namely, if* L *is a* C^∞*-lacuna, then* $\beta(x) = 0$ *for all* $x \in L$.

The first assertion of this theorem in the case of homogeneous $P = \bar{P}$ follows from the Herglotz-Petrovskij-Leray formula: at the point $x \in L$ all the partial derivatives $D^\nu E$ of the fundamental solution $E = E(P)$, where $|\nu| > \deg P - n$, are defined by the integrals of certain differential $(n-1)$-forms $\Lambda(x, \nu, P)$. These forms are regular in the domain $\mathbb{CP}_\xi^{n-1} - X^* - A^*$, taken over a cycle $t\beta(x)$ which lies in this domain and is obtained from $\beta(x)$ by a tubular operation $t: H_{n-2}(X^* - A^*) \to H_{n-1}(\mathbb{CP}_\xi^{n-1} - X^* - A^*)$. Therefore, when $\beta(x) = 0$, $E(P)$ is, in the restriction to L, a polynomial whose degree does not exceed $\deg P - n$.

The second assertion follows from the facts that a) $E|_L$ is a polynomial if L is a lacuna, b) in our case the tubular map is injective (in particular, $\beta(x) = 0$ if $t\beta(x) = 0$), and c) for sufficiently large N the classes of all possible differential forms $\Lambda(x, \nu, P)$, with $|\nu| = N$, generate the complete cohomology group $H_{n-1}(\mathbb{CP}_\xi^{n-1} - X^* - A^*)$ (see Atiyah, Bott and Gårding [1970, 1973]). The case of non-homogeneous P reduces to that of homogeneous P (see Atiyah, Bott and Gårding [1970, 1973; §4.5]).

Similar methods can be used to establish the following precise version of Statement C of Theorem 1.2.

Theorem 2.2 (see Atiyah, Bott and Gårding [1970, 1973]). *If the operator* P *is hyperbolic and the set* A^* *is smooth, then every point of the wave front* $W(P)$ *is a singular point of the fundamental solution* $E(P)$. *For almost every hyperbolic operator* P *with the given principal part* $\bar{P}$, *the support of* $E(P)$ *coincides with the convex hull of the wave front. If* $n = 3$, *the last statement holds for all hyperbolic operators without exception. (For example, for even* $n \geqslant 4$ *the wave operator is not covered by the term "almost every".)*

§ 3. The Local Petrovskij Criterion

3.1. It is possible for a component of the complement of the wave front to be both a local lacuna near some points of its boundary and, at the same time, a carrier of diffusion near other points. The question as to whether a component is a lacuna is equivalent to the question whether it is a local lacuna near the origin. To investigate the sharpness near the remaining points of the front, Atiyah, Bott and Gårding [1970, 1973] introduced a local version of the Petrovskij criterion, which we now describe.

Let $y \neq 0$ be a point of the front $W = W(P)$ and let l be the component of the complement of W near y. Also, let $Y^* \subset \mathbb{CP}_\xi^{n-1}$ be the projectivization of the plane $Y \subset \mathbb{C}_\xi^n$ which is orthogonal to y. If the point $x \in l$ is sufficiently close to y, then the projection $Y \to X$ defines a homomorphism

$$p_x: H_{n-2}(Y^* - A^*) \to H_{n-2}(X^* - A^*).$$

Definition 3.1. By the *local Petrovskij criterion* is meant the condition that

$$\beta(x) \in p_x(H_{n-2}(Y^* - A^*)).$$

3.2. Theorem 3.1 (see Atiyah, Bott and Gårding [1970, 1973]). *If for points of the local (near y) component l of the complement of the front the local Petrovskij criterion is fulfilled, then the fundamental solution E(P) is holomorphically sharp at the point y from the side l.*

It turns out that Theorem 3.1 is almost always reversible.

3.3. Theorem 3.2 (see Vasil'ev [1986]). 1. *If the surface A^* touches the plane Y^* near the set $\mathbb{RP}_\xi^{n-1}$ only at a finite number of points, then the holomorphic sharpness of E(P) at a point y from the side of the component l implies the validity of the local Petrovskij criterion for all $x \in l$.*

2. *For almost every hyperbolic operator the hypothesis of Part 1 of this theorem is satisfied at every point $y \neq 0$ of its front.*

3.4. A Counter example. In the most general case the converse of Theorem 3.1 is not valid. Indeed, let $P = \xi_1(\xi_1^2 - \xi_2^2 - \xi_3^2)$. Then the front $W(P)$ consists of a circular cone and a ray spanned over the vector (1, 0, 0). The sharpness near the ray follows from the Hartogs theorem on a removable singularity but, as can be calculated easily, the local Petrovskij criterion does not hold.

§4. Geometry of Lacunae Near Concrete Singularities of Fronts

4.1. The local Petrovskij condition and, hence, the sharpness near points of the front are determined by the local geometry of the front. This geometry can be most conveniently described in terms of the (projective) generating functions which we now define. To make the discussion simple, we consider only the case where the point $y \in W$ in question corresponds to a unique point a of the set Re A^*, that is, the plane Y orthogonal to the point touches the surface Re A along a unique line. In $\mathbb{RP}_\xi^{n-1}$ we choose an affine coordinate system $z_0, z_1, \ldots, z_{n-2}$ with centre a such that the plane Y^* is described by the equation $z_0 = 0$. Near a, the surface A^* is then described by the equation $z_0 = f(z_1, \ldots, z_{n-2})$. The function f defined in this manner is called a *projective generating function of the surface* A^* at the point a. Clearly, $f(0) = \text{grad}\, f(0) = 0$.

4.2. Investigation of Sharpness Near Smooth Points of the Front. At a general point of the front, the singularity of f is *Morse*, that is, its second differential is a non-degenerate quadratic form. Its inertia index will be denoted by $i_\pm(a)$. In this case, the front is a smooth manifold near the point and divides the neighbourhood of the point into two parts. One of the parts contains a point x such that the corresponding plane X^* is defined by the condition $z_0 = \varepsilon$, $\varepsilon > 0$. We denote this part by l^+ and the other by l^-.

Theorem 4.1 (see Borovikov [1959], Davydova [1945]). *Let a be a general point of A^*. If n and $i_\pm \equiv i_\pm(a)$ are even, then both the components $l^\pm$ are local lacunae. If n is even and $i_\pm$ is not, then there is no lacuna near y. Finally, if n is odd, then a local lacuna is only l^- when i_+ is even and only l^+ when i_+ is odd.*

4.3. Investigation of Sharpness Near an Edge of Return and Near a Swallow-tail. At the points of flattening of the surface Re A^* (that is, at the points where its curvature form degenerates) the generating function is no longer Morse but has a degenerate singularity. The classification of such singularities has been dealt with in numerous works (see, for example, Arnol'd, Varchenko and Gusejn-Zade [1982, 1984]). Near the simplest singularities in this classification, namely, the singularities of the types A_2 and A_3, the wave front is diffeomorphic (see Fig. 2a) to the product of a linear space with the semi-cubical parabola (respectively, with the *swallow-tail*, that is, with the surface depicted in Fig. 4; in this figure edges of return correspond to points of type A_2). The sharpness condition on the side of one or the other component of the complement to the front again depends on whether n and the index i_+ of the quadratic part of the generating function are even. Moreover, by multiplying z_0 by -1 if necessary, we can consider only the modification A_3^+ of the singularity A_3 (see Table 1 below).

Theorem 4.2 (see Gårding [1977]). *Near points of the type A_2, the front is sharp on the side of component 2 if n is odd and i_+ is even, and for other values of n and i_+ there is no sharpness. Near the singularities A_3, local lacunae are the following: domain 3 when i_+ is odd and n is arbitrary; domain 2 when i_+ is even and n is odd.*

4.4. In the case of more complicated singularities a similar description of all local lacunae becomes much more unwieldy, and therefore we merely cite the results concerning the number of these lacunae.

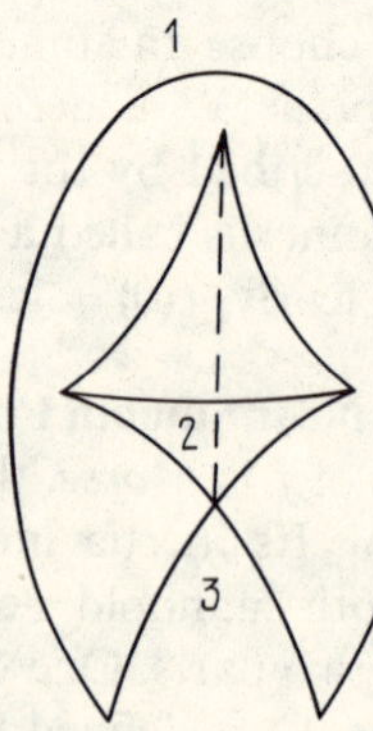

Fig. 4

Table 1

Type of singularity of the generating function	Normal form to which the generating function is reduced by diffeomorphism of arguments	Corank	$n \geqslant$	Number of lacunae			
				n even		n odd	
				i_+ even	i_+ odd	i_+ odd	i_+ even
A_1	Q	0	2	2	0	1	1
$A_{2k}, k \geqslant 1$	$Z_1^{2k+1} + Q$	1	$2k+1$	0	0	1	0
$A_{2k-1}^{\pm}, k \geqslant 2$	$\pm(Z_1^{2k} + Q)$	1	$2k$	0	1	1	1
D_4^-	$Z_1^2 Z_2 - Z_2^3 + Q$	2	5	0	3	1	1
$D_{2k}^+, k \geqslant 2$	$Z_1^2 Z_2 + Z_2^{2k-1} + Q$	2	$2k+1$	0	0	1	1
$D_{2k}^-, k \geqslant 3$	$Z_1^2 Z_2 - Z_2^{2k-1} + Q$	2	$2k+1$	0	2	1	1
$D_{2k+1}^{\pm}, k \geqslant 2$	$\pm(Z_1^2 Z_2 + Z_2^{2k} + Q)$	2	$2k+2$	0	0	1	1
$E_6^{\pm}$	$\pm(Z_1^3 + Z_2^4 + Q)$	2	7	0	0	1	1
E_7	$Z_1^3 + Z_1 Z_2^3 + Q$	2	8	0	0	1	1
E_8	$Z_1^3 + Z_2^5 + Q$	2	9	0	0	1	1
$X_9^{\pm}$	$\pm(Z_1^4 + \alpha Z_1^2 Z_2^2 + Z_2^4 + Q), \alpha > -2$	2	9	$\geqslant 1$	0	$\geqslant 2$	$\geqslant 0$
X_9^1	$Z_1 Z_2 (Z_1^2 + \alpha Z_1 Z_2 + Z_2^2) + Q, \lvert\alpha\rvert < 2$	2	9	0	0	0	0
X_9^2	$Z_1 Z_2 (Z_1 + Z_2)(Z_1 + \alpha Z_2) + Q, \alpha \in (0, 1)$	2	9	0	$\geqslant 2$	0	0
$X_{10}^{3\pm}$	$\pm(-Z_1^4 + Z_1^2 Z_2^2 + \alpha Z_2^5 + Q), \alpha > 0$	2	10	0	?	?	0
$X_{10}^{1\pm}$	$\pm(Z_1^4 + Z_1^2 Z_2^2 + \alpha Z_2^5 + Q), \alpha > 0$	2	10	?	0	?	0
J_{10}^3	$Z_1(Z_1 - Z_2^2)(Z_1 - \alpha Z_2^2) + Q, \alpha \in (0, 1)$	2	10	?	$\geqslant 1$	0	0
J_{10}^1	$Z_1(Z_1^2 + \alpha Z_1 Z_2^2 + Z_2^4) + Q, \lvert\alpha\rvert < 2$	2	10	?	?	0	0

Theorem 4.3. *Near any point at which the wave front of a general hyperbolic operator has a singularity of one of the types*[3] $A_k, D_k, E_k, X_9, X_{10}, J_{10}$ *the number of local lacunae is as given in Table 1 (or satisfies the inequality contained there). On the general fronts in* $\mathbb{R}_x^n$, *with* $n \leqslant 7$, *there may only be singularities belonging to one of the types mentioned in Table 1. (In this table* Q *denotes a non-degenerate quadratic form in the additional variables* $z_{r+1}, \ldots, z_n$, *where* r *is the corank of the singularity, that is, the defect of the curvature form of the surface* A^*: *it has been mentioned in the third column of the table;* i_+ *denotes the positive inertia index of* Q. *Then column "*$n \geqslant$*" indicates the minimum of those* n *such that the corresponding singularity appears on fronts in general in* $\mathbb{R}_x^n$.*).*

Conjecture. All the question marks in the table should be replaced by zeros.

The proof of this theorem concerning A_1 is given in Davydova [1945], and for A_2 and A_3 in Gårding [1977], while the proof for $A_k (k \geqslant 4)$, D_k and E_k is given in Vasil'ev [1986]. The assertions regarding other singularities have been

[3] The notation used here has been taken from the classification tables of Arnol'd, Varchenko and Gusejn-Zade [1982, 1984].

obtained with the aid of a computer. This computer experiment provides us with a very strong argument in favour of the above conjecture. The principal methodological remark in the proof of this theorem (and of Theorem 3.2) is that the problem of local lacunae is the problem of the local theory of singularities of smooth functions; the reader is referred to Arnol'd, Varchenko and Gusejn-Zade [1982, 1984] for this theory.

§ 5. Equations with Variable Coefficients

All the results of § 3 and § 4 have natural extension to the case of hyperbolic operators

$$P = \sum P_\alpha(x)(\partial/\partial x)^\alpha, \quad P_\alpha \in C^\infty(\mathbb{R}^n_x),$$

with variable coefficients (see Gårding [1977]). In this case also we can define, in particular, the notion of sharpness and local homological Petrovskij condition. We do not formulate these notions here and present only the fundamental result of this theory.

Theorem 5.1 (see Gårding [1977], Varchenko [1986]). *Near the points of the wave front of a strictly hyperbolic equation which correspond to the isolated singularities of generating functions, the local Petrovskij criterion implies the sharpness of the fundamental solution.*

This theorem was proved by Gårding [1977] for the case of analytic coefficients and near singularities of the types A_k as well as in the C^∞-case. This work also contains a hint to the proof of the theorem in the general formulation presented above that we realized by Varchenko [1986].

Bibliographical Comments

The theory of generalized functions has been dealt with at length in the books by Gel'fand and Shilov [1958a, 1958b], Schwartz [1961, 1966], Shilov [1965] and Vladimirov [1979]. These books also contain a number of useful exercises and illustrative examples.

The books by Ditkin and Prudnikov [1966], Doetsch [1967], and Titchmarsh [1937] discuss a wide range of problems connected with the theory and applications of the Fourier transformation and, in particular, questions connected with operational calculus.

The most effective methods of investigating general equations with constant coefficients suggested in Bernshtejn [1971], Bernshtejn and Gel'fand [1969], Björk [1979], Hörmander [1958], Lojasiewicz [1959], Palamodov [1960, 1967] are based on the ideas of algebraic geometry.

The works of Borovikov [1959], Gårding [1950], Gårding, Kotaké and Leray [1964], Hadamard [1932], Hörmander [1983, 1985], Leray [1953], and Riesz [1949] contain detailed information regarding the structure, singularities and asymptotics of fundamental solutions of hyperbolic equations while Atiyah, Bott and Gårding [1970, 1973], Davydova [1945], Gabriehlov [1986], Gårding [1977], Petrovskij [1945], Vasil'ev [1968] contain information regarding lacunae of these solutions.

The works of Petrovskij [1937a, 1937b, 1938] contain fundamental results on the well-posedness question of the Cauchy problem for hyperbolic equations while the work of Gel'fand and Shilov [1958c] contains fundamental results on uniqueness classes and well-posedness classes of the Cauchy problem for equations with constant coefficients. An exhaustive account on well-posedness of the boundary-value problems for parabolic systems (and equations) can be found in Ehjdel'man [1964].

Agranovich and Vishik [1964] present the theory of boundary-value problems in the Sobolev function spaces for equations with variable coefficients which is almost as effective as the one for equations with constant coefficients; Vishik and Ehskin [1965] treat this theory for pseudo-differential equations.

An account of the theory of Sobolev spaces and embedding theorems can be found in Egorov [1984], Hörmander [1963, 1983], Mikhlin [1977], Shubin [1978], Sobolev [1950], Volevich and Paneyakh [1965].

The works of Grushin [1963a] and Vajnberg [1966] discuss important questions of the stationary theory of scattering, connected with the Sommerfeld radiation conditions, for general equations while Tikonov and Samarskij [1951] treat these questions for the simplest concrete example.

Merzon [1973] obtained conditions for the solvability of equations in a cone (as well as in an angle).

The reader can become acquainted with the application of the Seidenberg-Tarski theorem from the works of Gorin [1961], Hörmander [1963, 1983, 1985], Shilov [1965], and Trèves [1961].

Those interested in the application of equations with constant coefficients to mathematical physics are referred to Bogolyubov and Shirkov [1973], Courant [1962], and Reed and Simon [1975, 1979].

The books by Bateman and Erdélyi [1953] and Whittaker and Watson [1927] contain vast information on special functions and their applications to partial differential equations.

A sufficiently complete account of the theory of analytic functions in several complex variables can be found in Gunning and Rossi [1965], Hörmander [1973], and Vladimirov [1964].

Agranovich [1961] has surveyed the theory of equations with constant coefficients while the book by Trèves [1961] can serve as a course book on these equations.

The most complete account of the modern theory of equations with constant coefficients can be found in the monographs of Hörmander [1983, 1985] and Palamodov [1967]. These monographs discuss such vital questions as the approximation to solutions, P-convexity and solvability of equations in domains, scattering theory, the WKB method (the transition from wave optics to geometric optics), systems of equations etc. which could not be discussed in the present survey paper due to lack of space.

For a detailed account of the modern theory of partial differential equations with variable coefficients the reader is referred to Egorov [1984], Hörmander [1983, 1985], and Shubin [1978].

The books by Courant [1962], Hörmander [1963], Mikhlin [1977], Petrovskij [1961], Tikhonov and Samarskij [1951], and Vladimirov [1981] can be used as general guide books on partial differential equations.

References*

Agranovich, M.S.
[1961] On partial differential equations with constant coefficients. Usp. Mat. Nauk *16*, No. 2, 27–93. English transl.: Russ. Math. Surv. *16*, No. 2, 23–90, Zbl.101,74

* For the convenience of the reader, references to reviews in Zentralblatt für Mathematik (Zbl.), compiled using the MATH database, and Jahrbuch über die Fortschritte der Mathematik (FdM.) have, as far as possible, been included in this bibliography.

Agranovich, M.S., Vishik, M.I.
[1964] Elliptic problems with a parameter and parabolic problems of general type. Usp. Mat. Nauk *19*, No. 3, 53–161. English transl.: Russ. Math. Surv. *19*, No. 3, 53–157, Zbl.127,296
Arnol'd, V.I., Varchenko, A.N., Gusejn-Zade, S.M.
[1982] Singularities of Differentiable Maps. Vol. 1. Moscow: Nauka, Zbl.513.58001. English transl.: Monographs in Mathematics *82*, Boston, MA: Birkhäuser 1985
[1984] Singularities of Differentiable Maps. Vol. 2. Moscow: Nauka, Zbl.545.58001. English transl.: Monographs in Mathematics *83*, Boston, MA: Birkhäuser 1988
Atiyah, M.F., Bott, R., Gårding, L.
[1970] Lacunas for hyperbolic differential operators with constant coefficients. I. Acta Math. *124*, 109–189, Zbl.191,112
[1973] Lacunas for hyperbolic differential operators with constant coefficients. II. ibid. *131*, 145–206, Zbl.266.35045
Bateman, H., Erdélyi, A.
[1953] Higher Transcendental Functions. Vol. 2. New York-Toronto-London: McGraw-Hill. Zbl.52,295
Björk, J.-E.
[1979] Rings of Differential Operators. Amsterdam: North-Holland Publ. Co., Math. Library Series *21*, Zbl.499.13009
Bernshtejn, I.N.
[1971] Modules over rings of differential operators. Investigation of fundamental solutions of equations with constant coefficients. Funkts. Anal. Prilozh. *5*, No. 2, 1–16. English transl.: Funct. Anal. Appl. *5*, 89–101, Zbl.233.47031
Bernshtejn, I.N., Gel'fand, S.I.
[1969] The meromorphicity of the function P^{λ}. Funkts. Anal. Prilozh. *3*, No. 1, 84–85. English transl.: Funct. Anal. Appl. *3*, 68–69, Zbl.208,152
Bogolyubov, N.N., Shirkov, D.V.
[1973] Introduction to Quantum Field Theory; 2nd rev. ed. Moscow: Nauka. English transl.: New York: Wiley, 1979
Borovikov, V.A.
[1959] Fundamental solutions of linear partial differential equations with constant coefficients. Tr. Mosk. Mat. O-va *8*, 199–257, Zbl.90,312. English Transl.: Transl., II. Ser., Am. Math. Soc. *25* (1963), 11–76
Courant, R.
[1962] Partial Differential Equations. New York, London: Interscience Publishers, Zbl.99,295
Davydova, A.M.
[1945] Sufficient Conditions for Lack of Lacuna. Moscow: Izdat. Mosk. Univ.
Dirac, P.
[1958] The Principles of Quantum Mechanics. 4th ed. Oxford: Clarendon Press, Zbl.80,220
Ditkin, V.A., Prudnikov, A.P.
[1966] Operational Calculus. Moscow: Vysshaya Shkola, Zbl.148,367
Doetsch, G.
[1967] Anleitung zum praktischen Gebrauch der Laplace-Transformation und z-Transformation. 3rd rev. ed., Müchen, Wien: R. Oldenburg, Zbl.166,100
Egorov, Yu.V.
[1984] Linear Differential Equations of Principal Type. Moscow: Nauka. English transl.: New York: Contemp. Sov. Math. 1986, Zbl.574.35001
Ehjdel'man, S.D. (= Eidel'man, S.D.)
[1964] Parabolic Systems. Moscow: Nauka. English transl.: Amsterdam: North-Holland 1969, Zbl.121,319
Ehrenpreis, L.
[1954] Solutions of some problems of division. I. Amer. J. Math. *76*, 883–903, Zbl.56,106
Gabriehlov, A.M.
[1986] Proof of Petrovskij's theorem. Supplement to the book: Petrovskij, I.G., Izbrannye Trudy

(Selected Works), Vol. 1. Moscow: Nauka, Zbl.603.01018

Gårding, L.

[1951] Linear hyperbolic partial differential equations with constant coefficients. Acta Math. *85*, No. 1, 1–62, Zbl.45,202

[1977] Sharp fronts of paired oscillatory integrals. Publ. Res. Inst. Math. Sci. Kyoto Univ. *12*, Suppl., 53–68, Zbl.369.35062

Gårding, L., Kotaké, T., Leray, J.

[1964] Uniformisation et développement asymptotique de la solution du problème de Cauchy linéaire, à données holomorphes; analogie avec la théorie des ondes asymptotiques et approachées. Bull. Soc. Math. Fr. *92*, 263–361, Zbl.147,81

Gel'fand, I.M., Shilov, G.E.

[1953] Fourier transforms of rapidly growing functions and the question of uniqueness of solution of the Cauchy problem. Usp. Mat. Nauk *8*, No. 4, 3–51, Zbl.52,116

[1958a] Generalized Functions. Vol. 1. Generalized Functions and Operations on Them. Moscow: Fizmatgiz. English transl.: New York: Academic Press 1964, Zbl.91,111

[1958b] Generalized Functions. Vol. 2. Spaces of Test and Generalized Functions. Moscow: Fizmatgiz. English transl.: New York: Academic Press 1968, Zbl.91,111

[1958c] Generalized Functions. Vol. 3. Some Questions of the Theory of Differential Equations. Moscow: Fizmatgiz. English transl.: New York: Academic Press 1967, Zbl.91,111

Gorin, E.A.

[1961] Asymptotic properties of polynomials and algebraic functions in several variables. Usp. Mat. Nauk *16*, No. 1, 91–118, Zbl.102,254. English transl.: Russ. Math. Surv. *16*, No. 1, 93–119

Grushin, V.V.

[1963a] On Sommerfeld type conditions for a class of partial differential equations. Mat. Sb. *61*, No. 2, 147–174. English transl.: Transl., II. Ser., Am. Math. Soc. *51*, 82–112 (1966)

[1963b] The extension of smoothness of solutions of differential equations of principal type. Dokl. Akad. Nauk SSSR *148*, No. 6, 1241–1244. English transl.: Sov. Math., Dokl. *4*, 248–252 (1963), Zbl.166,364

Gunning, R., Rossi, H.

[1965] Analytic Functions of Several Complex Variables. Englewood Cliffs, N.J.: Prentice-Hall Inc., Zbl.141,86

Hadamard, J.

[1932] Le problème de Cauchy et les équations aux dérivées partielles linéaires hyperboliques. Paris: Hermann, Zbl.6,205

Hörmander, L.

[1958] On the division of distributions by polynomials. Ark. Mat. *3*, 555–568, Zbl.131,119

[1963] Linear Partial Differential Operators. Berlin, Heidelberg, New York: Springer-Verlag, Zbl.108,93

[1973] An Introduction to Complex Analysis of Several Variables. 2nd ed. Amsterdam, London: North Holland Publ. Col., Zbl.271.32001, 3rd rev. ed. (1990)

[1983] The Analysis of Linear Partial Differential Operators. Vols. I and II. Berlin, Heidelberg, New York: Springer-Verlag, Zbl.521.35001 and Zbl.521.35002

[1985] The Analysis of Linear Partial Differential Operators. Vols. III and IV. Berlin, Heidelberg, New York: Springer-Verlag, Zbl.601.35001 and Zbl.612.35001

Komech, A.I.

[1974] Elliptic differential equations with constant coefficients in a cone. Vestn. Mosk. Univ., Ser. Mat. Mekh. *29*, No. 2, 14–20, Zbl.278.35032

Krejn, S.G.

[1967] Linear Differential Equations in Banach Space. Moscow: Nauka, Zbl.172,419. English transl.: Transl. Math. Monogr. *29*, Providence, R.I.: American Math. Soc. 1971

Leray, J.

[1954] Hyperbolic Differential Equations. Princeton: Inst. Adv. Study, Zbl.75,98

Lojasiewicz, S.
[1959] Sur le problème de division. Stud. Math. *18*, 87–136, Zbl.115,102
Lopatinskij, Ya.B.
[1953] On a device of reducing boundary-value problems for a system of elliptic differential equations to regular integral equations. Ukr. Mat. Zh. 5, No. 2, 123–151, Zbl.52,102. English transl.: Transl., II. Ser., Am. Math. Soc. *89*, 149–183 (1970)
Malgrange, B.
[1953] Equations aux dérivées partielles à coefficients constant. I: Solution élémentaire. CR Acad. Sci. Paris *237*, No. 25, 1620–1622, Zbl.52,342
Merzon, A.E.
[1973] On the solvability of differential equations with constant coefficients in a cone. Dokl. Akad. Nauk SSSR *211*, No. 2, 285–288. English transl.: Sov. Math. Dokl. *14* (1973), 1012–1015, Zbl.311.35014
Mikhlin, S.G.
[1977] Linear Partial Differential Equations. Moscow: Vysshaya Shkola
Nuij, W.
[1968] A note on hyperbolic polynomials. Math. Scand. *23*, No. 1, 69–72, Zbl.189,408
Palamodov, V.P.
[1960] On regularization and the problem of division. Dokl. Akad. Nauk SSSR *132*, No. 2, 295–298. English transl.: Sov. Math., Dokl. *1*, 560–564, Zbl.98,310
[1967] Linear Differential Operators with Constant Coefficients. Moscow: Nauka, Zbl.191,434. English transl.: New York, Berlin, Heidelberg: Springer-Verlag 1970
Paley, R., Wiener, N.
[1934] Fourier Transforms in the Complex Domain. Amer. Math. Soc. Colloqium Publication, Providence, R.I.: American Math. Soc., Zbl.11,16
Petrovskij, I.G.
[1937a] On the Cauchy problem in the domain of non-analytic functions. Usp. Mat. Nauk, *3*, 234–238
[1937b] Über das Cauchysche Problem für Systeme von partiellen Differentialgleichungen. Mat. Sb. *2*, No. 5, 815–868, Zbl.18,405
[1938] On the Cauchy problem for systems of partial differential equations in the domain of non-analytic functions. Bull. Mosk. Univ., Mat. Mekh. *1*, no. 7, 1–72, Zbl.24,37
[1945] On the diffusion of waves and the lacunas for hyperbolic equations. Mat. Sb. Nov. Ser. *17*, No. 2, 289–370, Zbl.61,213
[1961] Lectures on Partial Differential Equations. Moscow: Fizmatgiz. English transl.: New York: Wiley, 1955, Zbl.115,81
Reed, M., Simon, B.
[1975] Methods of Modern Mathematical Physics. Vol. 2. New York: Academic Press, Zbl.308.47002
[1979] Methods of Modern Mathematical Physics. Vol. 3. New York: Academic Press, Zbl.405.47007
Riesz, M.
[1949] L'intégral de Riemann-Liouville et le problème de Cauchy. Acta Math. *81*, 1–223, Zbl.33,276
Rudin, W.
[1964] Principles of Mathematical Analysis. New York: McGraw-Hill, Zbl.148,29
Schwartz, L.
[1961] Méthodes mathématiques pour les science physiques. Paris: Hermann, Zbl.101,413
[1966] Théorie des Distributions. Paris: Hermann, Zbl.149,95
Shilov, G.E.
[1965] Mathematical Analysis. Second Special Course. Moscow: Nauka. English transl.: Oxford: Pergamon Press, Zbl.137,262
[1968] Generalized Functions and Partial Differential Equations. New York: Gordon and Breach, Zbl.159,183

Shubin, M.A.
[1978] Pseudodifferential Operators and Spectral Theory. Moscow: Nauka. English transl.: Berlin, Heidelberg, New York: Springer-Verlag, 1987, Zbl.451.47064

Sobolev, S.L.
[1936] Méthode nouvelle à résoudre le problème de Cauchy pour les équations linéaires hyperboliques. Mat. Sb., Nov. Ser. *1*, No. 1, 39–71, Zbl.14.59
[1950] Applications of Functional Analysis in Mathematical Physics. Izdat. Leningrad. Gos. Univ. English transl.: Providence, R.I.: Amer. Math. Soc., 1963, Zbl.123,90

Tikhonov, A.N., Samarskij, A.A.
[1951] Equations of Mathematical Physics. 4th rev. ed. Moscow: Nauka. English transl.: Oxford: Pergamon Press, 1963, Zbl.111,290

Titchmarsh, E.
[1937] Introduction to the Theory of Fourier Integrals. Oxford: Clarendon Press, Zbl.17,404

Trèves, F.
[1961] Lectures on Linear Partial Differential Equations with Constant Coefficients. Rio de Janeiro: Notas de Mathematica, No. 27, Zbl.129,69, New York: Gordon and Breach 1966

Vajnberg, B.R.
[1966] The principles of radiation, limiting absorption and limiting amplitude in the general theory of partial differential equations. Usp. Mat. Nauk *21*, No. 3, 115–194. English transl.: Russ. Math. Surv. *21*, No. 3, 115–193, Zbl.127,137

Varchenko, A.N.
[1986] On normal forms of non-smoothness of solutions of hyperbolic equations. Izv. Akad. Nauk SSSR *51*, No. 3, 652–665. English transl.: Math. USSR, Izv. *30*, No. 3, 615–628 (1988), Zbl.646.35054

Vasil'ev, V.A.
[1986] Sharpness and the local Petrovskij condition for strictly hyperbolic operators with constant coefficients. Izv. Akad. Nauk SSSR *50*, No. 2, 242–283. English transl.: Math. USSR, Izv. *28*, 233–273 (1987), Zbl.615.35012

Vishik, M.I., Ehskin, G.I. (= Eskin, G.I.)
[1965] Convolution equations in a bounded domain. Usp. Mat. Nauk *20*, No. 3, 89–152. English transl.: Russ. Math. Surv. *20*, No. 3, 85–151, Zbl.152,342

Vladimirov, V.S.
[1964] Methods of the Theory of Functions in Several Complex Variables. Moscow: Nauka. English transl.: Cambridge, Mass.: M.I.T. Press, 1966, Zbl.125,319
[1979] Generalized Functions in Mathematical Physics. Moscow: Nauka. English transl.: Moscow: Mir Publishers, 1979, Zbl.515.46034
[1981] Equations of Mathematical Physics, 4th edition. Moscow: Nauka. English transl.: Moscow: Mir Publishers, 1984, Zbl.652.35002, 5th ed., compl. 1988

Volevich, L.R., Paneyakh, B.P. (= Paneah, B.P.)
[1965] Some spaces of generalized functions and embedding theorems. Usp. Mat. Nauk *20*, No. 1, 3–74. English transl.: Russ. Math. Surv. *20*, No. 1, 1–73, Zbl.135,165

Weyl, H.
[1940] The method of orthogonal projection in protential theory. Duke Math. J. 7, 411–444, Zbl.26,20

Whittaker, E.T., Watson, G.N.
[1927] A Course of Modern Analysis. 4th ed. Cambridge: Cambridge Univ. Press, FdM.53,180

Yosida, K.
[1965] Functional Analysis. Berlin, Heidelberg, New York: Springer-Verlag, 3rd ed., 1971, Zbl.126,115

Author Index

Subject Index

Encyclopaedia of Mathematical Sciences

Editor-in-Chief: R. V. Gamkrelidze

Analysis

Volume 13: **R. V. Gamkrelidze** (Ed.)

Analysis I

Integral Representations and Asymptotic Methods

1989. VII, 238 pp. 3 figs. ISBN 3-540-17008-1

Volume 14: **R. V. Gamkrelidze** (Ed.)

Analysis II

Convex Analysis and Approximation Theory

1990. VII, 255 pp. 21 figs. ISBN 3-540-18179-2

Volume 26: **S. M. Nikol'skij** (Ed.)

Analysis III

Spaces of Differentiable Functions

1991. VII, 221 pp. 22 figs. ISBN 3-540-51866-5

Volume 27: **V. G. Maz'ya, S. M. Nikol'skij** (Eds.)

Analysis IV

Linear and Boundary Integral Equations

1991. VII, 233 pp. 4 figs. ISBN 3-540-51997-1

Volume 19: **N. K. Nikol'skij** (Ed.)

Functional Analysis I

Linear Functional Analysis

1992. V, 283 pp. ISBN 3-540-50584-9

Volume 20: **A. L. Onishchik** (Ed.)

Lie Groups and Lie Algebras I

Foundations of Lie Theory. Lie Transformation Groups

1993. VII, 235 pp. 4 tabs. ISBN 3-540-18697-2

Volume 41: **A. L. Onishchik, E. B. Vinberg** (Eds.)

Lie Groups and Lie Algebras III

Structure of Lie Groups and Lie Algebras

1994. V, 248 pp. ISBN 3-540-54683-9

Several Complex Variables

Volume 7: **A. G. Vitushkin** (Ed.)

Several Complex Variables I

Introduction to Complex Analysis

1990. VII, 248 pp. ISBN 3-540-17004-9

Volume 8: **A. G. Vitushkin, G. M. Khenkin** (Eds.)

Several Complex Variables II

Function Theory in Classical Domains. Complex Potential Theory

1994. VII, 260 pp. 19 figs. ISBN 3-540-18175-X

Volume 9: **G. M. Khenkin** (Ed.)

Several Complex Variables III

Geometric Function Theory

1989. VII, 261 pp. ISBN 3-540-17005-7

Volume 10: **S. G. Gindikin, G. M. Khenkin** (Eds.)

Several Complex Variables IV

Algebraic Aspects of Complex Analysis

1990. VII, 251 pp. ISBN 3-540-18174-1

Volume 54: **G. M. Khenkin** (Ed.)

Several Complex Variables V

Complex Analysis in Partial Differential Equations and Mathematical Physics

1993. VII, 286 pp. ISBN 3-540-54451-8

Volume 69: **W. Barth, R. Narasimhan** (Eds.)

Several Complex Variables VI

Complex Manifolds

1990. IX, 310 pp. 4 figs. ISBN 3-540-52788-5

Volume 74: **H. Grauert, T. Peternell, R. Remmert** (Eds.)

Several Complex Variables VII

Sheaf-Theoretical Methods in Complex Analysis

1994. VIII, 369 pp. ISBN 3-540-56259-1

Printing: Mercedesdruck, Berlin
Binding: Buchbinderei Lüderitz & Bauer, Berlin